Residential Construction Problem Solver

Construction Books from McGraw-Hill

Bianchina: Forms & Documents for the Builder
Bolt: Roofing the Right Way, 3/e
Bynum, Woodward, Rubino: Handbook of Alternative Materials in Residential Construction
Clark: Retrofitting for Energy Conservation
Domel: Basic Engineering Calculations for Contractors
Feldman, Feldman: Construction & Computers
Frechette: Accessible Housing
Gerhart: Everyday Math for the Building Trades
Harris: Noise Control Manual for Residential Buildings
Hacker: Residential Steel Design and Construction
Hutchings: National Building Codes Handbook
Jahn: Practical Cost-Saving Techniques For Housing Construction
Jahn: McGraw-Hill's Best Practices For Housing Construction
Jahn, Dettenmaier: Offsite Construction
Korejwo: Bathroom Installation, Design and Remodeling
Korejwo: Kitchen Installation, Design and Remodeling
Miller, Baker: Carpentry and Construction, 2/e
Philbin: The Illustrated Dictionary of Building Terms
Philbin: Painting, Staining and Refinishing
Powers: Kitchens: Professional's Illustrated Design and Remodeling Guide
Powers: Bathrooms: Professional's Illustrated Design and Remodeling Guide
Scharff and the Staff of Roofer Magazine: Roofing Handbook
Scharff and the Staff of Walls & Ceilings Magazine: Drywall Construction Handbook
Shuster: Structural Steel Fabrication Practices
Skimin: The Technician's Guide to HVAC Systems
Trellis: Documents, Contracts and Worksheets for Home Builders
Vernon: Professional Surveyor's Manual
Woodson: Be a Successful Building Contractor, 2/e

Dodge Cost Books from McGraw-Hill

Marshall & Swift: Dodge Unit Cost Book
Marshall & Swift: Dodge Repair & Remodel Cost Book
Marshall & Swift: Dodge Heavy Construction Unit Cost Book
Marshall & Swift: Dodge Electrical Cost Book

Residential Construction Problem Solver

Bart Jahn

McGraw-Hill
New York San Francisco Washington, D.C. Auckland Bogotá Caracas Lisbon London
Madrid Mexico City Milan Montreal New Delhi San Juan Singapore Sydney Tokyo Toronto

McGraw-Hill
A Division of The **McGraw·Hill** Companies

Library of Congress Cataloging-in-Publication Data

Jahn, Bart
  Residential construction problem solver/Bart Jahn.
    p. cm.
  ISBN 0-07-032962-1 (hardcover).—ISBN 0-07-032961-3 (pbk.)
  1. House construction—Handbooks, manuals, etc. 2. Building failures—Prevention—Handbooks, manuals, etc. 3. Accidents—Prevention—Handbooks, manuals, etc. I. Title.
TH4813.B37 1998
690'.837—dc21   98-15167
                                                                                                            CIP

Copyright © 1998 by The McGraw-Hill Companies, Inc. All rights reserved. Printed in the United States of America. Except as permitted under the United States Copyright Act of 1976, no part of this publication may be reproduced or distributed in any form or by any means, or stored in a data base or retrieval system, without the prior written permission of the publisher.

1 2 3 4 5 6 7 8 9 0   KGP/KGP   9 0 3 2 1 0 9 8

ISBN 0-07-032962-1 (H)
       0-07-032961-3 (P)

The sponsoring editor for this book was Zoe G. Foundotos, the editing supervisor was Penny Linskey, and the production supervisor was Clare Stanley. Interior design by Jaclyn J. Boone. Composition by Jaclyn J. Boone and Michele Pridmore of McGraw-Hill's Professional Book Group in Hightstown, New Jersey.

Printed and bound by Quebecor/Kingsport.

McGraw-Hill books are available at special quantity discounts to use as premiums and sales promotions, or for use in corporate training programs. For more information, please write to the Director of Special Sales, McGraw-Hill, 11 West 19th Street, New York, NY 10011. Or contact your local bookstore.

This book is printed on recycled, acid-free paper containing a minimum of 50% recycled, de-inked paper

---

Information contained in this work has been obtained by The McGraw-Hill Companies, Inc. ("McGraw-Hill") from sources believed to be reliable. However, neither McGraw-Hill nor its authors guarantee the accuracy, or completeness of any information published herein, and neither McGraw-Hill nor its authors shall be responsible for any errors, omissions, or damages arising out of use of this information. This work is published with the understanding that McGraw-Hill and its authors are supplying information, but are not attempting to render engineering or other professional services. If such services are required, the assistance of an appropriate professional should be sought.

*I would like to graciously dedicate this book to my nephews and nieces:
Robert, Carolyn, Liana, Raymond, Heidi, Brian, Brooke,
Rhonda, Kathryn, Stephanie, Madison, and Nicolas.*

CONTENTS

Introduction ix

1. Rough Grading, Underground Utilities, and Streets *1*
2. Concrete Foundations and Slabs *24*
3. Masonry Block Retaining Walls *37*
4. Layout and Wall Framing *45*
5. Framing Floors and Ceilings *59*
6. Layout and Framing for Doors *73*
7. Layout and Framing for Windows *83*
8. Framing Stairs *96*
9. Shear Panel *104*
10. Fireplace Framing *108*
11. Roof Framing *113*
12. Moment Frames *121*
13. Exterior Framing *125*
14. Miscellaneous Framing *130*
15. Framing for Bathtubs *134*
16. Backing *137*
17. Straightedge *144*
18. Rough Plumbing, Electrical, and HVAC *150*
19. Drywall *190*
20. Cabinets *208*
21. Finish Carpentry *227*
22. Ceramic Tile *249*
23. Insulation, Lath, and Stucco *271*
24. Roofing *279*
25. Painting *282*
26. Hardware *291*
27. Finish Plumbing, Electrical, and HVAC *304*
28. Miscellaneous Finish *313*
29. Flooring *331*
30. Concrete Walkways and Driveways *345*
31. Walls, Fences, and Gates *379*
32. Miscellaneous Exterior *409*
33. Landscaping *414*
34. Construction Management *433*
35. The Superintendent, the Construction Trailer, and Field Paperwork *437*
36. The Final Preparation Phase *445*
37. Customer Service *448*
38. Sales Models *454*

Index 459

INTRODUCTION

ONE OF THE FUNDAMENTAL PROBLEMS IN HOUSING CONSTRUCTION IS THAT builders and contractors do not send memos back and forth regarding mistake prevention. Every new housing construction project struggles with assembly-line type problems that were solved months and years ago on other projects. New housing construction projects are so isolated in terms of communicating debugging information that two similar projects going up side by side, built by different companies, can each be making the same costly mistakes without either one knowing about or being able to benefit from the other's experience. The result is that hundreds of thousands of people working in housing construction find themselves at different points on the uphill slope of the learning curve, repeating many of the same hard-earned lessons.

A second factor that contributes to the debugging problem in housing construction is the lack of feedback from the construction to the architect. Over the years, builders have reduced the traditional oversight role played by the architect during the building construction in order to save money. After the building plans are complete, many builders feel their own supervisory staff is capable of resolving most of the design and construction problems and questions that may arise. The added cost of involving the architect in an oversight, supervisory role during the construction is seen by builders as a redundant and unnecessary expense to the project budget. One result of this cost-saving approach is that architects have fewer opportunities to get out in the field and observe the problems that occur during the construction. Without direct participation in the construction, and without debriefing sessions after the construction, architects are kept unaware of construction difficulties, and the opportunity for future prevention through improved building plans is missed.

This book is an illustrated checklist of problems and mistakes that actually occurred on real housing construction projects, taken from my previous three books—*Practical Cost-Saving Techniques for Housing Construction, McGraw-Hill's Best Practices for Housing Construction,* and *Offsite Construction.* There are also about 250 new topics that have been added to this checklist that were observed and recorded after the writing of these books.

The goal of this book is to forewarn the reader of costly housing construction problems and mistakes before they occur. Each checklist topic is a miniature case-history of a problem or mistake that resulted in a loss of time, money, or quality during the construction. All of these problems and mistakes can be avoided by first knowing about them in advance, and then taking the appropriate precautions during the design and construction phases of the project.

The value of this information lies in its benefit to cost ratio. There are literally tens of thousands of man-hours, and hundreds of thousands of dollars represented in the construction problems and mistakes discussed in the following pages. Yet builders, contractors, and architects need only avoid one or two of these construction problems to recoup the cost of this book.

Debugging housing construction is the last area of information remaining to complete the technology of housing construction. Because of the uniqueness of every housing construction project, and the lack of communication in the building industry regarding mistake prevention, the only way to achieve progress in this area is to observe and record problems and mistakes one-by-one as they occur, and then pass along this information. The intent of this book therefore is to provide builders, contractors, architects, home inspection consultants and construction students with an exhaustive checklist identifying many of these debugging problems, issues, and mistakes, so that housing construction can move closer to the assembly-line efficiency typical of other manufacturing industries.

The checklist information is arranged in each chapter so that the design issues come first, followed by the field construction issues. Each chapter therefore starts out with problems and mistakes that can be resolved upfront on the building plans. The latter half of each chapter then covers problems and mistakes that must be anticipated and resolved during the field construction.

One of the best features of this checklist are the illustrations. Because one picture is worth a thousand words, the illustrations enable the reader to scan each page and quickly determine which checklist topics apply to the construction. Not only does this make the checklist easier and quicker to use for builders, architects, purchasing agents and inspectors working in the industry, but also makes the technical information more accessible and user-friendly for beginning architecture and construction students, assistant superintendents, and homeowners.

The main point of this introduction is to say that there is a large group of problems and mistakes that remain in housing construction, and that most of them can be prevented simply by knowing about them in advance. Although problems and mistakes can be resolved after-the-fact on an individual basis on most housing construction projects, it is much easier and more cost-effective to remove these problems upfront on the design plans and during the course of the construction, before they occur. The challenge is to identify the problems and mistakes that occur on individual projects, and then communicate this information to the industry at large. That is the goal of this book.

Bart Jahn

Thanks to the following contributors: George Carter, Matt Ford, Jeff Ford, Robert Jahn, Brian Jahn, Barry Jahn, John P. Kreitzer, Bruce Lewis, John Montoya, Rodger More, and Dennis Post.

Special thanks to Paul Dettenmaier, long-time friend, for his many contributions and insightful editing of each of my books. Paul is co-author of my third book, *Offsite Construction*.

CHAPTER 1

Rough Grading, Underground Utilities, and Streets

☐ *1*

To better analyze the various subcontractors' bids for rough grading, request that the lump-sum bids be broken down into the categories listed below. This allows the builder to analyze the yardage figures used in each bid and identify bids that have built-in extra costs as a result of unrealistic yardage estimates.

1. Mobilization
2. Clear and grub, and demolition work
3. Site preparation
4. Removal and compaction
5. Raw or design cut
6. Finish

☐ *2*

When dirt is exported off the site, check that the amount of soil in each loader bucket scoop results in the correct yardage of dirt being placed inside the export truck bed; the total amount of export yardage usually is calculated by counting the number of export truck cycles and multiplying that figure by the yards of dirt in each truck. If there is less dirt in each truck bed, fictitious yardage is being exported off the site, costing the builder more money than was anticipated.

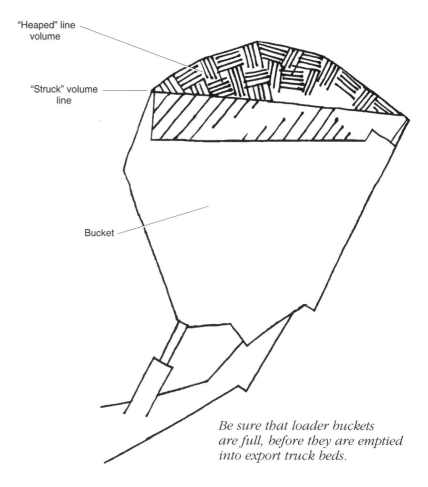

Be sure that loader buckets are full, before they are emptied into export truck beds.

☐ *3*

During the rough grading, consider using the dirt for activities that span the duration of project construction. In determining the export or import yardage, the goal should be to have a zero balance of dirt on the site at the end of construction, with no further exporting or importing required.

☐ *4*

For projects that must export dirt off the site during the rough grading design phase, consider the option of raising the elevation of the building pads or the entire project, with the goal of balancing the existing dirt onsite without having to export any dirt.

☐ **5**

To save on soil export costs, designate some building pads on the rough grading plans to be "as built" instead of showing a precise elevation number. If some pads can be raised 1 or 2 feet, this will provide a "safety valve" on the project to absorb extra dirt during the rough grading and reduce some of the export costs.

☐ **6**

Do not use the civil engineer's cut and fill map exclusively in estimating the import and export requirements for a project. Construction activities that can generate excess dirt and are not included on the cut and fill map include the following:
1. Trenching for sewers, storm drains, and underground utilities
2. Planting of trees and landscaping
3. Design of street sections
4. Installation of site walls
5. Permanent slopes

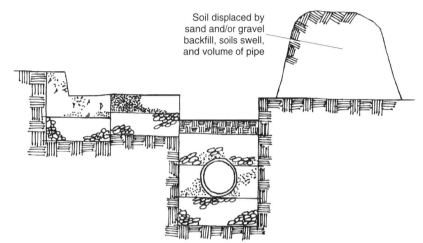

Include within the soils import/export, the calculation of the extra soils generated from underground utilities.

☐ **7**

Before the start of rough grading for hillside view lots, do a line-of-sight study to determine whether the elevations of the building pads as designed maximize the view potential of the lots. You don't want to add the expense of raising some of the pads 3 to 5 feet to improve the views after the rough grading has been completed.

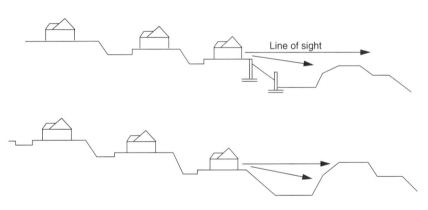

Check elevations of view lots before starting rough grading.

☐ **8**

For large projects, only finish the grade and certify the number of building pads in the project that are immediately ready for construction. Building lots that sit untouched beyond the typical 6-month rough grading certification time must be finish processed and recertified a second time.

☐ 9

For subsurface conditions that may include rock, do not economize on the number of geologic tests and concentrate the tests in the areas of the project that require the deepest cuts.

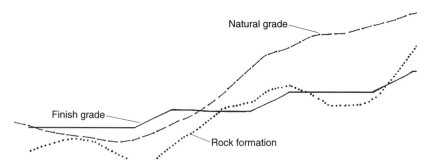

Take enough soil tests in the correct locations to determine accurately subgrade conditions.

☐ 10

Discuss ahead of time with the soil engineer and civil engineer and include in the rough grading subcontract the maximum height of vertical cuts for mass grading projects. This helps prevent vertically cut slopes from collapsing and adding extra cost to the project budget for remedial rough grading slope repairs.

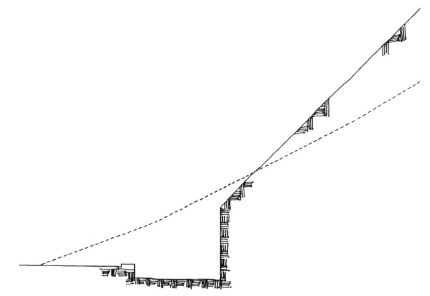

Vertical cut into hillside for a retaining wall. Broken line shows a slope failure.

☐ *11*

For high-density condominium and apartment projects with entry doors one floor level above ground level, consider placing the dirt fingers between the buildings during the rough grading phase rather than after the buildings are in place. It may be faster, easier, and cheaper to build the dirt fingers during the rough grading operation.

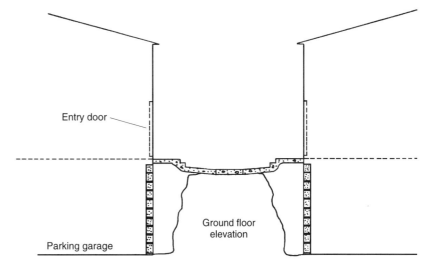

Entry doors and concrete walkways at second floor level.

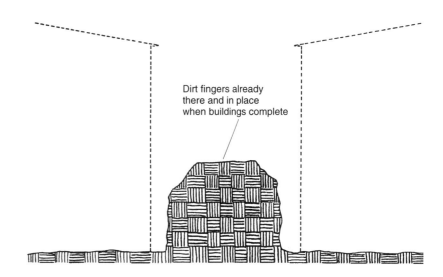

Build dirt fingers during rough grading operation.

☐ **12**
For split-level houses, cut back the top edge of the vertical cut at 1:1 or 1½:1 (height to vertical) during the rough grading operation. This provides a safer working area between the vertical slope and the split-level retaining wall.

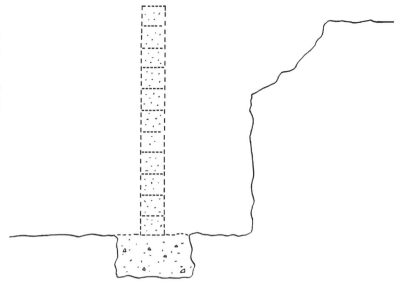

Vertical cut in slope angled back at 45 degrees.

☐ **13**
For split-level lots, include on the rough grading plans any temporary ramps that must be cut through the vertical slopes to provide access to the upper and lower levels for the building construction. This prevents the rough grading work from becoming an afterthought extra by placing it on the plans and in the rough grading subcontract. It also places the temporary dirt ramps in the official design documents and therefore within the scope of work of the civil engineer, the soil engineer, and the city and/or county inspector.

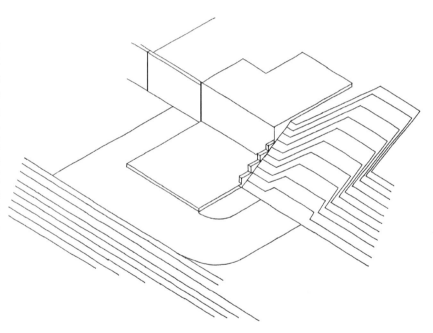

Side view of temporary ramp cut into the slope at the end of a row of split-level building pads.

☐ **14**
For projects that must export or import dirt, consider the nature of the surrounding community, the length of the haul routes, and the city and/or county requirements in determining the dust-control measures that will be used. If extensive dust-control measures are required either by the city and/or county or by the surrounding community, include them in the rough grading bidding instructions and the subcontract.

☐ **15**
Include the rough grading subcontractor in the discussions with the civil engineer and the field surveying crew regarding the rough grading staking plan in terms of the number, locations, and setbacks of the grading stakes.

☐ **16**
For hillside lots with finish grade elevation differences of only a few feet, don't spend time and money to finish grade 2:1 slopes between the building pads when masonry block retaining walls will be built within the first weeks of construction that will destroy those finished graded slopes. Have the rough grading plans show 24-inch-high vertical cuts, for example.

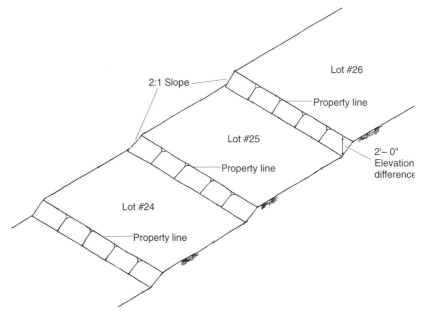

Don't complete rough grading work on two-to-one slopes that will be destroyed by short retaining walls just a few weeks later.

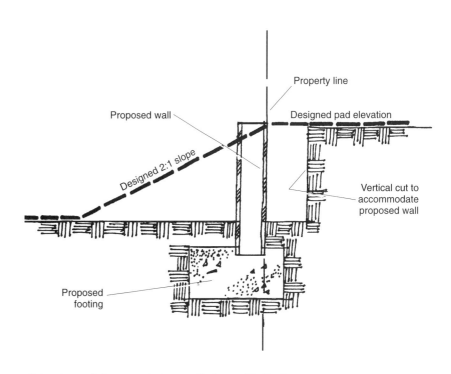

Side view of shot retaining wall that will displace temporary two-to-one slope.

Continued on next page

Rough Grading, Underground Utilities, and Streets

Continued from previous page

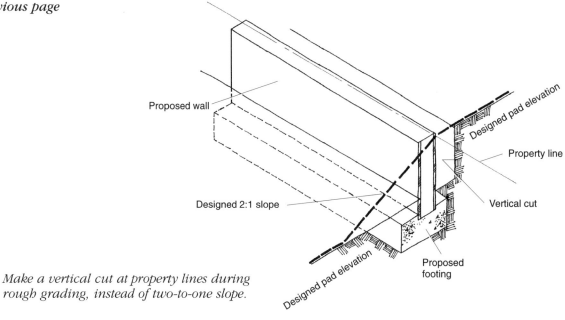

Make a vertical cut at property lines during rough grading, instead of two-to-one slope.

☐ **17**

For projects where tree removal is required, look for companies or individuals who will cut down and chop up the trees for resale as mulch, fuel, or firewood at little or no cost to the project instead of including tree removal in the scope of work of the rough grading subcontract.

☐ **18**

During the clear and grub phase of the rough grading operation, consider that a small percentage of the on-site vegetation can be acceptably incorporated into the rough grading of the project, saving some of the cost of hauling and dumping during the rough grading.

☐ **19**

For projects with expansive soils that require presaturation and when there is time before the start of building construction, consider building small dirt berms around the perimeter of the building pads and presaturating the entire pads for a 2- to 3-week period before construction to achieve the depth and extent of presaturation required. You can save time by getting the presaturation done ahead of the construction schedule.

☐ **20**

For projects with city and/or county street curbs and gutters already in place, determine whether an encroachment permit and bond are required and obtain them before the job site preconstruction meeting with the city and/or county inspector, the civil engineer, the soil engineer, and the rough grading subcontractor. Failure to obtain the encroachment permit and pay the bond can delay the start of the rough grading.

☐ **21**

When the installation of underground utilities in the streets will ruin the finish rough grading of curb benches and sidewalk parkways, do not spend time and money to build the various tier-leveled benches at the street curbs during the rough grading. Instead, concentrate on getting the street subgrade elevations correct so that the amount of dirt is balanced and then build the required tier-leveled benches after the underground utilities are in.

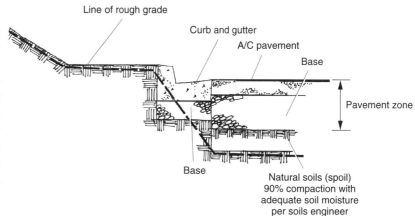

Rough grade streets along dark broken line; then build benches and vertical after underground utilities are complete.

☐ **22**

When a project has a mix of public and private streets or parking areas, don't use the "construction notes and estimates of quantities" figures given on the title pages of the engineering improvement plans as a basis for estimating the cost of the street improvements. These figures are calculated for the purpose of determining the dollar amount for city and/or county improvement bonds and may not include private parking areas.

☐ **23**

To avoid the added cost of having to rebuild and test the dirt bench that goes underneath a street curb as a result of deterioration over the interval between the rough grading and the curb and gutter installation, consider leaving the curb bench "fat" by 6 to 12 inches and cutting the bench to the correct elevation when the curb and gutters are installed.

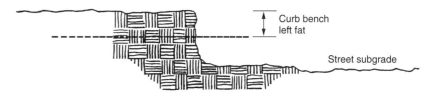

Fat curb bench will be cut to correct elevation later.

☐ **24**

In consultation with the soil engineer, make a reminder checklist that specifies when to call for soil tests during the rough grading and during the trenching and backfilling of underground utilities if the project cannot afford a full-time soil technician on-site.

☐ *25*

For large projects with a superintendent and assistant superintendents, have the soil engineering firm give one of the assistant superintendents a "crash course" on soil testing. The assistant superintendent can then watch the various backfilling and compaction operations and test subgrades for driveways, walkways, landscape irrigation, and drainage pipe trenches for 90 percent compaction by using a steel probe bar at a time during construction when a soil engineering technician would not normally be on the job site.

☐ *26*

For large mass grading projects, request that the soil engineering firm staff the project with an experienced soil technician. An experienced soil technician knows how and when to move safely in and out of the rough grading operation to make the required tests and observations.

☐ *27*

When there is a range for the optimum moisture content of the soil during the rough grading operations, for example, from 14 to 17 percent, before the start of the rough grading work reach a consensus between the soil engineer and the rough grading subcontractor regarding which level of moisture content is preferred. This helps keep the rough grading operation consistent even if soil technicians are changed partway through the rough grading. You don't want a new soil technician insisting on mixing more water into the soil during the processing to reach a 17 percent moisture content when the rough grading up to that point has been acceptable at 14 percent.

☐ *28*

Resolve up front in the subcontracts the issue of who will bear the extra cost for the soil technician to perform retests in areas that did not pass compaction tests. The cost of retests should be charged to the rough grading subcontractor. If the builder establishes this policy, this relieves the soil engineer of the burden of being the "bad guy" for charging for retests.

☐ *29*

Check for recommendations in the soil report that are more restrictive than the city and/or county rough grading and off-site construction standards and make sure the various subcontractors understand that the soil report takes precedence over the public standards. You don't want a subcontractor backfilling and compacting trenches in 8-foot lifts according to the city and/or county standards when the soil report recommends backfilling and compaction in 2-foot lifts, for example.

☐ *30*

Look for a provision in the soil engineering contract proposal for reimbursement for the travel expenses of a soil technician coming from another project to your project simply because it is more convenient to make your project the second or third stop each day. The project that is the first stop does not pay for commuting time for the technician, and so the second or third project should not pay for the travel time from the previous project. Work out a compromise with the soil engineering firm on this issue.

☐ *31*

Put a provision in the soil engineering contract that allows you to increase the linear footage intervals for soil tests when the test results consistently are higher than the required 90 percent compaction. This helps prevent a situation where the rough grading subcontractor speeds up the operation to get the quality down closer to 90 percent and the soil technician cannot keep up with the faster pace. The soil engineering firm then adds a second soil technician, costing the project more money. When the soil tests performed by the soil technician are all passing (90 percent or above), and the process of mixing and placing the soil by the grading contractor remains the same, there should be flexibility in the provisions of the soil engineering contract to increase the test linear foot intervals from every 25 feet to every 100 feet, for example.

☐ *32*

On mass grading projects, the soil engineer will structure the proposal and staff the job site in accordance with the proposed project schedule. An additional reason to monitor the progress of the work and avoid slippage on the construction schedule is that the soil engineer will justifiably ask for more money to cover the extra cost of maintaining the job site staff when the rough grading operation runs beyond the scheduled completion date.

☐ *33*

To simplify the finding of subdrain pipes that may be buried a few inches below the surface of a hillside slope for a rough grading buttress fill, wrap steel wire around the pipe when it is installed and use a metal detector to find the buried pipe ends later.

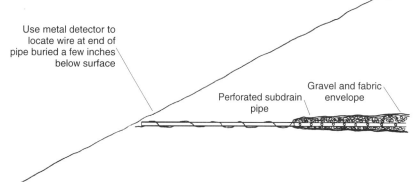

Subdrain in hillside slope encased in gravel and fabric envelope. Steel wire wrapped around subdrain pipe helps locate end of pipe.

☐ *34*

On a mass grading project where the conditions of approval, based on the environmental impact report, require that the project provide an on-site archaeologist and/or paleontologist to search for fossils or ancient artifacts during the rough grading operation, choose a company specializing in this area that has a reputation for allowing builders to perform their work while the rough grading proceeds.

☐ **35**
Overlay the improvement plans and the precise grading plans to determine the crossing points and elevations of the various underground utilities so that you can discover any conflicts before the start of the off-site construction.

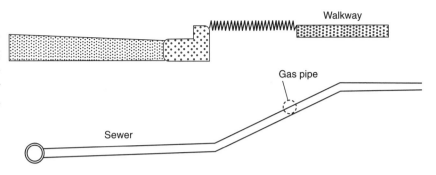

Conflict between underground sewer lateral and gas pipe.

☐ **36**
When the off-site construction and the construction of the sales models occur simultaneously, analyze and if necessary change the staking offset dimensions for the underground utilities and other off-site construction activities in order to place the surveyor's stakes out of the way of the sales models' building construction; restaking can be expensive.

☐ **37**
For a high-density project where the street curbs and gutters are not poured until late in the building construction, have the surveyors stake not only the line and grade for the utilities but also the lateral terminus points for utilities that extend past the curb and gutter, such as a fire hydrant. This gets these utilities correctly placed behind the curb instead of in the street.

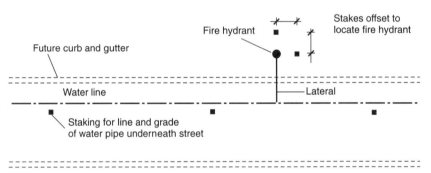

Staking for underground utility.

☐ **38**
For a high-density project using reclaimed water for landscape irrigation and/or fire sprinklers, check the city and/or county requirements for the minimum separation between the underground reclaimed water lines and the domestic water lines. At locations where the two lines cross, encase the reclaimed water pipe inside a larger-diameter sleeve pipe that extends on each side of the crossing point a length equal to the required separation (usually 10 feet).

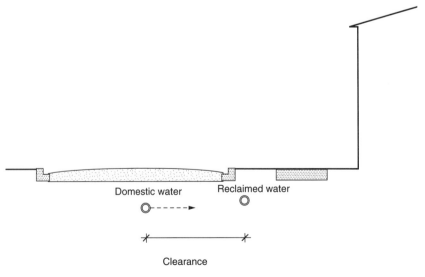

Clearance between domestic and reclaimed water.

12   **Chapter 1**

☐ **39**

Obtain copies of the engineering "cut sheets" showing the line and grade of the underground sewers from the underground pipeline subcontractor. A cut sheet is an as-built record of the line and grade of the sewers at sewer structures, cleanouts, lateral connection points, and 20-foot intervals. Cut sheets help locate lateral ends later for the service connections and locate the underground sewer lines if there is a problem.

☐ **40**

To simplify the installation of underground utility services during building construction, consider using conduit pipe instead of direct burial cable. This allows the conduit pipe for all the utilities to be installed at the same time, shortening the period during which trenches are open on the job site. The utility companies later pull their cables through the conduit pipe at their convenience, eliminating scheduling and coordination problems.

☐ **41**

Check that the location of the electrical transformer in relation to surrounding exterior stairways, garden walls, and landscaping trees provides the clearances required by the electrical utility company and that the transformer is installed with the correct orientation so that the door opening is facing the proper direction for clearance and access. This allows a serviceperson to turn off the high-voltage breaker switches, using an 8-foot-long "hot stick," from a safe distance.

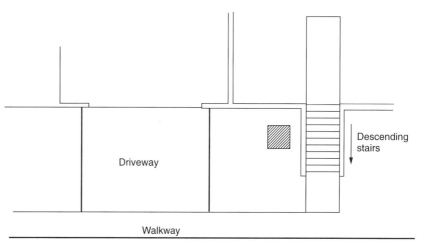

Plan view of electrical transformer box placed within landscaped area.

☐ **42**

Choose locations within the project for electrical transformers that are out of view and inconspicuous and soften their visual impact with garden walls and landscaping.

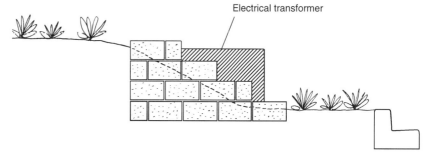

Masonry block wall hides portion for electrical transformer box.

Rough Grading, Underground Utilities, and Streets

☐ **43**

For projects that use conduit pipe for underground "dry" utilities such as electricity, telephone, and television, angle the conduit pipe services to a utility meter room at a 45-degree angle. This saves 45 degrees out of the recommended maximum total of 225 degrees of angle bends (two 90-degree bends plus one 45-degree bend equals 225 degrees).

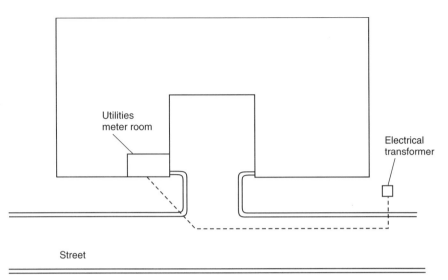

Route of underground utility services from electrical transformer to meter room.

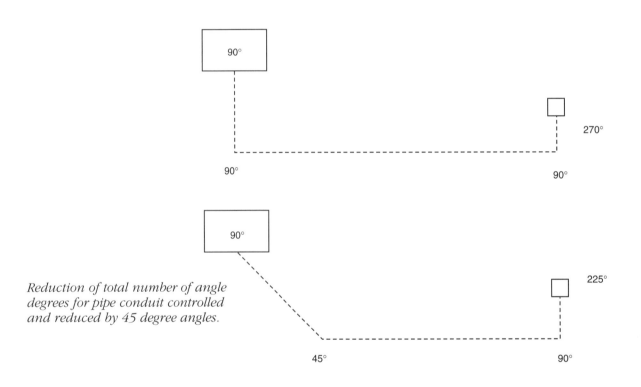

Reduction of total number of angle degrees for pipe conduit controlled and reduced by 45 degree angles.

☐ **44**

When direct burial cable is installed by the utility companies for electrical, telephone, and television services to the houses, schedule the trenching and installation of those services after the concrete slabs are poured and the site is cleaned but before the framing lumber is delivered and the framing starts. This provides a job site that is clear and free of obstacles for faster trenching, installation, and backfilling, preventing the disruptions that occur when the trenching work is done simultaneously with the building construction.

☐ **45**
Place a scrap piece of 2x4 wood at the end of a sewer lateral located at the street curb to make the sewer lateral end easier to find during trenching for the service connection to the house.

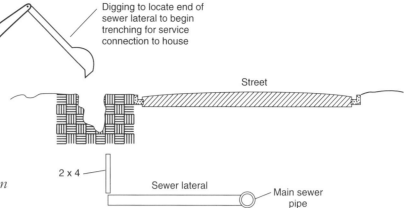

Short piece of wood marking location of sewer lateral end.

☐ **46**
Extend the sewer lateral, which is below the street curb, a few feet beyond the other utility laterals. This allows the backhoe operator to dig up and locate the end of the sewer lateral to make the service connection without hitting and damaging the other utilities.

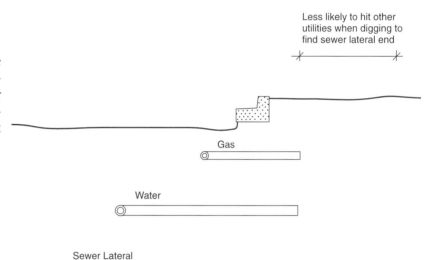

Sewer lateral extending further than other utilities.

☐ **47**
Provide at least 4 inches of clearance between the top of the concrete sidewalk or rough grade and the bottom flange of a fire hydrant to provide working space for the installation of the breakaway bolts. This may require staking the top of the street curb elevation to get the fire hydrant riser and flange at the correct elevation. The bolts usually are installed from the bottom upward so that the nuts are on the top.

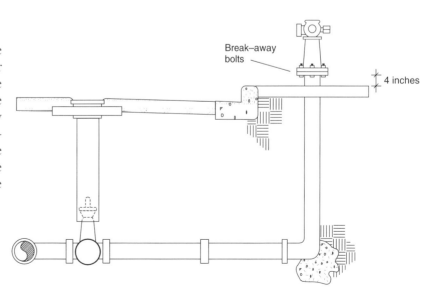

Typical fire hydrant installation with adequate clearance for installation of break-away bolts.

Rough Grading, Underground Utilities, and Streets

☐ **48**
Take photographs of each phase of the off-site construction to serve as a record for future reference.

☐ **49**
Have the soil engineering technician check that all the trenches on the project, in both public and private areas, are backfilled and compacted uniformly in compliance with the soil engineer's recommendations.

☐ **50**
On projects with nonpermeable, expansive clay soils, check with the soil engineer to see whether water jetting is permissible as a method to obtain compaction in sand-bedded joint utility trenches. Introducing water into a trench that will not drain because of the characteristics of the soil can cause the backfill material in the trench to have too high a moisture content, resulting in an unstable soil condition in an underground utility trench in the street, for example.

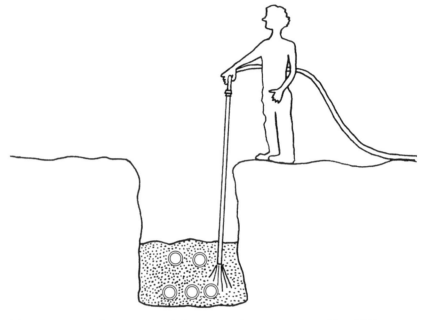

Water jetting achieves consolidation and compaction of backfill in joint utility trenches.

☐ **51**
When trenches for underground utilities are dug in the wrong locations, check that the backfilling and compaction for those trenches are done with the same amount of care as are those for the trenches with utilities in them.

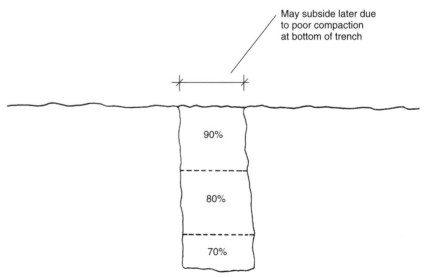

Trench dug in wrong location and incorrectly backfilled.

☐ **52**
Obtain a valve key or wrench from the underground utility pipeline subcontractor for the turning on and shutting off of the main water line valves located below grade in the streets.

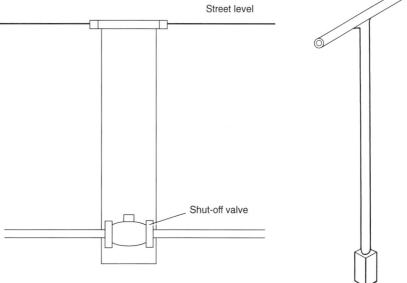

Water main shut-off valve in street.

Valve key or wrench.

☐ **53**
For underground utility crossings underneath street curbs where it is difficult to obtain 90 percent compaction of the backfill because the street curb and gutter are in the way, consider using a two-stack cement and sand slurry mix. This will prevent paving settlement failures at these locations.

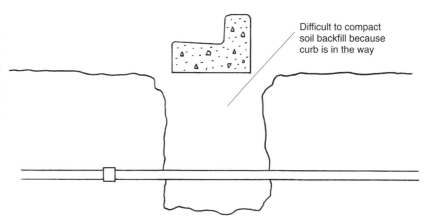

Utility pipe crossing underneath street curb.

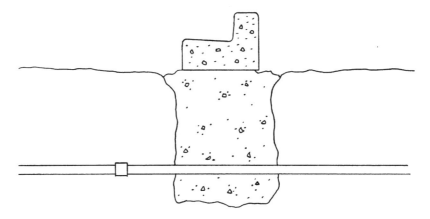

Concrete slurry mix is used to backfill trench area underneath street curb.

Rough Grading, Underground Utilities, and Streets   **17**

☐ **54**
Consider placing extra empty conduit pipes in the utility trenches that cross the streets in case they are needed later for landscape lighting, landscape irrigation control wiring, and monitored fire alarm systems.

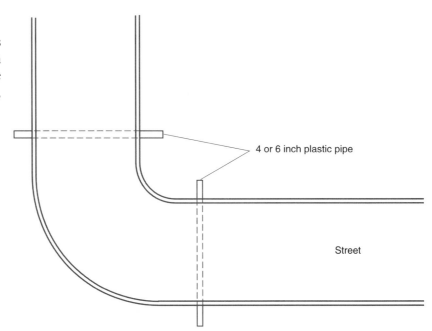

Extra sleeve pipes placed underneath street in anticipation of future underground utilities.

☐ **55**
After each phase of the underground utilities is completed, spray-paint and photograph newly chipped, cracked, or broken sections of the street concrete curbs and gutters so that each new subcontractor will not be responsible for earlier damage and can bear the costs for the repair and replacement of the damaged curb and gutter sections it is responsible for. Each subcontractor should then walk the job site with the builder to visually inspect the spray-painted curb areas before the start of its work.

☐ **56**
For flag lots, consider the following three issues: guest parking spaces, turnarounds for fire trucks, and the proximity of fire hydrants.

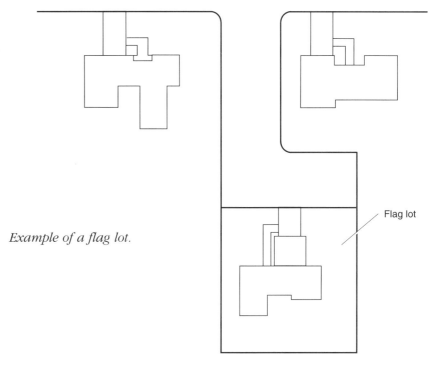

Example of a flag lot.

☐ **57**
In dividing up a project into phases, to eliminate built-in confusion during building construction, do not include within one phase of the construction duplicate lot numbers from a previous phase. For example, do not include within the last 49 houses under construction on a large project lot numbers 21 through 26 in phase 4 and lot numbers 1 through 43 in phase 5; if you do this, lot numbers 21 through 26 will be duplicated.

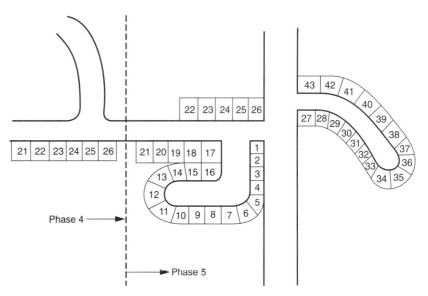

Duplicate lot numbers from previous phase are included in the construction of another phase.

Rough Grading, Underground Utilities, and Streets

☐ **58**

For high-density condominium projects, to separate the incoming and outgoing automobile traffic for the homeowner-occupied units from the building construction traffic, consider using a temporary chain-link fence that has a colored fabric windscreen with steel posts embedded in automobile tires filled with concrete. This is easy to assemble and disassemble each time a phase of the construction is completed and you will move on to the next phase in the project.

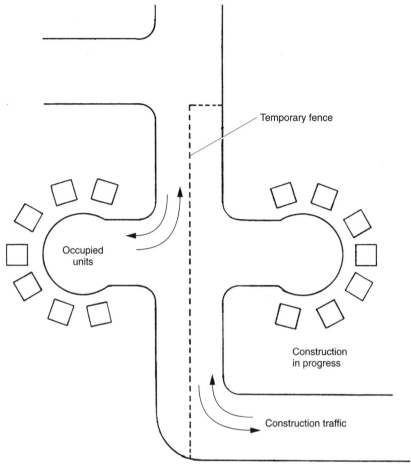

Plan view of temporary fence in street separating occupied units from new construction.

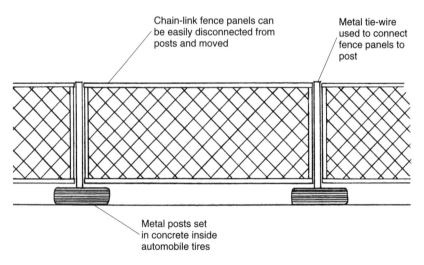

Temporary fencing in center of streets to separate construction traffic from occupied units.

☐ **59**

For the final city and/or county punch list inspection of the new public streets within a project for the purpose of getting the streets accepted by the public entity and getting the street improvement bond released, request that the city and/or county person who makes the inspection have the actual authority to accept the streets and release the bond. This eliminates the situation where people make preliminary inspections and generate punch lists that are not final, causing subcontractors to make two or three trips to the project to make repairs.

☐ **60**

For projects that have private streets, install the light bulbs in the streetlights while they are lying horizontally on the ground, before they are stood up vertically. This prevents having to have someone climb a ladder later to install the streetlight bulbs.

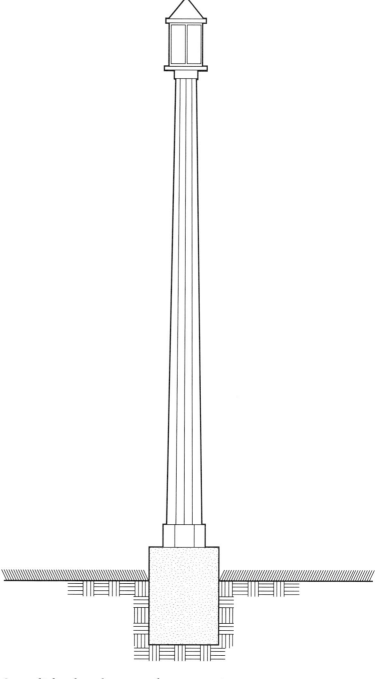

Street light placed on top of concrete pier.

Rough Grading, Underground Utilities, and Streets

☐ **61**

Analyze the locations of paired water meters at the street curbs to shorten the distances for water services to the houses. One location for the water meters shown in the example drawings has the water services taking a longer route underneath driveways. Some plumbing subcontracts are written to provide a minimum water service run per house, such as 35 feet, and then charge an extra dollar amount for each linear foot beyond the minimum contractual length. You don't want 100-foot-long water service runs when they could be reduced to 40 feet long by having a more favorable layout.

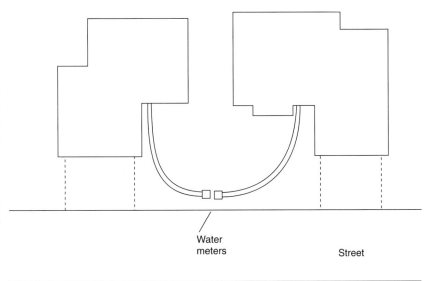

Water meters paired together at location which results in shorter runs.

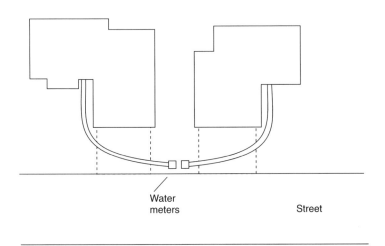

Water services running underneath driveways.

☐ **62**

Place red-painted wood or rebar stakes around water meters at the street curb to prevent forklifts and other vehicles from mistakenly driving over the water meters, resulting in water leaks that must be repaired by the water utility company and will be charged to the builder.

☐ **63**

Have the water company install the vault boxes around the water meters at the street curb, with adequate clearance around the shutoff valve. This provides working space to turn on or shut off the water to a house without scraping and skinning one's hands and knuckles.

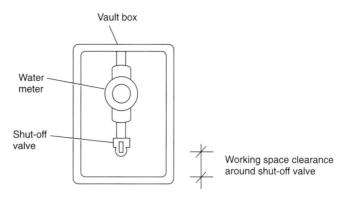

Water meter valve box placed to provide clearance around shut-off.

☐ **64**

Sketch as-built drawings showing the locations of underground utility services to each house, with the dimensions, while the utility trenches are still open. This provides information regarding the locations of the underground utility services so that they are not damaged by future construction activities such as walls, fences, trees, and landscaping.

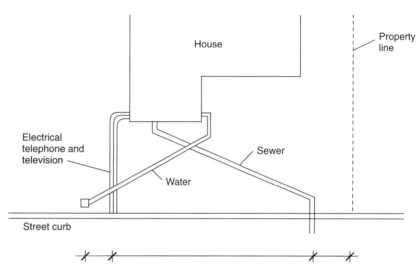

Typical as-built drawing for detached single-family home.

Rough Grading, Underground Utilities, and Streets

CHAPTER 2

Concrete Foundations and Slabs

☐ **65**
Get plumbing trench layout sheets from the plumber before laying out the building footings for trenching. At this early stage in the work the plumber is not always on the job site to indicate the plumbing trench locations.

☐ **66**
Make simplified cheat sheets showing the dimensions of the building footprints to make the concrete formboards easier to check.

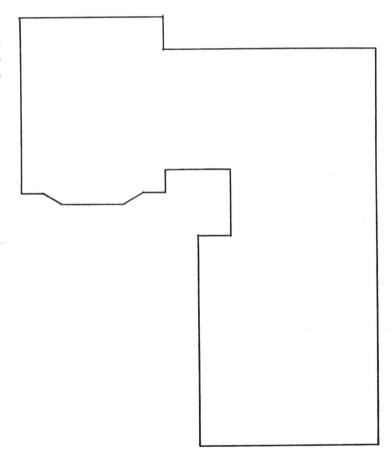

☐ **67**
Minimize variation in the types and sizes of hold-downs, post anchors, anchor bolt spacings, and shear panel nailing patterns to simplify the construction.

☐ **68**
Make sure that the offset dimension from the edge of the concrete for the grade beam rebar does not conflict with the required locations of anchor bolts for hold-downs and structural steel columns that will be installed later.

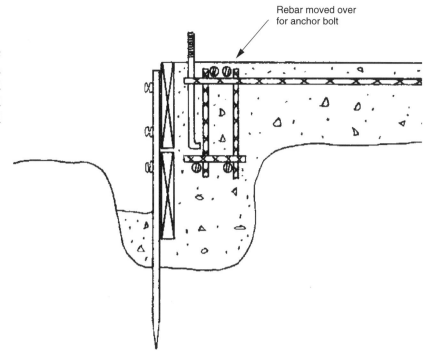

☐ **69**
For party wall foundations, check that there is enough space for the plumbing pipes and the horizontal rebar. If the rebar must be heated and bent around the plumbing, the builder may need a cut-sheet detail from the structural engineer that shows this change to the plans in order to pass the foundation inspection.

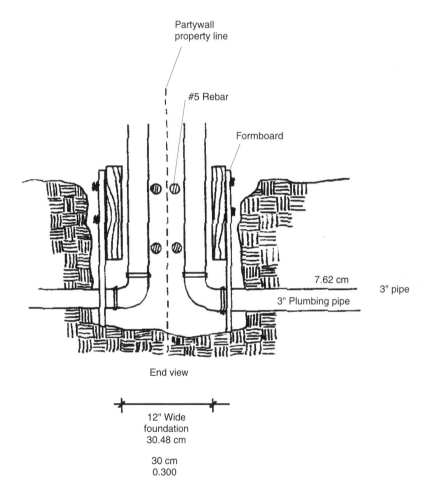

Concrete Foundations and Slabs

☐ **70**
Check for conflicts between the locations of anchor bolts and the installation of future wood posts that are 4 inches by 4 inches or larger.

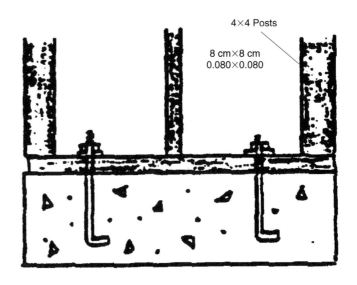

☐ **71**
Analyze the layout of anchor bolts so that the location and orientation of hold-downs result in bolt heads, washers, and nuts that end up inside the walls. This eliminates the need to furr out walls that have bolts and nuts that project beyond the wall framing.

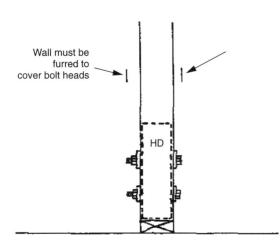

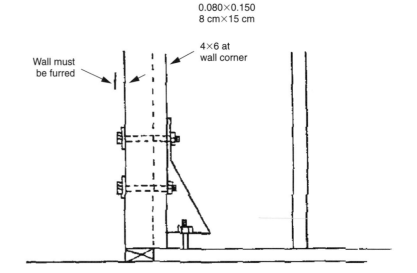

26 Chapter 2

☐ **72**
Check that a block wall footing turned inward toward the building does not interfere with anchor bolts for the building concrete slab that is poured later.

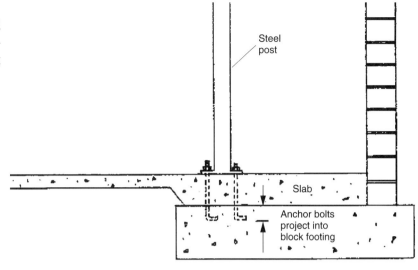

☐ **73**
Make sure there is enough dimension on plans to provide adequate concrete coverage around anchor bolts at the front of the garage for steel moment frames and columns.

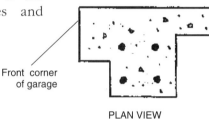
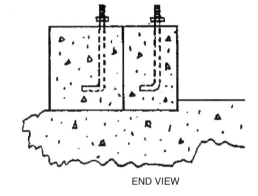

☐ **74**
Set anchor bolts in concrete for steel columns and posts so that the bolt threads extend into and below the surface of the concrete. This allows the bottom nut underneath the base plate to be adjusted all the way down to the concrete if necessary.

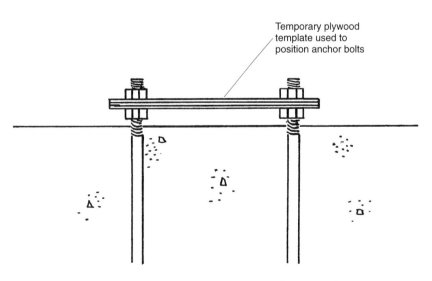

Continued on next page

Concrete Foundations and Slabs

Continued from previous page

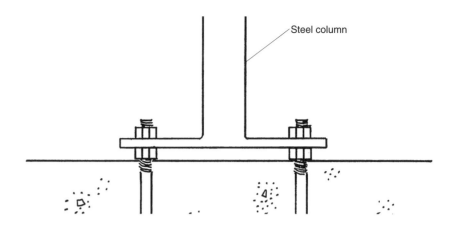

☐ **75**
A 3-inch plumbing pipe in a 3½-inch wall will result in a wall with a bow or bulge in the drywall in any area where there is a coupling. Walls with 3-inch plumbing pipes should be 2×6 walls.

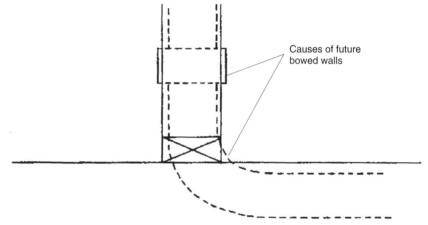

☐ **76**
Check for utility services that may pass too close to a concrete footing for a structural post. The post footing should be equal to or lower in elevation than the adjacent underground utility.

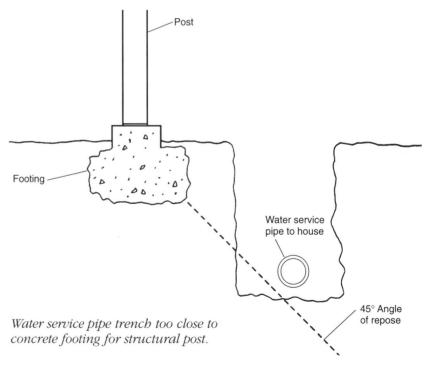

Water service pipe trench too close to concrete footing for structural post.

28 **Chapter 2**

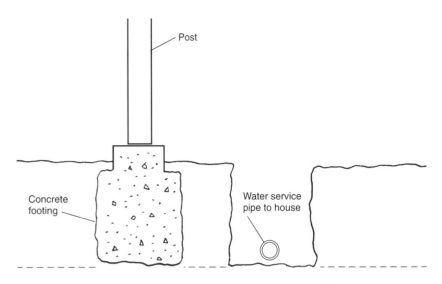

Concrete footing lowered to same elevation as trench for water service.

☐ **77**
At a split-level slab, provide a return at the step at a wall corner for the drywall to die into. Otherwise the edge of the drywall will be exposed and must somehow be finished.

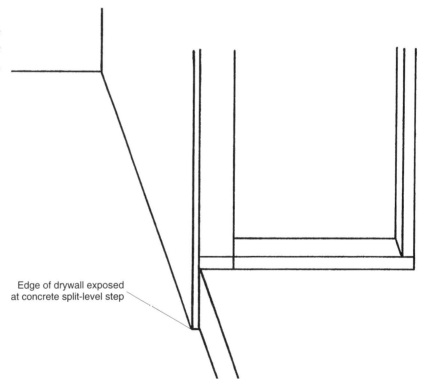

Continued on next page

Continued from previous page

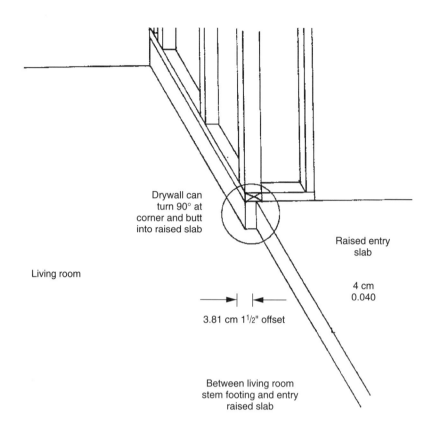

☐ **78**
Schedule survey staking for building foundation layout for a weekday other than Friday so that work can begin on the same day or the following day, since the project will be accessible to children over the weekend. This will help prevent children from pulling up the expensive staking to use as spears and javelins.

☐ **79**
Check that there is enough working clearance from adjacent slopes and verticals for the trenching equipment, especially on high-density projects with zero lot lines.

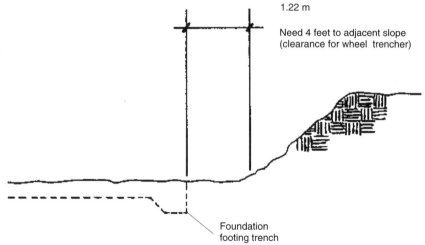

30 Chapter 2

☐ **80**
Block off the ends of building footings by using formboards in areas where excess concrete slough might interfere later with a masonry block wall footing rebar.

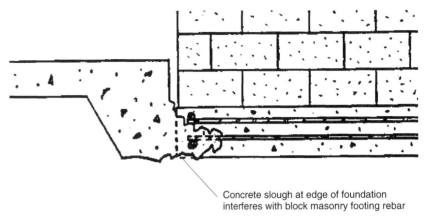
Concrete slough at edge of foundation interferes with block masonry footing rebar

☐ **81**
Use a custom-fabricated steel jig to position the anchor bolts in the concrete for steel moment frames and columns installed later on large multiunit projects. This gets the anchor bolts in exactly the right locations for easy structural steel erection later.

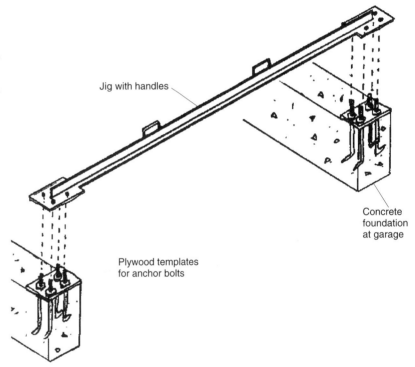

☐ **82**
Check that the ends of dryer vents contained within concrete slabs actually touch the surfaces of the formboards during the pouring of the concrete. This allows the ends of the dryer vents to be clearly visible after the formboards are removed and easily found later in the construction process.

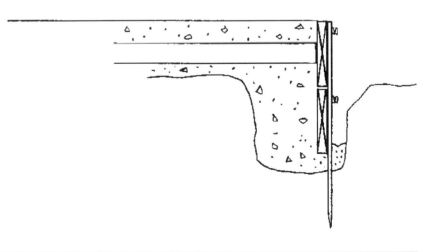

Concrete Foundations and Slabs

☐ **83**

When the sides of garages have wood siding or masonry veneer returns, the utility sleeves in the foundation concrete must be laid out and installed beyond the width of the returns so that the utility panels will be outside the siding or masonry.

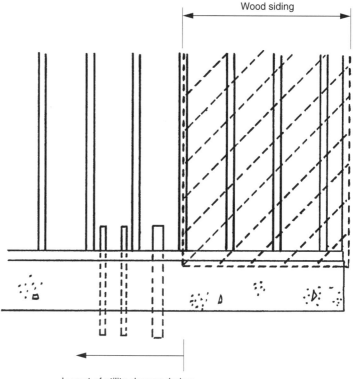

Layout of utility sleeves during concrete form setting allows utility panel boxes to fall outside of siding or masonry

☐ **84**

Use straight conduit for utility sleeves rather than using conduit with elbow bends at the bottom. You don't have to match straight conduit with the elevations of the various utilities within a joint trench.

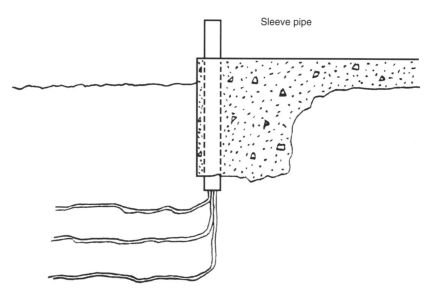

Underground cable entering utility sleeve pipes without 90 degree sweeps.

☐ **85**
Use the coupling or female end of conduit at the edge of concrete for easy insertion of conduit later.

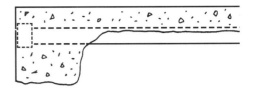

Pipe coupling embedded within concrete.

☐ **86**
Prevent low spots on garage slabs that create water puddles. Some homeowners will view this as a major construction defect, causing the builder to jackhammer out and replace sections of the garage concrete slab.

☐ **87**
Include the straightening and wire brush cleaning of anchor bolts and the grinding smooth of the outside perimeter edges of concrete slabs as part of the concrete subcontractor's activities.

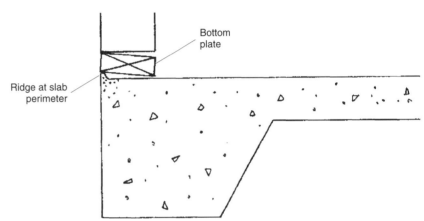

☐ **88**
Camber the edges of wood sleepers embedded in concrete foundations used as knockouts for plumbing pipe channels for easier removal after the concrete has hardened and the formboards have been removed.

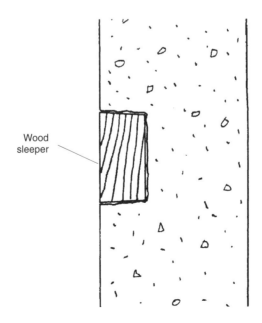

Concrete Foundations and Slabs 33

☐ **89**
Install cambered wood sleepers all the way down to the bottom and side of the foundation trench to displace the concrete and create an open channel that is free of excess concrete slough that otherwise would have to be broken out and removed later.

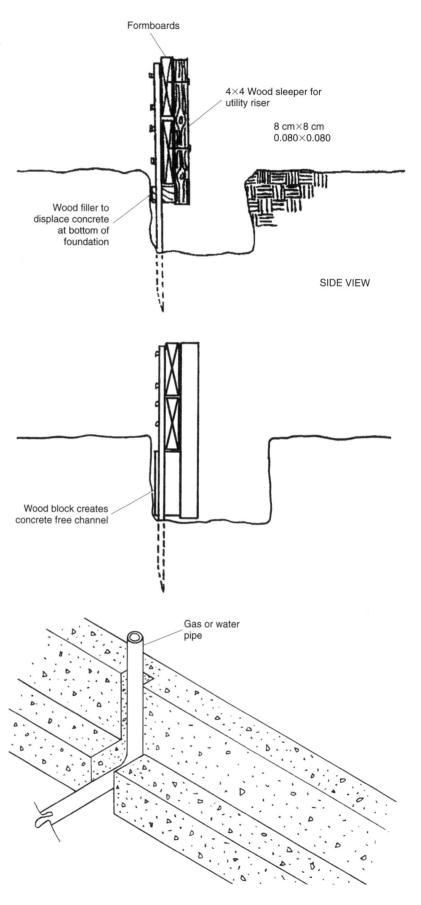

34   **Chapter 2**

☐ **90**
If lightweight concrete is scheduled to be poured on the upper floor levels after the drywall has been installed, schedule the pouring of preliminary lightweight concrete before the installation of drywall in areas that will be enclosed by the drywall and thus made inaccessible later; otherwise holes must be cut through the drywall, requiring patching later.

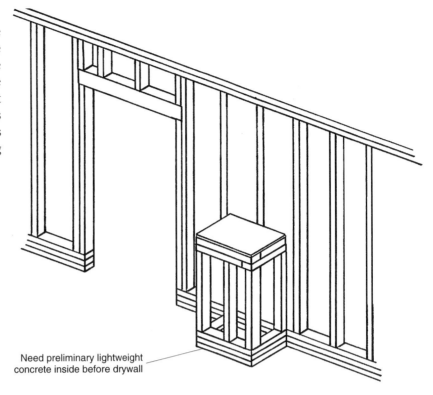

Need preliminary lightweight concrete inside before drywall

☐ **91**
Collect the anchor bolt nuts and washers directly from the concrete subcontractor and keep them in storage until the rough framing starts. If this issue is not addressed up front, the concrete subcontractor may simply leave the gunnysack bags or plastic bags containing the nuts and washers on the ground around one of the slabs, which can then be accidentally broken by other tradespersons, resulting in the nuts and washers being spread out on, mixed into, and lost in the loose dirt.

☐ **92**
For hold-down hardware for a post located at a split-level floor, use the correct hold-down type and set the anchor bolt at the correct elevation above the slab so that the bolts through the post are high enough up on the post to achieve the intent of the structural design.

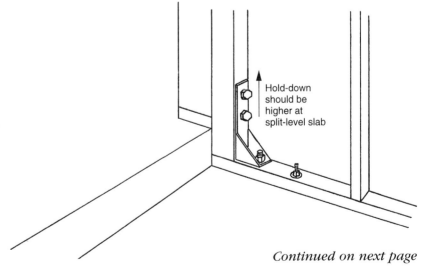

Hold-down should be higher at split-level slab

Continued on next page

Concrete Foundations and Slabs

Continued from previous page

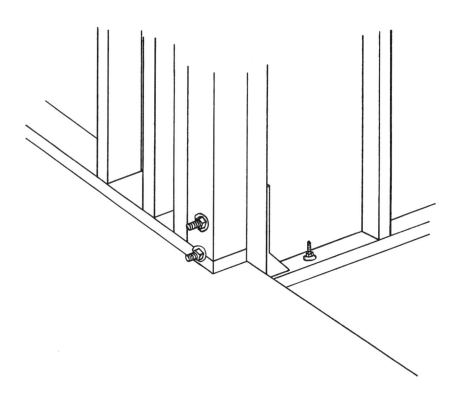

CHAPTER 3

Masonry Block Retaining Walls

☐ **93**
Make sure block retaining wall footings do not encroach into building footings that are poured later.

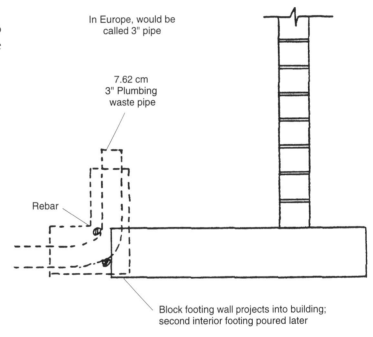

☐ **94**
Check that block retaining wall footings do not encroach into garage grade beam footings that are poured later.

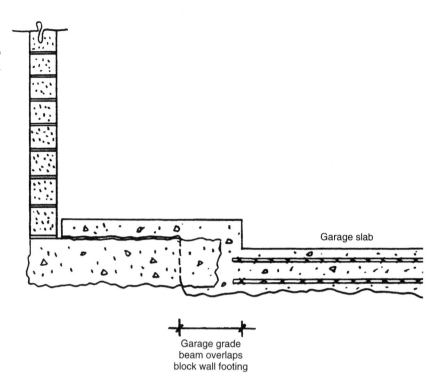

☐ **95**
Check that block retaining wall footings are low enough in elevation not to interfere with future underground plumbing pipes.

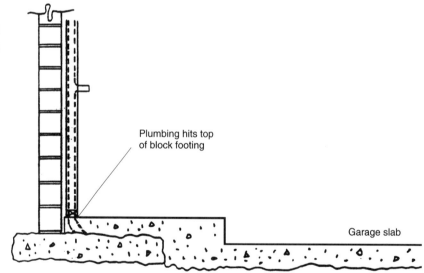

☐ **96**
Check that the block retaining wall footings as shown on the plans are turned away from landscaping.

Masonry block retaining wall footing turned away from planter.

☐ **97**
Make sure that the top-of-footing elevation for block walls provides enough depth of soil for landscape planting.

Masonry block retaining wall footing depth interferes with landscaping.

Masonry Block Retaining Walls

☐ **98**
Check that the block wall footings are designed and built at the correct elevation to coordinate with utility room conduit sweeps.

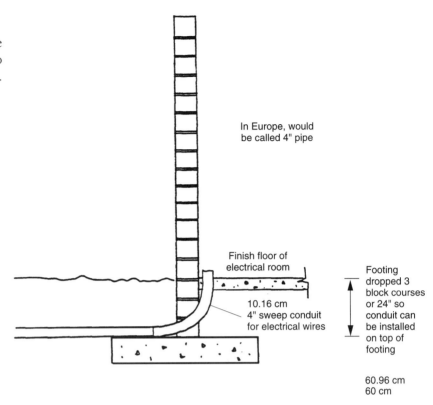

☐ **99**
In the structural details in the plans, coordinate the top of split-level block wall with the bottom elevation of the upper-level concrete slab to avoid having to cut the top row of masonry blocks.

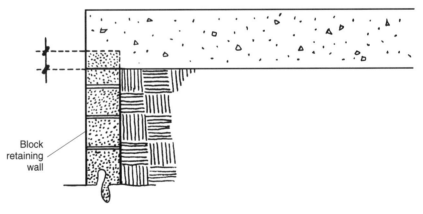

40   **Chapter 3**

☐ **100**
Design and install a good waterproofing system at block retaining walls.

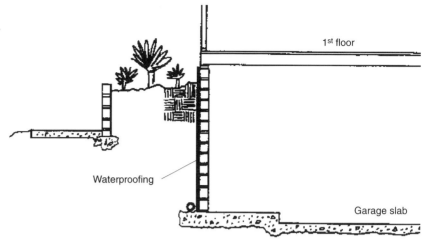

☐ **101**
During the staking and layout phase, before trenching, ensure that there will be enough working space behind the block wall for waterproofing and subdrain pipe installation.

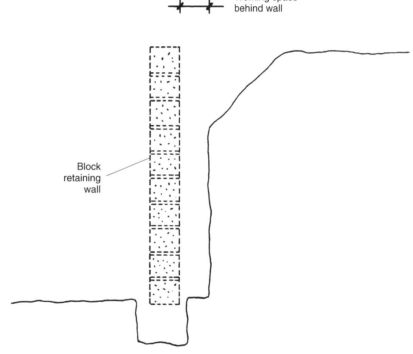

☐ **102**
Accelerate the masonry block construction schedule when design conditions require special inspection. This saves money by shortening the amount of time during which the special inspector is required on the job site.

☐ *103*
Coordinate masonry block work so that grouting occurs in the morning, when special inspection is required. This eliminates expensive time-and-a-half overtime for the special inspector for afternoon grouting that runs beyond quitting time but must be completed.

☐ *104*
Lay out the locations for the top of block wall anchor bolts so that the bolts miss large wood beams that are installed later.

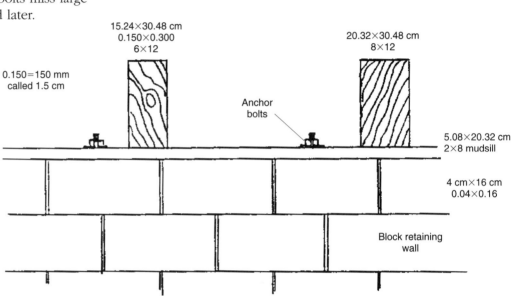

☐ *105*
Check that the grouting of the block wall cells does not result in grout spilling over the backside of the wall and on top of the block wall footing. This can adversely affect the installation and slope of the block wall subdrain pipe. Have the masonry subcontractor break off and remove this accidental spillage as part of the block wall construction.

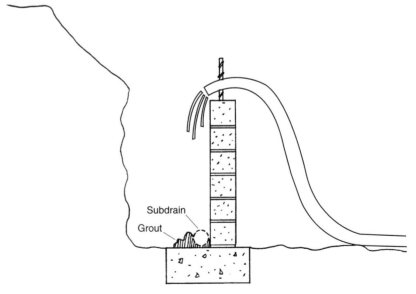

Masonry block grout spillage on back side of wall interferes with subdrain installation.

☐ **106**
Check that backfill soil is processed and compacted in accordance with the soil engineer's recommendations so that landscaping, concrete walkways, and A/C condensers do not settle and subside later.

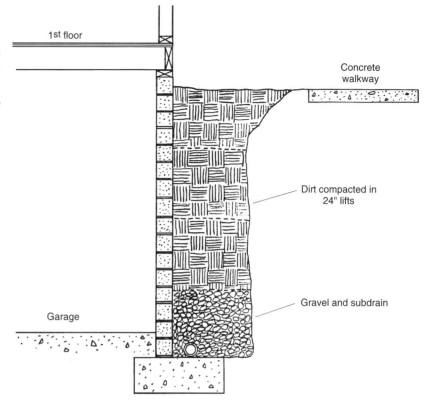

Backfill between building retaining wall and vertical slope.

☐ **107**
For masonry block walls that retain expansive soils, consider backfilling the walls using a granular type of soil, such as sand, to remove the expansive soil factor. Sometimes this allows the design of the block wall to be reduced, but one must consider this option early in the design phase to have it included in the engineering calculations.

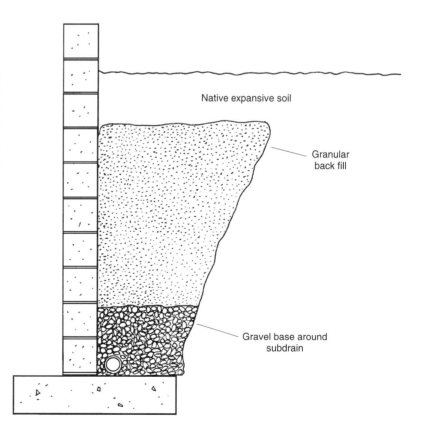

Use granular backfill (such as sand) instead of expansive soil behind retaining wall.

Masonry Block Retaining Walls

☐ *108*
For block wall subdrain plastic pipes that daylight out through a slope, use plastic caps over the ends of the pipes and drill holes through the caps to allow water to escape. This prevents rodents such as gophers from entering the subdrain pipes and blocking off the drainage flow with their nests.

CHAPTER 4

Layout and Wall Framing

☐ *109*
Laundry closets should have a minimum inside finish dimension of 34 inches when louvered bifold doors are used at their fronts. This provides enough clearance for the metal guides that align the bottom corners of the bifold doors.

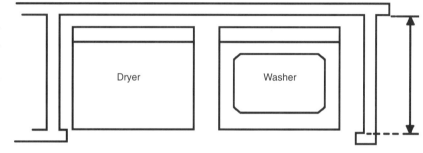

☐ *110*
Check the forced-air-unit furnace manufacturer's specifications for the minimum required clearances in front of and behind the furnace against the forced-air-unit closet dimensions shown on the plans.

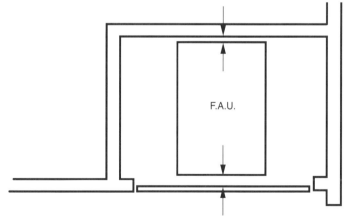

☐ **111**
For a bathroom with 45-degree corners, check that the toilet fits close enough to the back wall to engage the closet ring.

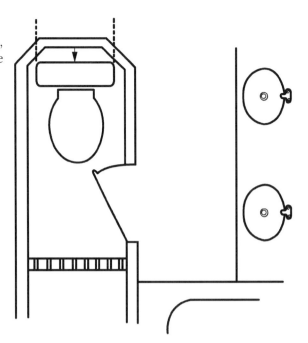

☐ **112**
The vertical chase in a walk-in closet should be deeper than the closet shelf and pole; otherwise the clothes pole rosette holders may overlap. If they are not, the rosettes must be cut to fit around one another or the poles must be moved closer to the closet walls, reducing the width of the space for hanging clothes.

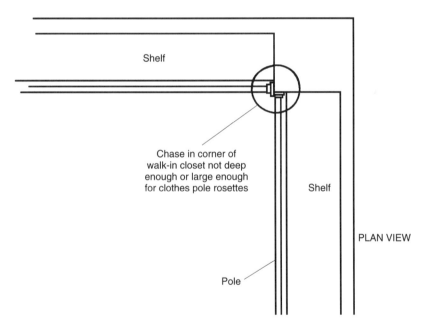

☐ **113**
Lay out a trash chute door inside a closet to avoid interference with the door leading into the closet. The trash room door and the trash chute should both be openable at the same time.

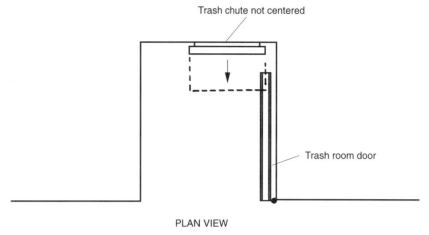

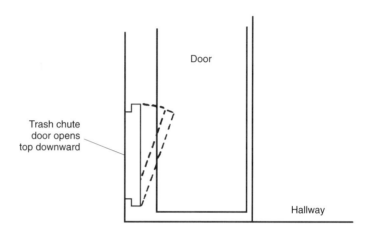

☐ **114**
When no plumbing pipes are involved, delete the pony wall at a kitchen countertop and use a cabinet with a finished back instead. This eliminates the need to frame the pony wall at the correct height for the cabinet countertop as well as the framing, drywall, taping, and painting costs.

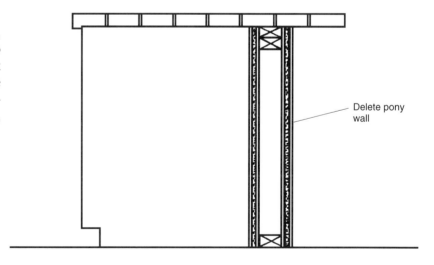

Layout and Wall Framing 47

☐ **115**
Check that a second floor hallway pony wall is high enough above the linen cabinet to provide a leftover wall reveal that is sufficiently wide to paint easily. You don't want a slim horizontal band of 1/8-inch or 1/4-inch wide drywall left between the pony wall's wood cap and trim and the top of the cabinet.

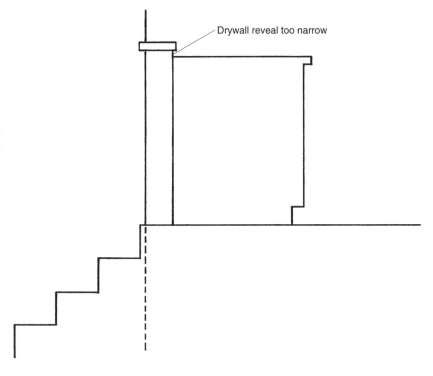

☐ **116**
Frame openings for bathtubs by adding a 1/2-inch extra allowance for installation and considering the thickness of the shear panel when applicable.

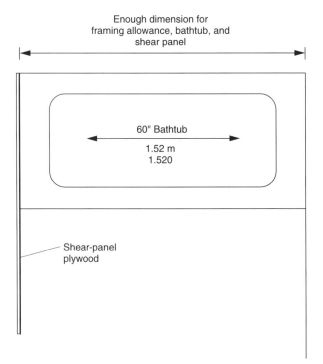

☐ *117*
Provide enough clearance in electrical meter closets for the meter glass covers and the plywood closet doors to fit.

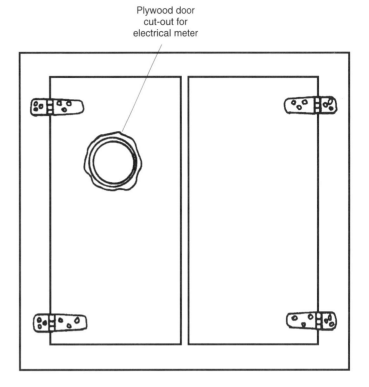

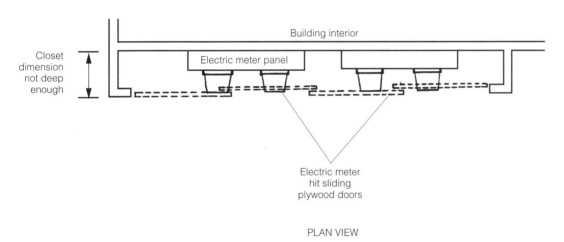

PLAN VIEW

☐ **118**

Have the architect or structural engineer use grid lines on the structural plans to help locate and align structural members from the concrete footings up through the roof.

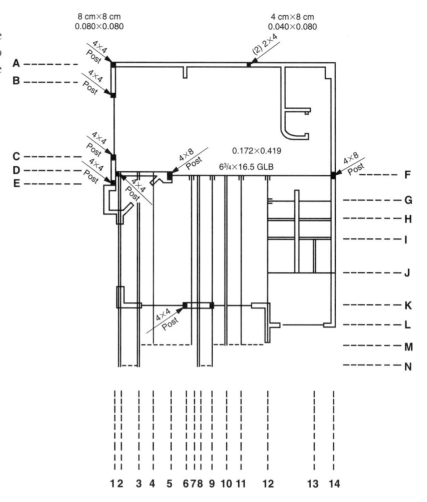

☐ **119**

Check for and revise metal strap details for wood posts that place the nailing on one side of the strap too close to the edge of the post because of the surrounding conditions.

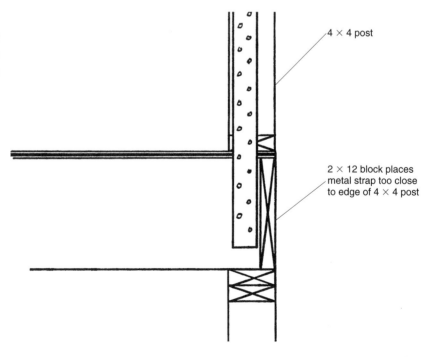

50   Chapter 4

☐ *120*
Check that the structural plans reflect changes in the architectural plans for standard and reverse floor plans, such as shear panel plywood at windows that change locations and sizes.

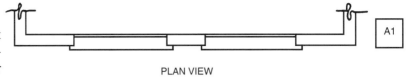
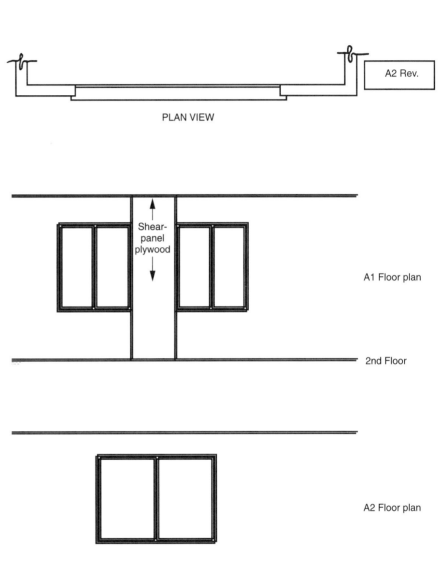

☐ *121*

Cut cripples uniformly on a radial-arm cutoff saw, not individually by hand with a portable electric saw merely because that is how it is included in the framing piecework price. Uniformly cut cripples of equal length help provide accurate rough framed window openings with uniform window frame drywall reveals.

☐ *122*

Straighten crooked hold-down bolts before standing up the wall framing, while there is still working space around the anchor bolts. A steel pipe can be used to bend the anchor bolts straight in areas where there is working space in all directions.

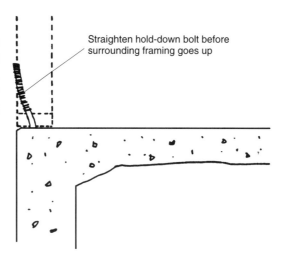

☐ *123*

Look for plumbing pipes that will end up slightly outside the wall framing during the snapping of the layout chalk lines and shift the walls over, if possible, to avoid bowed walls.

☐ *124*

Frame openings for medicine cabinets at least 3 inches away from a bathroom inside wall corner to provide clearance for swinging the medicine cabinet's door.

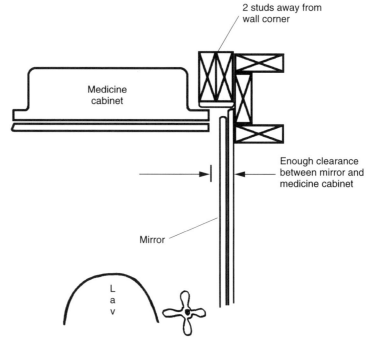

☐ **125**

Lay out the studs on the back wall of a wardrobe closet so that the shelf and pole support brackets can be centered within the closet.

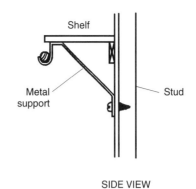

SIDE VIEW

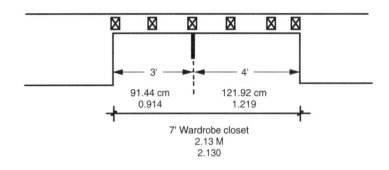

7' Wardrobe closet
2.13 M
2.130

PLAN VIEW

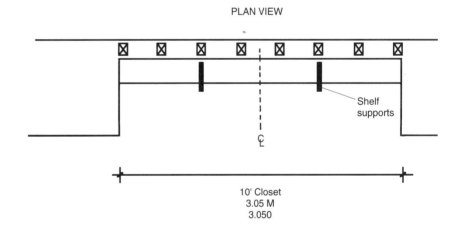

10' Closet
3.05 M
3.050

☐ **126**
Lay out the studs on the side walls of pantry closets to provide backing for the shelf support cleats.

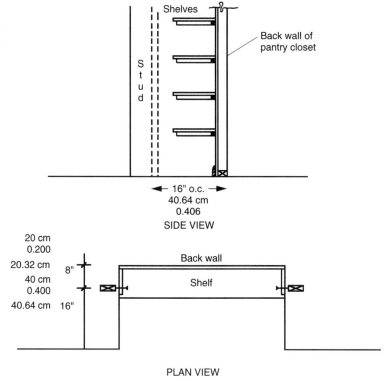

☐ **127**
Lay out studs on the bathroom wall to provide an open bay centered behind a pedestal sink to hide the plumbing.

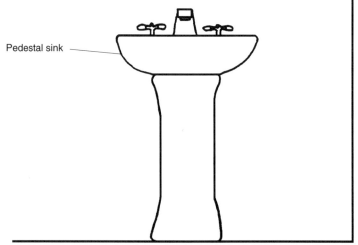

54   Chapter 4

☐ **128**
Design bases and ceiling soffits with enough dimensional width to provide adequate clearances around prefabricated columns for caulking and painting.

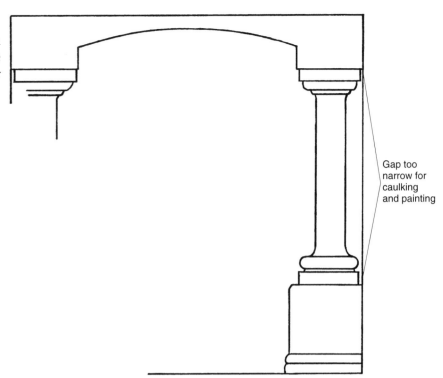

Gap too narrow for caulking and painting

☐ **129**
Consider the aesthetic appearance of uniformly spaced handrail spindles in designing the dimensions of columns and the leftover open spaces between columns. Small adjustments in the sizes of the columns can result in spindles that are all spaced 4 inches apart instead of 4 inches at the center and only 3 or 3^1/$_2$ inches at each end.

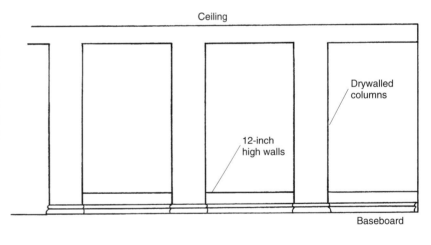

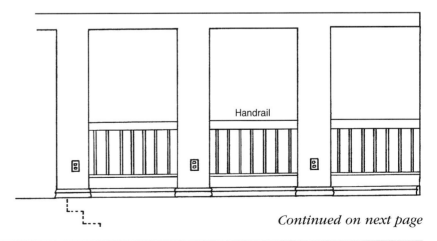

Continued on next page

Layout and Wall Framing 55

Continued from previous page

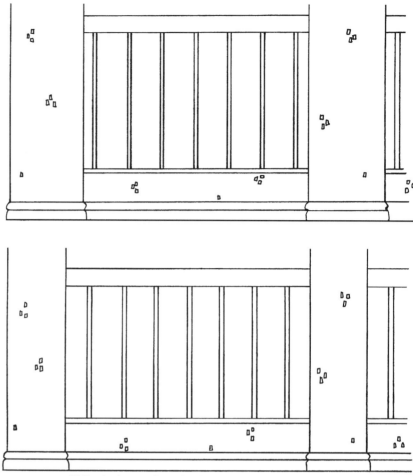

Width of columns adjusted, or handrail spindle spacing changed, to achieve uniform spindle spacing

☐ *130*
Check that the bathroom wall framing is square and that the bathtubs are installed square with the opposite wall. Bathtubs that are out of square with the bathroom walls become immediately apparent after square-patterned vinyl flooring or floor tiles are installed.

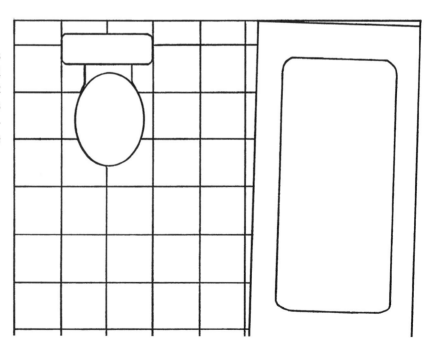

☐ **131**
Get the framing square, level, and plumb for an opening through a kitchen wall for a pass-through countertop. The surrounding cabinets that are installed later will expose drywall reveals that are out of square, level, or plumb.

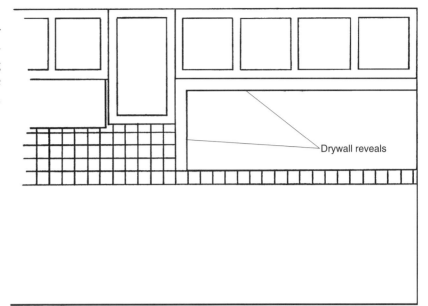

☐ **132**
Don't frame let-in braces at the same locations as 3-inch plumbing pipes, as both a 3/4-inch-thick let-in brace and a 3-inch pipe will not fit within a 3 1/2-inch wall.

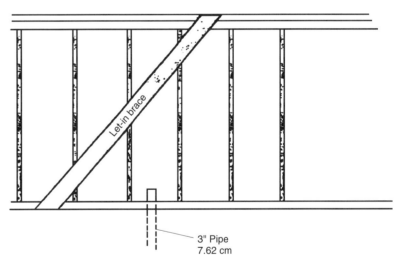

☐ **133**
Don't frame let-in braces at the same locations as electrical breaker panels installed later, during the rough electrical phase. The plastic sleeve conduits coming up through the garage foundation concrete identify the locations of breaker panels during wall framing.

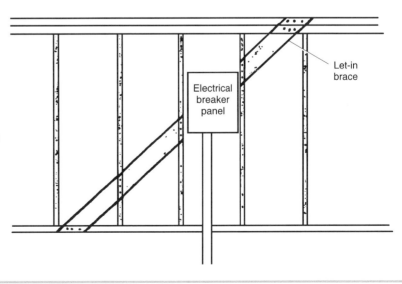

☐ *134*

When preliminary insulation is to be installed during the rough framing phase, order a few bundles of insulation and have them delivered to the job site to be installed by the framer. This is easier than trying to fit the insulation subcontractor into the rough framing schedule for a few hours of preliminary insulation work each week.

☐ *135*

Check that a bath door opens fully against a bathroom wall and does not hit the side of the bathtub during the last 6 inches of its swing, for example. If the bathroom wall and the door are about equal in width, the door should open into the wall, not the bathtub.

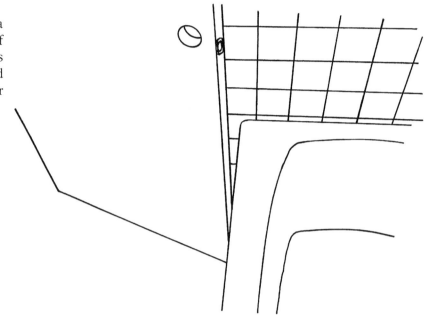

☐ *136*

Provide a method for identifying the lot numbers and floor plans of the houses in a way that is noticeable from the street. Spray-paint the lot numbers and floor plans on the garage door headers and the entry door concrete slabs.

CHAPTER 5

Framing Floors and Ceilings

☐ **137**
Don't mix cambered and noncambered structural members in the same floor system. This will result in floor humps.

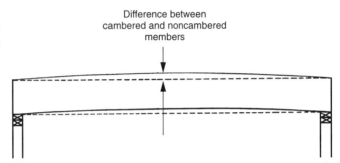

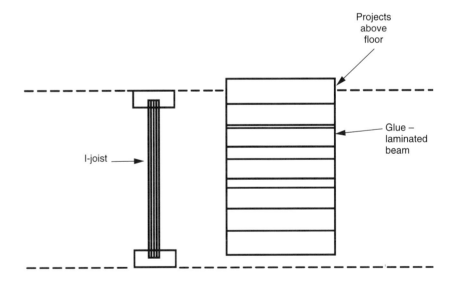

☐ **138**
Check for fulcrum-type deflection problems in areas where downward settlement at one location creates upward movement elsewhere, affecting a balcony deck or roofline, for example.

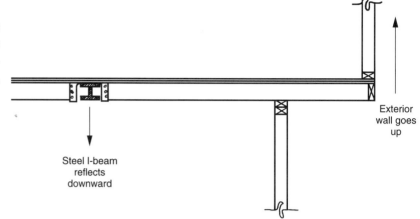

☐ **139**
Use the correct type of glue-laminated beam when the ends are cantilevered to support a balcony deck or another wall.

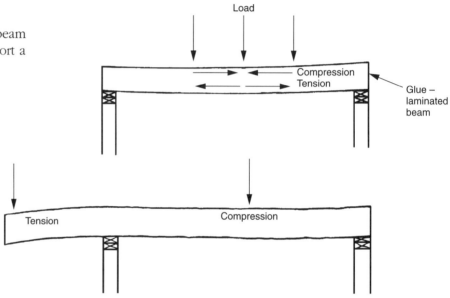

☐ **140**
Avoid the use of large-dimension lumber in the floor system. It may twist, causing a floor hump.

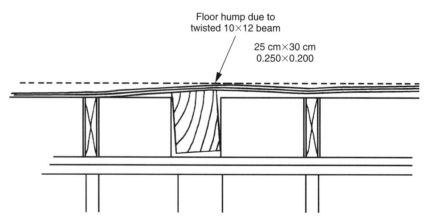

☐ **141**
Don't use large-dimension lumber for beams that will be covered with lath and stucco. When the beams twist over time, stucco cracks will result. Use glue laminated beams, laminated veneer lumber, or structural steel instead.

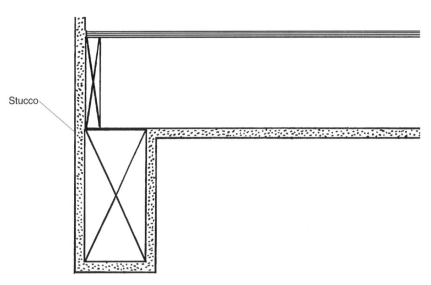

☐ **142**
Consider using drop beams instead of flush beams at certain locations to save the cost of joist hanger hardware.

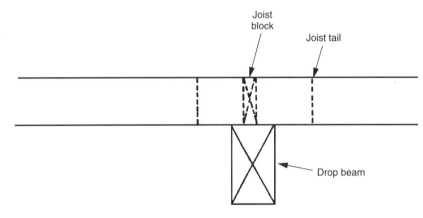

☐ **143**
Design the orientation of steel saddles and buckets so that the bolts and nuts end up inside the walls whenever possible. This prevents having to furr out walls to cover the nut and bolt ends.

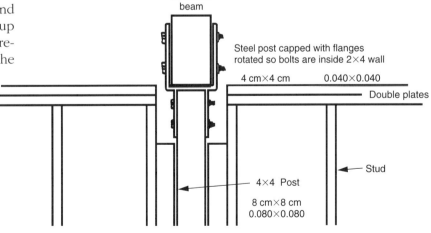

Framing Floors and Ceilings

☐ **144**
Check that steel hangers, buckets, and saddles for structural posts and beams that would otherwise be covered over when the exterior walls are stuccoed do not have their nuts and bolt ends exposed when the exterior wall covering is wood bevel siding, for example.

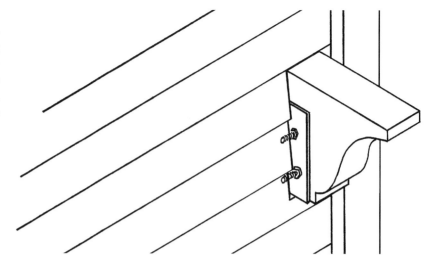

☐ **145**
Get the prefabricated wood I–joist design and layout from the manufacturer and obtain review and approval from the project structural engineer before steel I-beam fabrication with attached hangers and buckets. This prevents expensive field welding of additional saddles and buckets on the job site.

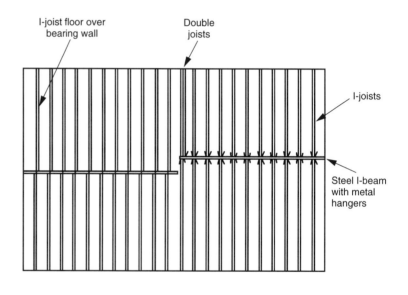

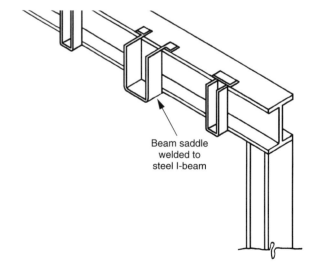

PLAN VIEW

☐ **146**
Allow enough time for an engineering review of shop drawings of structural steel and truss joist designs before the start of construction so that items with longer procurement lead times can be ordered.

☐ **147**
For 9-foot or higher ceilings with a mixture of assorted structural members, such as standard joists, double joists, glue laminated beams, laminated veneer lumber, and solid wood beams, consider using a metal resilient channel on the undersides of the ceilings as a means to straighten the ceilings.

☐ **148**
Lay out or head out ceiling joists for flush lights.

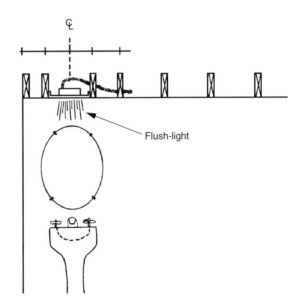

☐ **149**
Thick metal joist hangers cause bulges on the undersides of ceilings. To avoid bulges, the backsides of the ceiling drywall sheets must be cut out at the joist hangers or the undersides of the ceiling joists must be furred out. Include this activity in the scope of work sections in the subcontracts to prevent this from becoming "extra work."

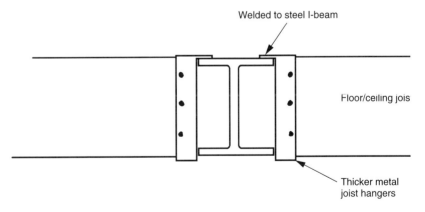

Continued on next page

Framing Floors and Ceilings 63

Continued from previous Page

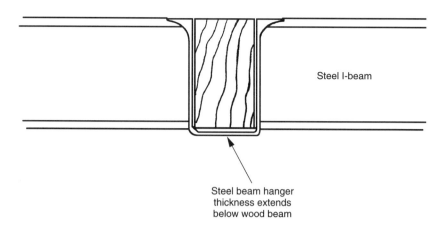

☐ 150

Frame laminated veneer lumber beams used as garage door headers at the correct height so that the exposed bottom "ply" edges can be covered with a finish wood trim.

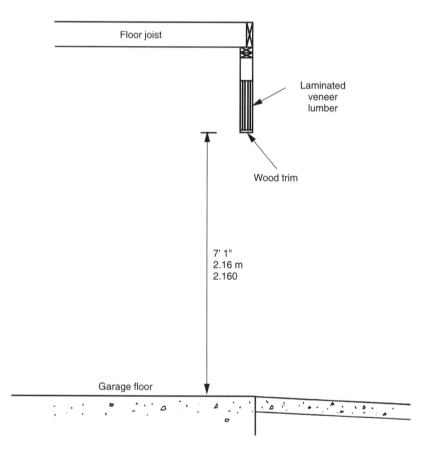

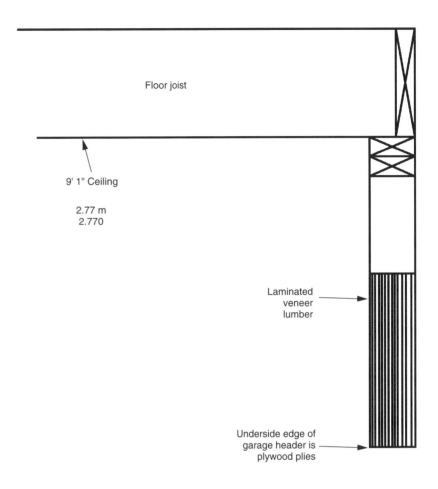

☐ *151*
Bolt up two or three laminated veneer lumber beams combined into one composite group before installation when the working space after installation will be tight.

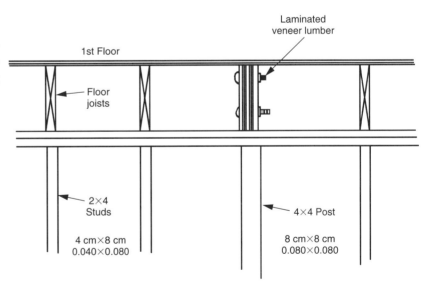

Framing Floors and Ceilings 65

☐ **152**

Make the slot cut at the end of the wood beam for connection with the steel I-beam flange the same dimensions as the thickness of the flange; otherwise the wood may break when the bolts are tightened.

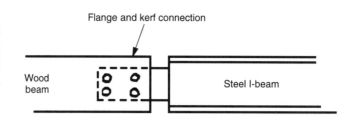

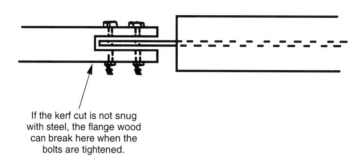

☐ **153**

Use a template to mark and cut out the notches that provide slope to the tops of balcony deck floor joists, measuring from the bottom upward to ensure uniformity. If they are always measured from the bottom upward, joists that vary in depth from $11^{1}/_{8}$ inches to $11^{1}/_{2}$ inches will not affect the straightness of the balcony deck floor. The amount cut off may vary, but the leftover depths of the joists will uniformly match the template.

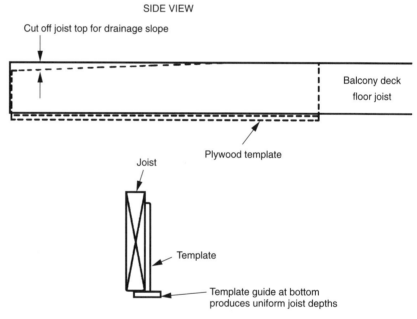

☐ **154**
For subfloors covered with lightweight concrete, cut out the second upper bottom plate in interior doorways before pouring the lightweight concrete so that the lightweight concrete can be finished flush with the remaining permanent bottom plate.

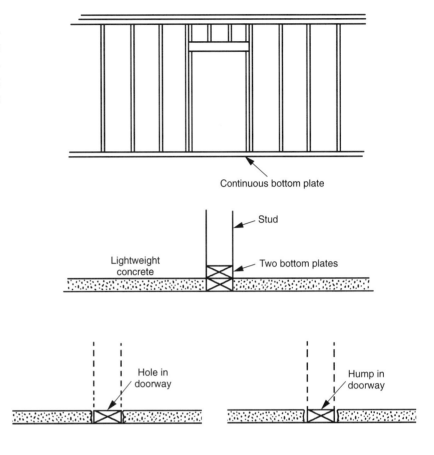

☐ **155**
Select aesthetically pleasing wood beams for exposed garage door headers.

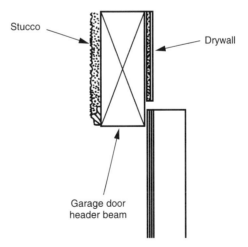

Framing Floors and Ceilings 67

☐ **156**
Get the twist portion of metal twist straps above the drywall ceiling line; otherwise the twist may project out beyond the drywall.

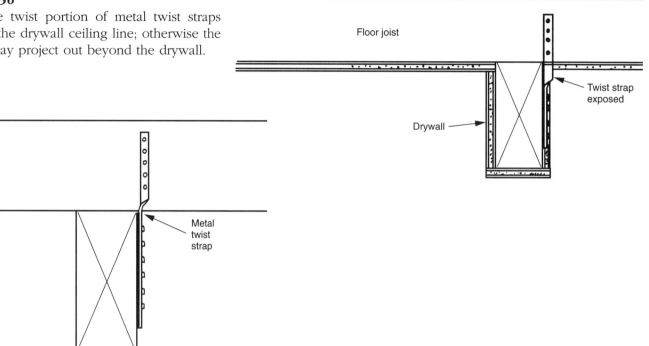

☐ **157**
Check the tops of floor joists and beams for straightness before installing plywood floor sheathing.

☐ **158**
Provide drip kerfs on exposed exterior wood beams.

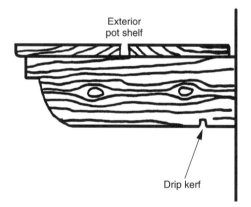

☐ **159**
For large-size second floor balcony decks, install area drains in the centers of the decks to allow for any downward deflection that may occur. If the drainage design is planned around scuppers at one perimeter side of the deck, the deflection in the floor framing may create low points in the deck, resulting in water puddles that cannot reach the scuppers.

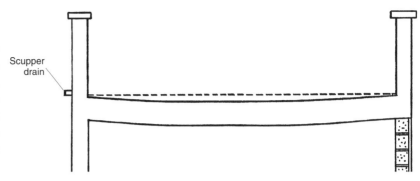

☐ **160**
Design and build the widths of kitchen ceiling dropped soffits to match the dimensions of the upper cabinets so that the drywall reveals around the cabinets are uniform.

Continued on next page

Framing Floors and Ceilings 69

Continued from previous page

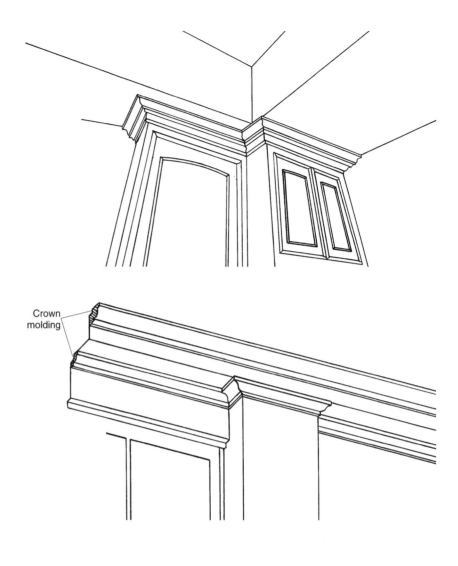

□ *161*
When a dropped ceiling soffit butts into a wall corner at a 45-degree angle, center the soffit with the corner.

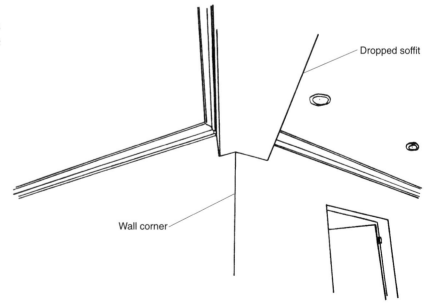

☐ **162**
When several sloping ceilings and dropped soffits converge at one location, analyze the various elevations while the project is in the design phase or early during the rough framing phase in terms of the aesthetics of the ceiling. If the aesthetics of the ceiling are not good, structural beam sizes or types can be changed or drop beams can be changed to flush beams to raise the beams above the ceiling line.

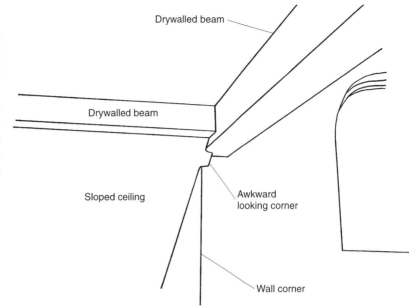

☐ **163**
To end up with equal drywall reveals on both sides of a prefabricated tapered column, consider the top and bottom dimensions of the column, the thickness of the drywall and the baseboard, and an additional amount for clearance between the baseboard and the column during the rough framing of the ceiling soffits.

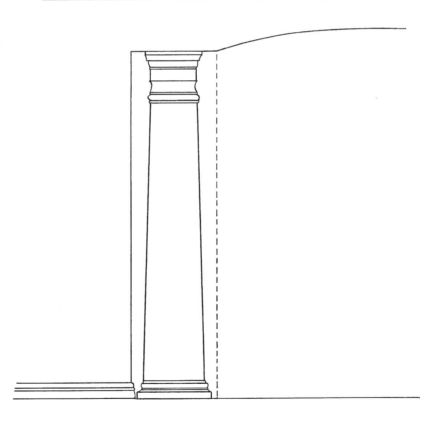

Framing Floors and Ceilings

☐ *164*
Check that arched soffits framed above interior doorways will have adequate clearance for the door casing.

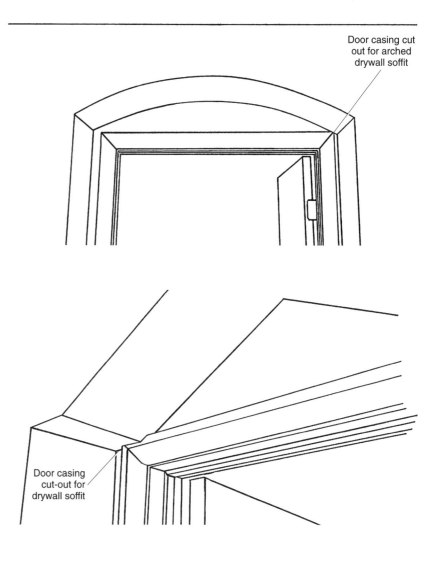

☐ *165*
When houses on hillsides have both concrete slab subfloors and conventionally framed plywood subfloors, make sure the elevations of the two floor systems will end up at the exact same height. This will prevent floor humps underneath carpeting at the transition between the two subfloors.

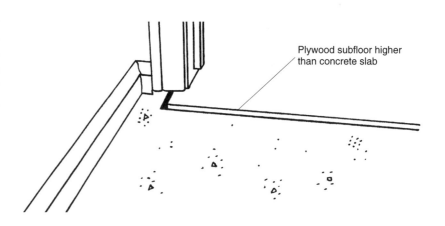

CHAPTER 6

Layout and Framing for Doors

☐ **166**
Keep the width variation of interior door sizes to a minimum to simplify ordering, spreading, and installation. For production tract housing, for example, coat closet doors that vary between 1-8, 1-10, 2-0, and 2-2 for different floor plans make it difficult to borrow the required missing door sizes from units out ahead to complete the units in sequence when deliveries of materials have shortages.

☐ **167**
For condominium and apartment projects with 8-foot-high ceilings in the hallways, stairways, and unit interiors, do not use floor-to-ceiling doors. They are too difficult to spread and install without hitting the ceilings and causing scratches or gouges in the drywall.

☐ **168**
Check hallways and closets to make sure there is enough room for the width of the door and the door casing.

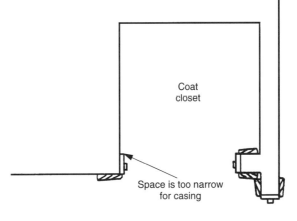

PLAN VIEW

☐ **169**
Check that there is enough wall room for a cabinet, scribe mold, and door casing.

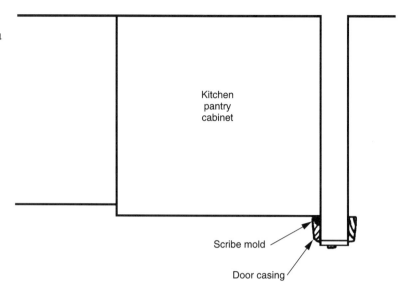

PLAN VIEW

☐ **170**
Check that there is enough wall room for a bathroom cabinet, countertop overhang, and door casing.

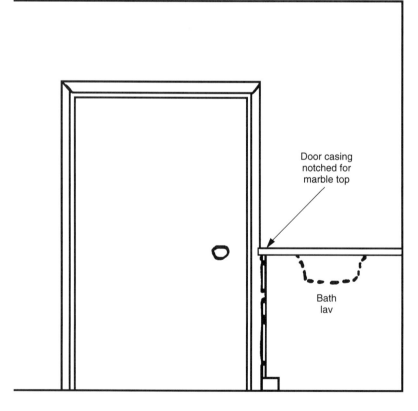

☐ **171**
Check that there is enough clearance between a toilet and a swinging bathroom door.

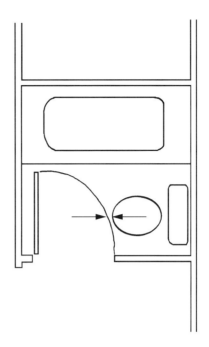

☐ **172**
Check that door openings are framed far enough away from interior wall corners to allow clearance for the door casing.

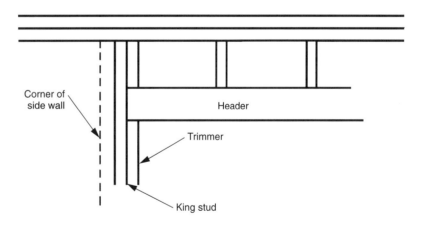

Layout and Framing for Doors 75

☐ *173*
Design roof overhang protection for stairway bulkhead doors that open onto the flat roofs of condominium or apartment projects; otherwise the door weather stripping will be the only waterproofing protection against rainfall hitting the door and leaking into the interior stairway.

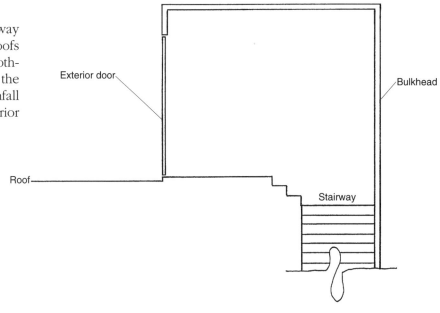

☐ *174*
Check plumb and alignment for single and double doors.

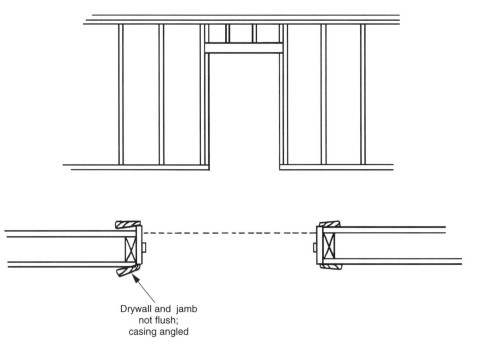

76   **Chapter 6**

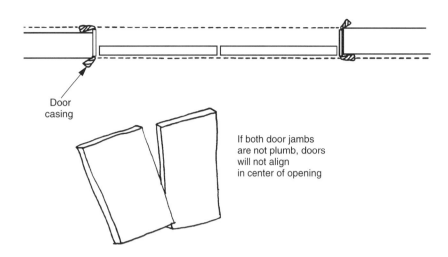

☐ **175**
Check that condominium and apartment corridor fire door jambs are plumb so that the door does not rub on the top of the carpet as the door is opened out into the corridor.

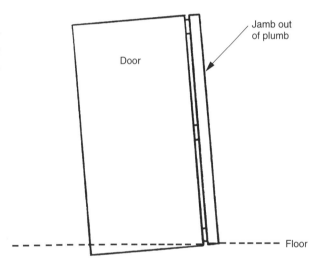

☐ **176**
Cut out bottom plates within interior doorways, using a handsaw, so that the ends of the bottom plates are flush and even with the sides of the trimmers.

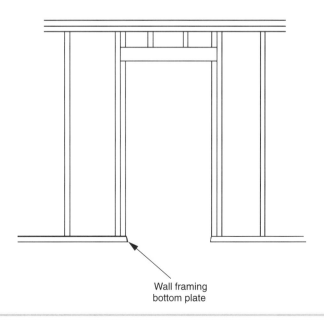

Layout and Framing for Doors 77

☐ **177**
Shim gaps between headers and trimmers at windows and doors by using small wood wedges.

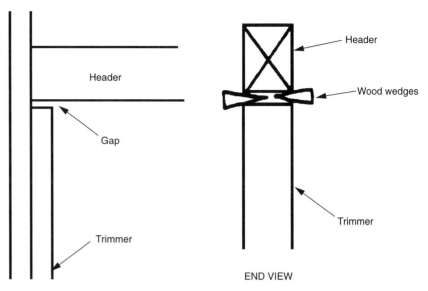

☐ **178**
Check that pocket door jamb sides are plumb so that the door slides at the same elevation in and out of the pocket channel and therefore does not rub against or bind the plastic bottom guide.

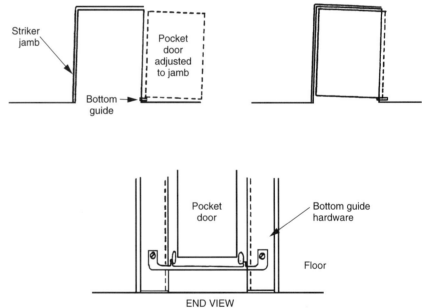

☐ *179*
The sides of laundry closets must be plumb for louvered bifold doors to open fully against each side without binding at the top.

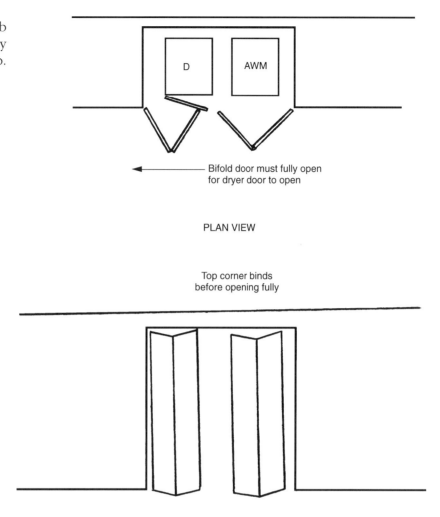

☐ *180*
Do not deliver wood jambs to the job site too early unless there is a protected place, such as a metal container bin or an empty garage, for storage until the jambs can be installed. If they are placed inside the garages of houses under construction, the jambs may get damaged.

☐ *181*
Install wood jambs as late in the construction as possible to minimize the opportunities for the jambs to be damaged by other workers carrying materials through the doorways.

☐ *182*

Shim the gap between the doorjamb and the trimmer or between the trimmer and the king stud at the location of deadbolt locks on exterior doors. This keeps the long deadbolt striker plate screws from pulling the jamb closer to the trimmer and widening the gap at the door.

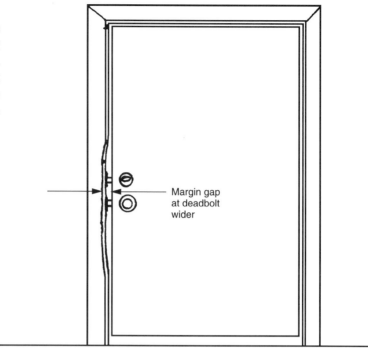

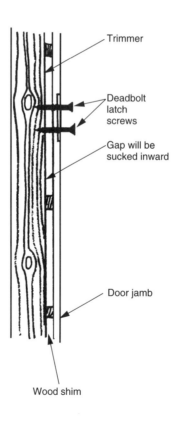

☐ **183**
Include in the concrete subcontract a provision for filling in with concrete the gaps between the backsides of the doorjambs and the raised foundation stem walls at garage openings. This provides a better appearance from inside the garage compared to looking at the exposed backside of lathing paper.

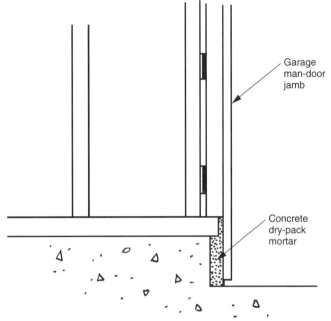

☐ **184**
Provide enough clearance at inside wall corners for the full width of detailed door casing trim in door openings that are next to one another.

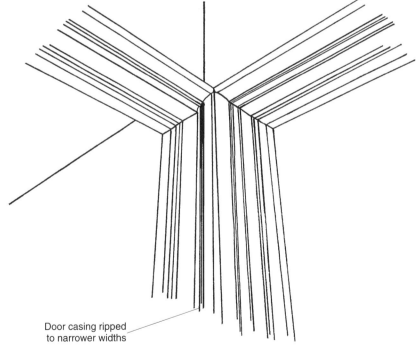

Layout and Framing for Doors

☐ **185**

For interior doorways with bull-nose drywall corner bead instead of door casing, frame the door opening in the center between the two side walls so that the baseboard returns on each side are equal in length. You don't want one side to have a 1- or 2-inch-wide baseboard return and the other side to have no baseboard. This difference is visually apparent with bull-nose corner bead at doorways that otherwise would not be noticeable when door openings have wood casing trim.

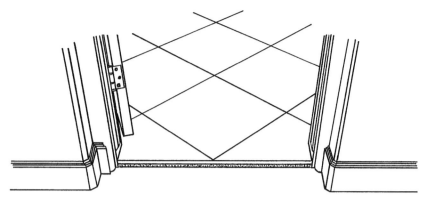

CHAPTER 7

Layout and Framing for Windows

☐ **186**
Keep variation in window sizes to a minimum to simplify construction.

☐ **187**
Whenever possible, avoid 8-foot-high dual-glazed sliding glass doors. They are too heavy to spread and install in large numbers in production tract housing for second or third floor balcony decks.

☐ **188**
If one tempered glass window size is duplicated elsewhere in the building, order all the other similar-sized windows to be tempered glass also to prevent a nontempered glass window from being mistakenly installed at a location that should have tempered glass.

☐ **189**
The leftover wall space reveals between windows above kitchen sinks and adjacent upper cabinets should be equal after the cabinets are installed.

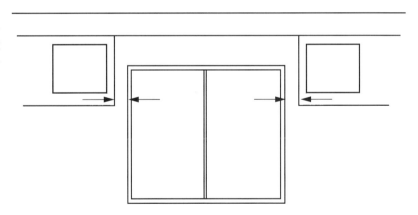

☐ *190*
Check that a window is truly centered below a gable roof peak when the floor plans show a dimension from the outside face of an exterior wall to the inside face of an interior wall for a bedroom, for example.

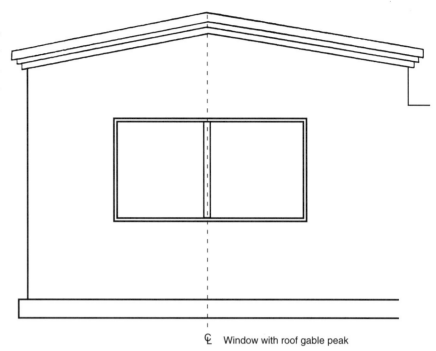

₡ Window with roof gable peak

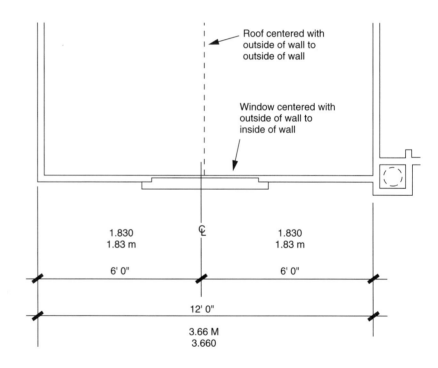

☐ **191**
Separate two adjacent windows at an inside corner to reduce the number of intersecting corners and thus simplify the construction.

☐ **192**
Check that there is enough dimensional width on short-length walls for the width of the window plus some wall reveal on both sides.

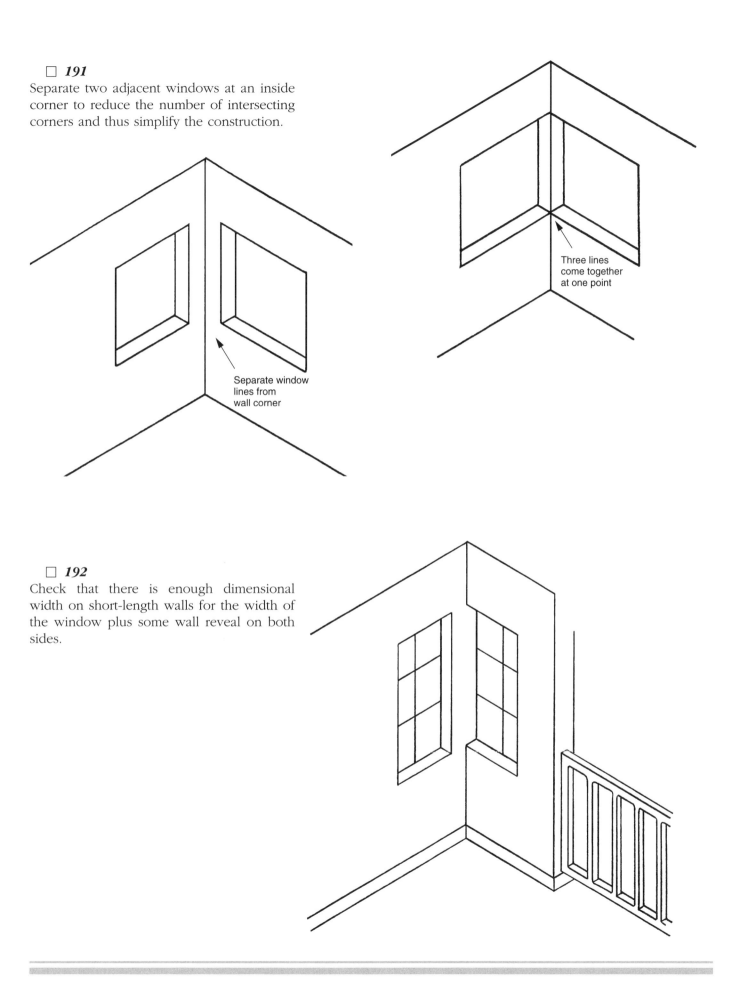

Layout and Framing for Windows 85

☐ **193**

Check that the top elevation of S-shaped or "barrel" roof tiles at the top of a sloping roof does not cause them to project up above the bottom of a window.

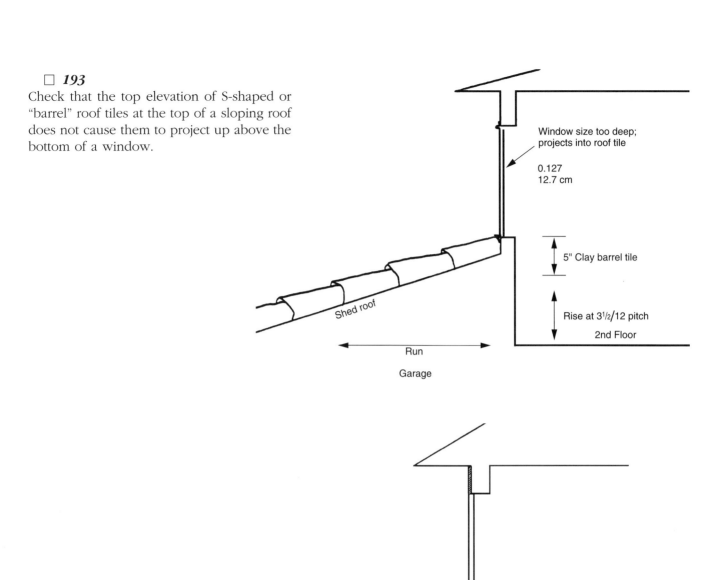

☐ **194**
Check that a cabinet countertop overhang does not project into an adjacent window. The countertop overhang dimension is not always shown on the plans.

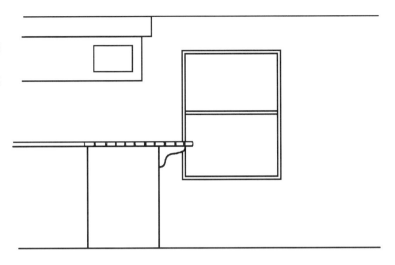

☐ **195**
Check that the architectural floor plans match the elevation drawings for header heights when 8-0 and 6-8 sliding glass doors are shown on the same walls. The header height for the window should be 8-0 or 6-8 but not both.

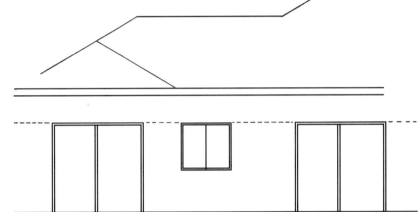

☐ **196**
Design and frame exterior windowsill pot shelves with a downward slope away from the windows to prevent water puddles on the sills that create the potential for leaks.

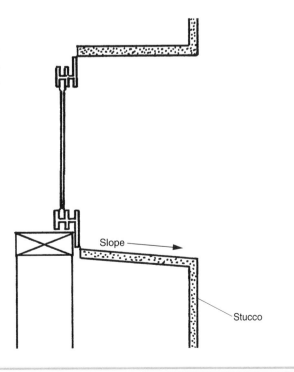

Layout and Framing for Windows

☐ **197**

Inspect window frames and/or sashes when they are delivered to the job site and off-loaded to check for dents, cracks, and scratches; identify these defects on the paperwork and send the paperwork back with the delivery truck. This prevents the window company from later claiming that the defective windows were damaged on the job site.

☐ **198**

Consider storing window frames and sashes in a metal container bin or the empty garage of a completed but unsold house until they are spread and installed. If they are placed inside the garages of houses under construction for too long a period before installation, the windows can become damaged as a result of the surrounding construction activity.

☐ **199**

Insist on complete window deliveries with no back orders of a particular size or type to be delivered later. Window shortages can disrupt the construction, as some activities, such as exterior wood siding or lathing, cannot be completed until the back ordered windows are delivered.

☐ **200**

When metal window frames are installed without the glass sash in place, check the installation for squareness before the exterior wall finish material, such as stucco or wood siding, is applied.

☐ **201**

When marble or Corian® is added to a kitchen windowsill, for example, the reveal at the window flange must be the correct size or the marble will extend up above the edge of the flange, not allowing the window to be removed.

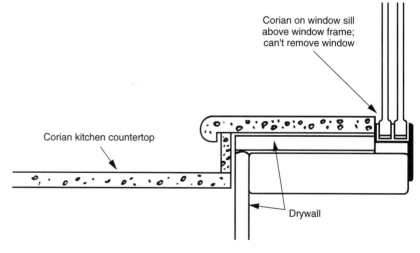

☐ **202**
Make sure the spaces around window frames during the rough framing will result in uniformly equal window frame reveals after the drywall is added later.

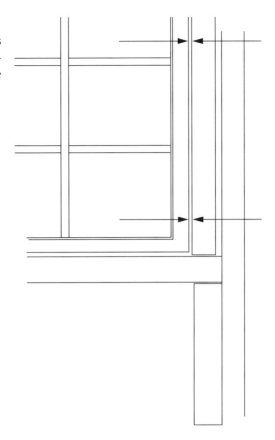

☐ **203**
Avoid small window frame reveals. They make it difficult to caulk the drywall-to-flange joint with an 1/8- or 1/4-inch window frame reveal.

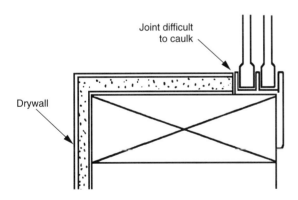
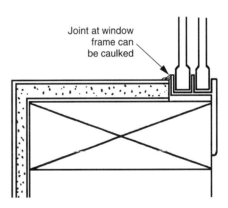

Layout and Framing for Windows

☐ **204**

Two adjacent windows at an inside corner that will be finished with stool and apron wood trim must be installed in line so that the bottom reveals will be uniformly equal after the stool is added to the windowsill; the adjacent ends of the stool must be at the same elevation.

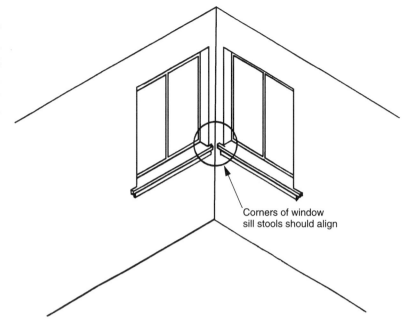

Corners of window sill stools should align

☐ **205**

When windows share plantons or casing, the window frames must be installed in line horizontally and vertically so that the window frame reveals will be uniformly equal.

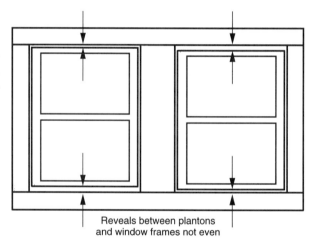

Reveals between plantons and window frames not even

☐ **206**
Get a window's lower header roughly framed at the correct elevation for kitchen countertop tile that extends onto the windowsill.

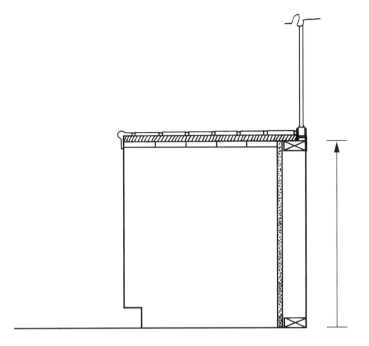

☐ **207**
Use one-piece plywood at window arches by eliminating cripples and raising the header up to the top plates. This provides more uniform reveals than does a two-piece plywood arch.

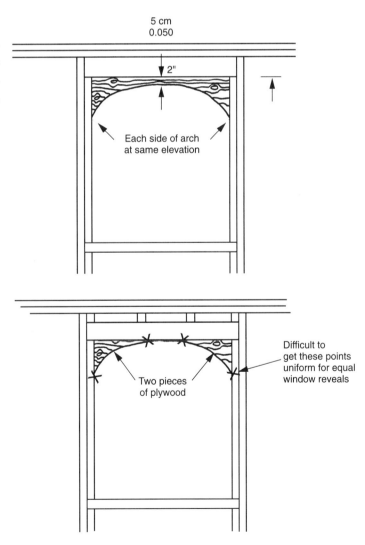

☐ *208*

Use plywood scraps to prevent canopy windows from being broken by falling objects during the course of construction.

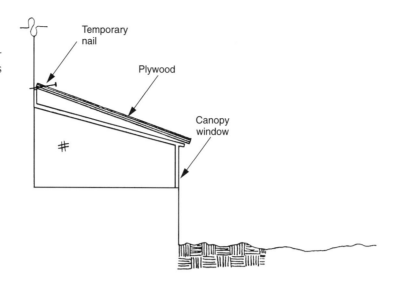

☐ *209*

When greenhouse canopy windows located at the kitchen sink come with midheight wood shelves and metal support clips, keep the shelves and clips stored in a safe place until the end of construction and then install the shelves along with the other move-in materials, such as showerheads and faucet aerators. This prevents the shelves from being stolen if they are left inside the greenhouse windows during the course of construction.

☐ *210*

For skylights on the roof of a house next to a highway or freeway, purchase skylights that are rated for sound insulation. It is no good to have dual-glazed windows for sound insulation and then have loud traffic noise come through the skylights into a master bathroom, for example.

☐ *211*

For windows with wood surrounds, design the windows with enough clearance from the adjacent wall corners that the wood surround trim can be full width around each window.

☐ *212*
Check that a window is not designed too close to an adjacent midheight roof overhang, resulting in the roof tiles extending a few inches into the window opening.

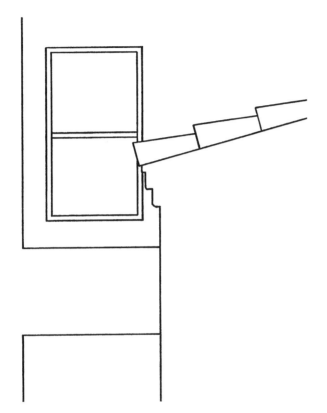
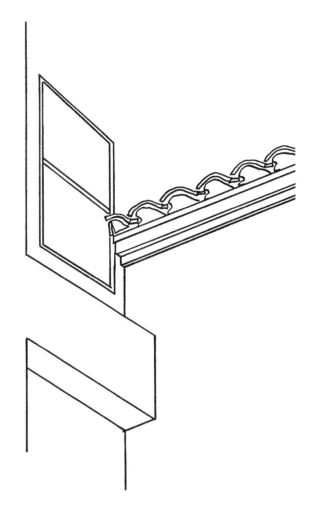

☐ *213*
For a small bathroom, check that a window that will be close to an adjacent mirror is installed perfectly plumb and level. The close relationship between the window and the mirror makes any difference between the two immediately apparent.

Layout and Framing for Windows 93

☐ **214**
For windows and sliding glass doors with wide, detailed wood casing trim, check that there is some clearance between the casing and a drywalled column pop-out. If the gap between the casing and the pop-out is 1/2 inch or less, it is difficult to caulk and paint the edge of the casing and drywall cleanly. A narrow gap like this also gives the appearance of an unplanned design and construction mistake.

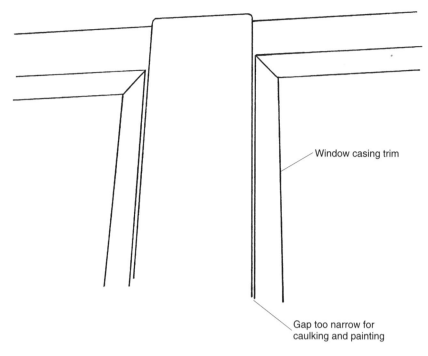

☐ **215**
For windows with wood casing trim, check that there is enough clearance from the countertop splash tile.

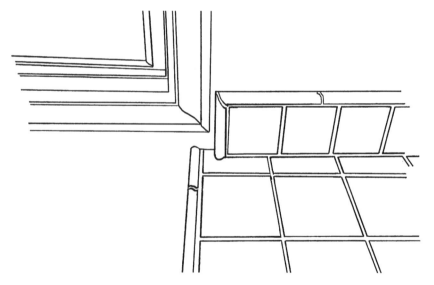

☐ *216*

For windows with wood casing trim, check that the casing will not project into the ceramic tile for a bathtub and/or shower. On the floor plans the window may look like it will miss the bathtub, but after a full-height vertical row of tile is added to the side of the bathtub and wood casing is added to the window opening, the tile and casing may overlap.

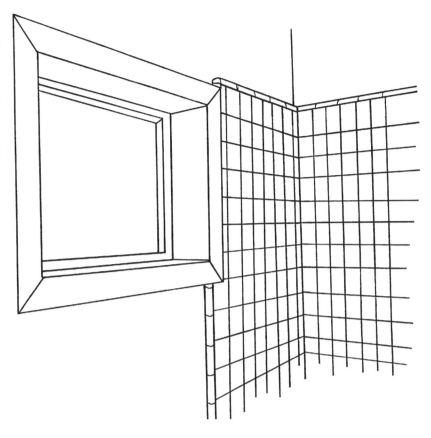

CHAPTER 8

Framing Stairs

☐ **217**
Avoid curved inside corners at stair landings, as they are difficult to frame, drywall, and tape. The wall should continue straight across yet still have pie-shaped steps.

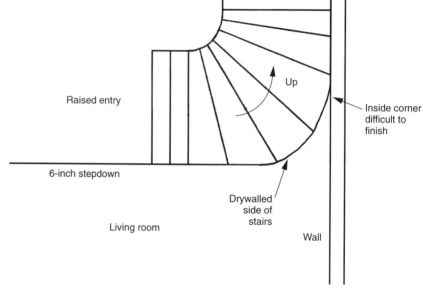

☐ **218**
Check for overspan of stair stringers that also function as ceiling joists for 1-hour-rated drywall used for the undersides of stairways that are exposed.

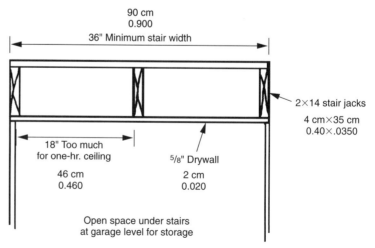

☐ **219**
Analyze the locations and height clearances of structural beams in ceilings above stairs in relation to the stair steps below. It is difficult to change the location or size of a structural beam during construction if it is discovered that headroom clearances are not adequate.

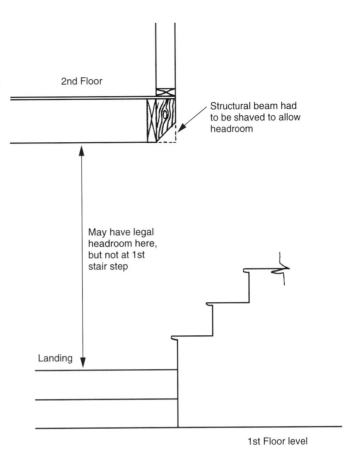

☐ **220**
Coordinate the arched ceiling height in the hallway at the top of a stairway with the stairway ceiling height to obtain an arch that is centered in the hallway. You cannot have a light fixture centered with an arch that itself is not centered between the two side walls.

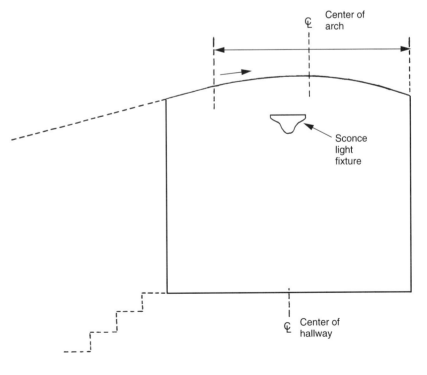

Framing Stairs 97

☐ **221**
Check that the stair handrail pony wall is high enough for the handrail. The pony wall must be framed at the correct height for the wood cap, apron, and handrail to fit.

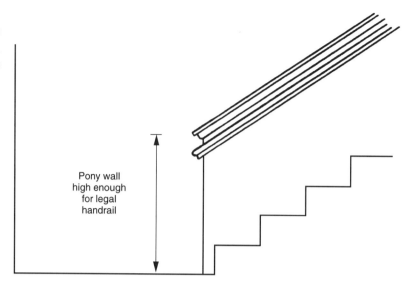

☐ **222**
When stairs get skirtboard, fur out the stair stringer or jack 1¹/₂ inches from the wall framing to provide an open channel for the skirtboard to slip down into without having to be saw cut to match the steps.

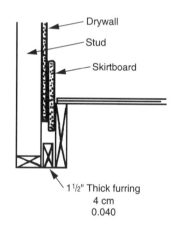

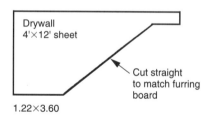

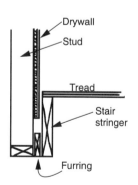

98 **Chapter 8**

☐ **223**
Use thicker plywood at pie-shaped stair treads at overspan areas or add blocking to the stair framing to reduce the spans in order to eliminate weak and spongy plywood areas.

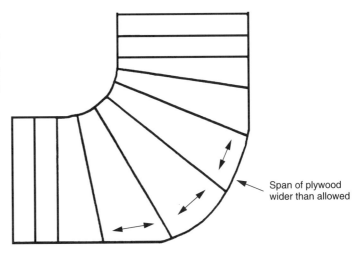

Span of plywood wider than allowed

☐ **224**
Adequately attach and/or bolt handrail pony wall posts to the underfloor joists at the floor level and the stair landing to provide stability to the pony wall.

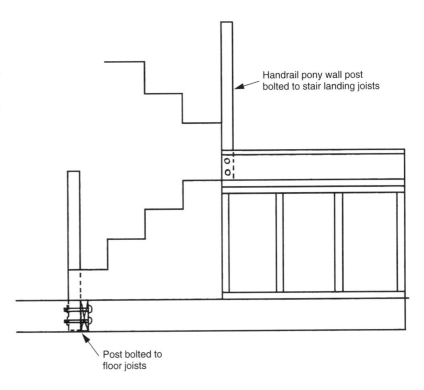

Handrail pony wall post bolted to stair landing joists

Post bolted to floor joists

Framing Stairs

☐ **225**

Stair plywood treads and risers should not extend into the skirtboard channel and thus interfere with the future skirtboard installation.

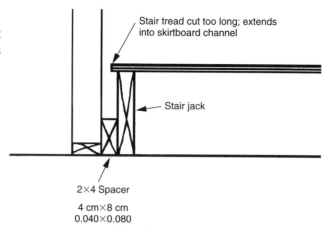

☐ **226**

Design and build the stairs so that the bottom step ends up several inches short of the wall corners. That way the carpet that wraps around the stair steps also ends up short of the wall corners and does not extend into the hallway, for example.

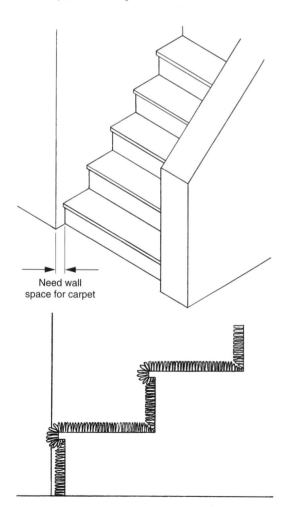

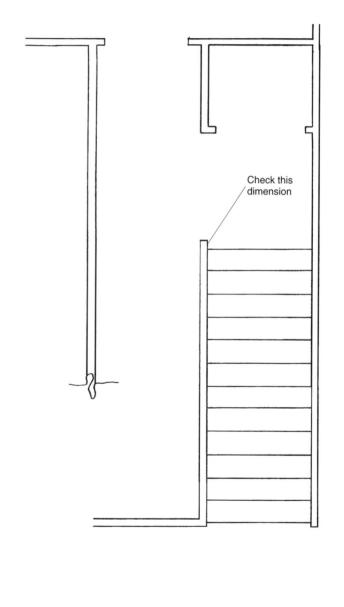

Chapter 8

☐ **227**
Check the vertical rise and horizontal run of a stairway to determine whether the bottom step actually will end up in the location shown on the plans. If the stairway will keep half of an entry double door from fully opening, discover this during the design phase so that changes can be made on the plans.

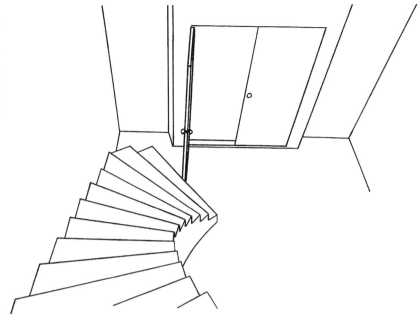

☐ **228**
Check the vertical rise and horizontal run of a stairway to determine whether the top step actually will end up where it is shown on the plans and will not extend partially into a sliding glass door opening for a second floor balcony deck, for example.

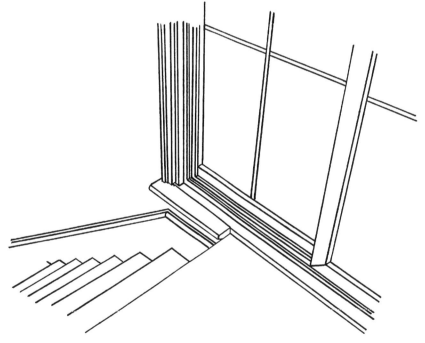

☐ **229**
When the stair steps separating a split-level floor get wood flooring, plan to locate the riser at a 45-degree wall corner to anticipate the projection of the tread nosing. Hold back the first step riser several inches from the wall corner during the framing phase to keep the nosing from ending up centered on the wall corner, which will look like a mistake.

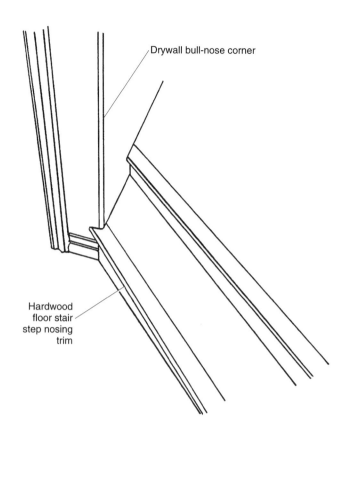
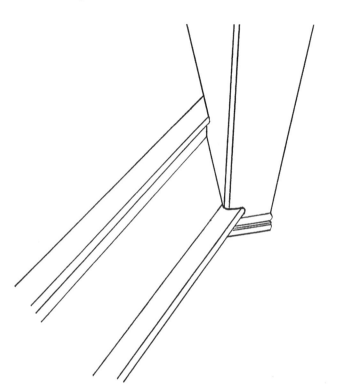

☐ **230**
Don't frame the riser for a stair step at a split-level floor in line with the adjacent pony wall. This will provide no wall space for the carpeting to tuck into.

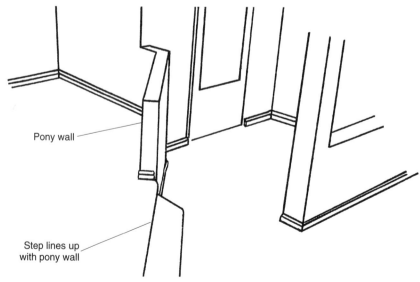

☐ **231**
At a transition at a stairway from a full-height wall to an open area, check that the stair steps will work out so that a hardwood tread nosing will not extend beyond the corner of the full-height wall, exposing the end grain of the hardwood nosing.

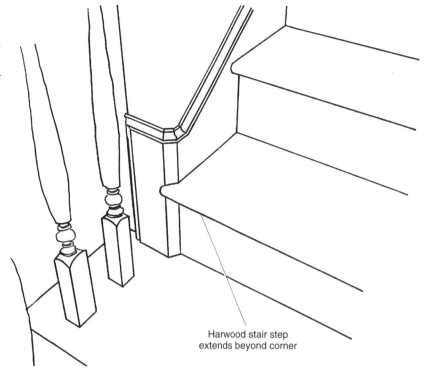
Harwood stair step extends beyond corner

☐ **232**
Check that the bottom stair step at a mid-height landing or at the bottom of the stairs is held back far enough from the wall corner that a hardwood border piece will not project beyond the corner.

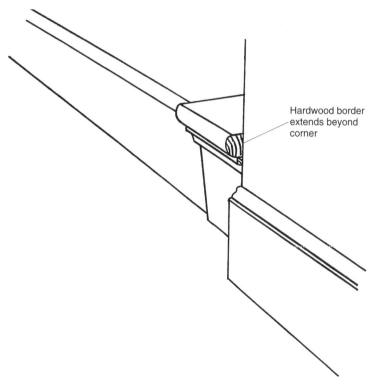
Hardwood border extends beyond corner

Framing Stairs 103

CHAPTER 9

Shear Panel

☐ **233**
Consider conditions where the second floor wall framing can be cantilevered slightly beyond a first floor wall to end up flush with the first floor's shear panel plywood. This eliminates the need to furr out the wall framing on the second floor.

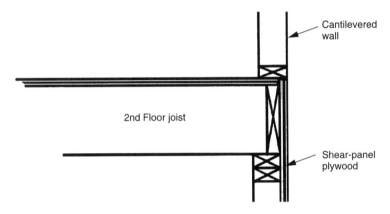

☐ **234**
Analyze the possibility of upgrading the shear panel design to reduce the overall lengths of shear walls at locations where furring can then be eliminated. Upgrading the thickness of the plywood and the nail spacing pattern may allow the shear panel to stop at a wall intersection and not have to continue partway into a hallway, for example.

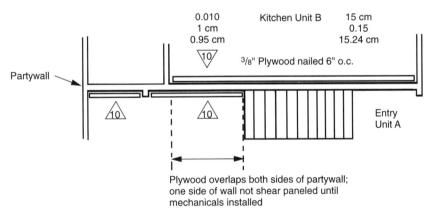

Continued on next page

Continued from previous page

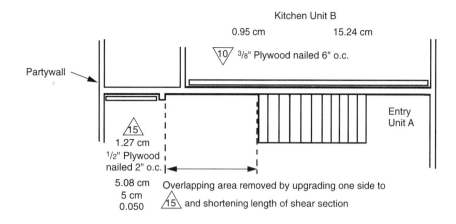

☐ **235**
Analyze the locations of attic access panels in the ceilings when the structural design calls for shear panel plywood sections in the roof attic space that block off access to areas in the attic. When this occurs, get the structural engineer to design approved openings through the shear panel attic walls or look for additional attic access locations in the ceiling.

☐ **236**
Minimize or eliminate 2-inch and 3-inch shear panel nailing spacings that require 3-inch-thick mudsills, anchor bolts placed higher, and 3×4 studs at the plywood vertical joints. Keeping shear panel nail spacings at 4 inches or more simplifies the construction.

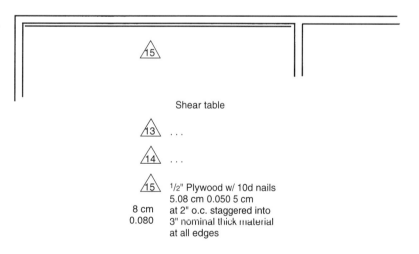

Shear Panel 105

☐ **237**

Avoid conditions where shear panel plywood extends behind items such as prefabricated fireplaces. This makes it more difficult to coordinate the installation, nailing, and inspection of the shear panel with the delivery and installation of the fireplaces, especially with party wall 1-hour drywall fire walls behind fireplaces, for example.

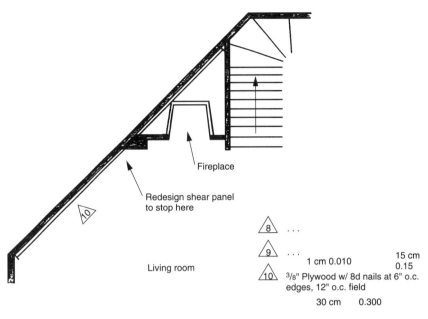

☐ **238**

Change the wall framing stud layout to accommodate shear panel so that the first sheet of plywood can be nailed up without having to be cut to a narrower width. If the same floor plan or condition is repetitive, this can save a lot of time and material on a large housing tract.

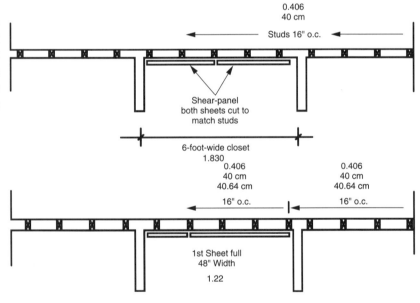

☐ **239**

Don't cover both sides of a wall (usually at the garage front) with shear panel plywood until the interior anchor bolt spacing and/or hold-downs have been inspected by the building inspector. You must divide the shear panel for two installations and two inspections or get the hold-downs and anchor bolts inspected separately during an earlier inspection.

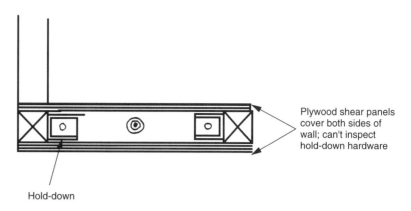

☐ **240**

Plan in advance with the building inspector to divide shear panel inspections when necessary to improve the coordination of the work. Draw in new signature lines on the back of the inspection card to document those inspections.

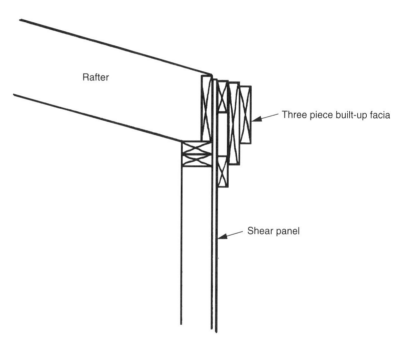

CHAPTER 10

Fireplace Framing

☐ **241**
Make sure the required manufacturer's clearances around the fireplace boxes and flues match the architectural dimensions provided on the plans. This is especially critical for fireplaces that are open on two sides and separate two living areas, such as a family room on one side and a living room on the other side.

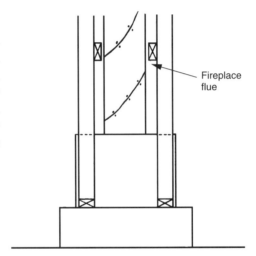

☐ **242**
When the vertical chase for a fireplace flue is offset horizontally from the location of the fireplace because of the floor plan of a second and/or third floor above, check the manufacturer's specifications to determine whether the sheet metal flue has the number of angle pieces required to reach the chase.

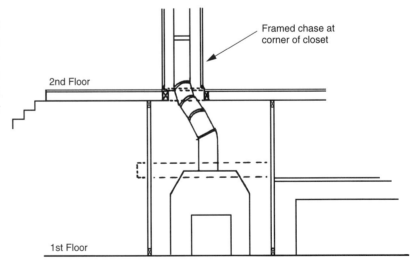

☐ **243**
Analyze the required locations for wood backing at corners for decorative wood faces that have plantons wider than a standard three-stud corner. A 2×12 planton places the edge of the wood siding beyond the width of a three-stud corner.

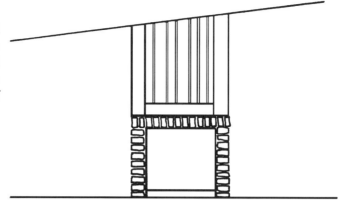

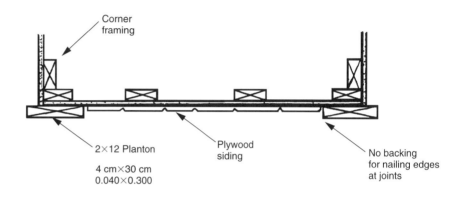

PLAN VIEW

☐ **244**
Analyze the fireplace mantel and pop-out dimensions in relation to the sizes of the decorative tile and/or marble inserts and firebox reveals to prevent the occurrence of exposed cut tile edges.

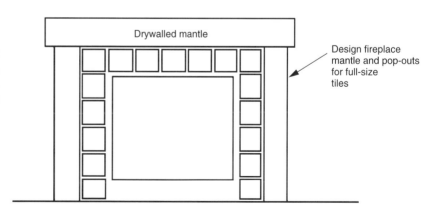

Fireplace Framing *109*

☐ **245**

Check that the faces of prefabricated metal fireplaces will end up flush with the drywall so that there will not be wide caulking joints after the tile and/or marble is installed over the fireplace faces.

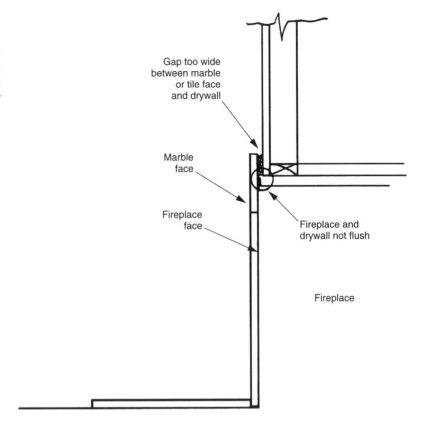

☐ **246**

Support second floor fireplaces with floor joists so that the front faces are plumb. Fireplaces should not be supported by rim joists and plywood alone.

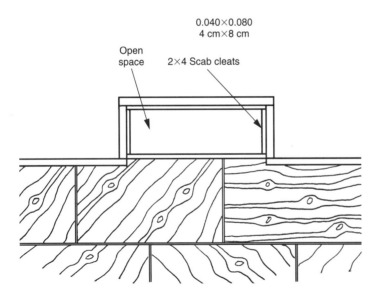

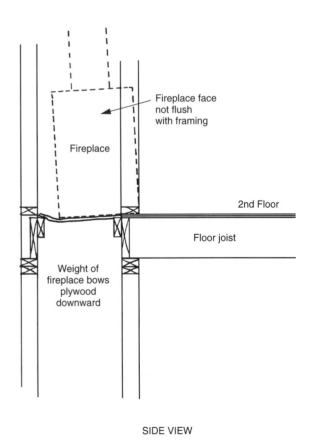

☐ **247**
Add plywood to the top of a sloped area at a fireplace chimney for structural support so that workers using this area as a step to reach higher wall areas do not damage the lath paper or stucco.

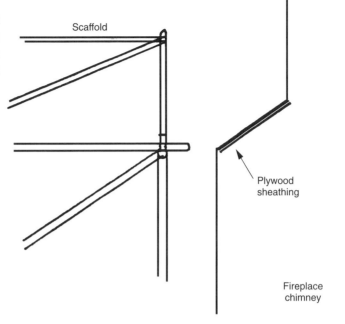

☐ **248**
For the bracing of fireplace chimneys, use plywood scrap pieces instead of 1×6-inch let-in braces to keep the chimney framing rigid and plumb until the stucco plastering solidifies the chimney; plywood is stronger than let-in braces for this application.

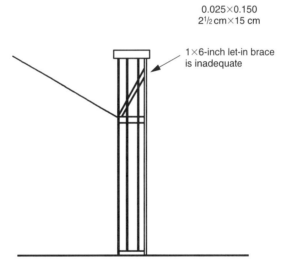

Fireplace Framing *111*

☐ *249*
Design and frame the elevation of the top of the fireplace chimney stack on the roof so that the top piece of the sheet metal flue works out to be a standard-length piece with a factory joint. This prevents makeshift joints at the top of the fireplace flue that can leave holes open to sparks that could cause a fire.

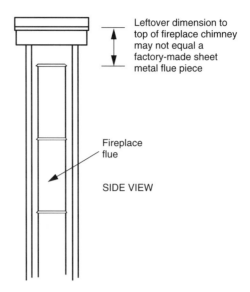

☐ *250*
Consider covering the firebox openings of fireplaces with plastic masking as soon as they are installed during the rough framing phase to keep out dirt, dust, debris, and paint overspray from the inside of the fireplace.

☐ *251*
Use the correct type of caulking to seal around the opening through the fireplace firebox for the gas pipe's nipple. The correct specialized type of caulking hardens with heat and prevents sparks and cinders from exiting the firebox through the gap around the gas pipe.

CHAPTER 11

Roof Framing

☐ **252**
Calculate the total amount of rise from the top of a supporting beam to the bottom of a roof ridge, for example, to determine whether there is enough room for the stirrups of a metal cap on the bottom and a bucket on the top to fit on the wood post.

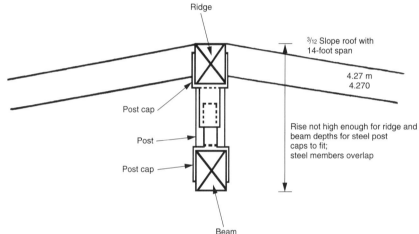

☐ **253**
Check whether the structural detail for a ridge to the hip steel hanger also provides a stirrup for the king rafter. Otherwise the king rafter will butt into the steel hanger without any means of attachment.

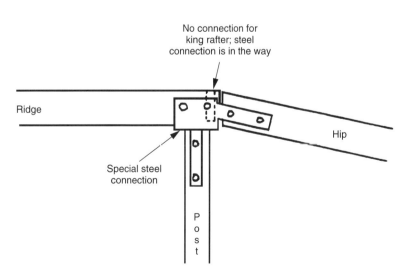

Continued on next page

Continued from previous page

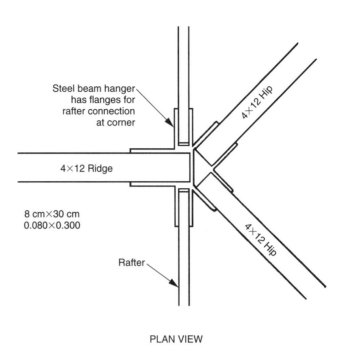

PLAN VIEW

□ **254**
Check the elevation and orientation of steel post attachment flanges so that they do not interfere with floor joist and roof rafter installation. This will help you avoid having to cut out wood members for metal flanges that can be placed elsewhere and kept out of the way.

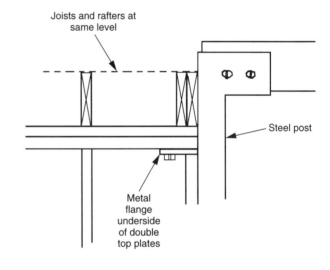

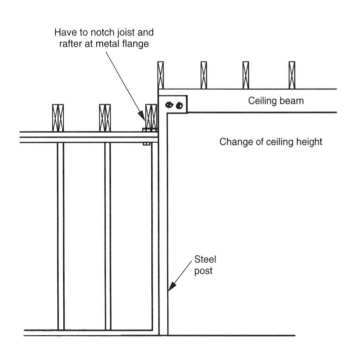

☐ **255**
For houses with restrictions on roof ridge height, consider the option of cutting a 3½ by 12 pitch roof at 3¼ by 12 instead, for example, to gain an additional margin of error without noticeably changing the appearance of the roof. Another approach is to cut the rafter tail bird mouths a little deeper at the double top plates.

☐ **256**
Consider ahead of time the layout of roof rafters and ceiling joists to miss ceiling flush lights that will be installed later, during the rough electrical phase.

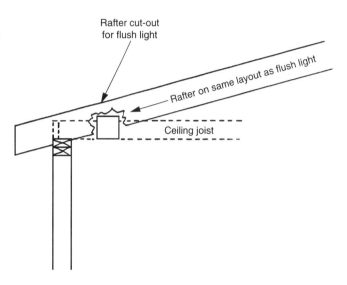

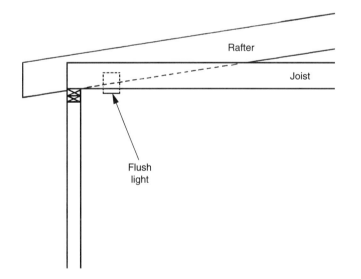

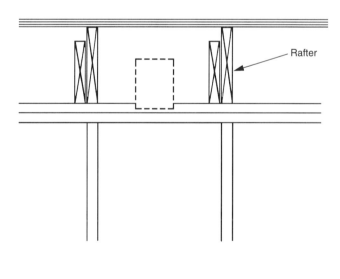

Roof Framing 115

☐ *257*

Avoid narrow spaces between a roof rafter tail and a fireplace chimney that are difficult to finish because of a lack of room to work. Add a false rafter tail or a solid block at these locations.

☐ *258*

Provide an installation method for raising thick steel hangers above the ceiling line to prevent drywall ceiling bulges or include in the drywall subcontract a provision to cut out the drywall around the hanger and finish tape the hole.

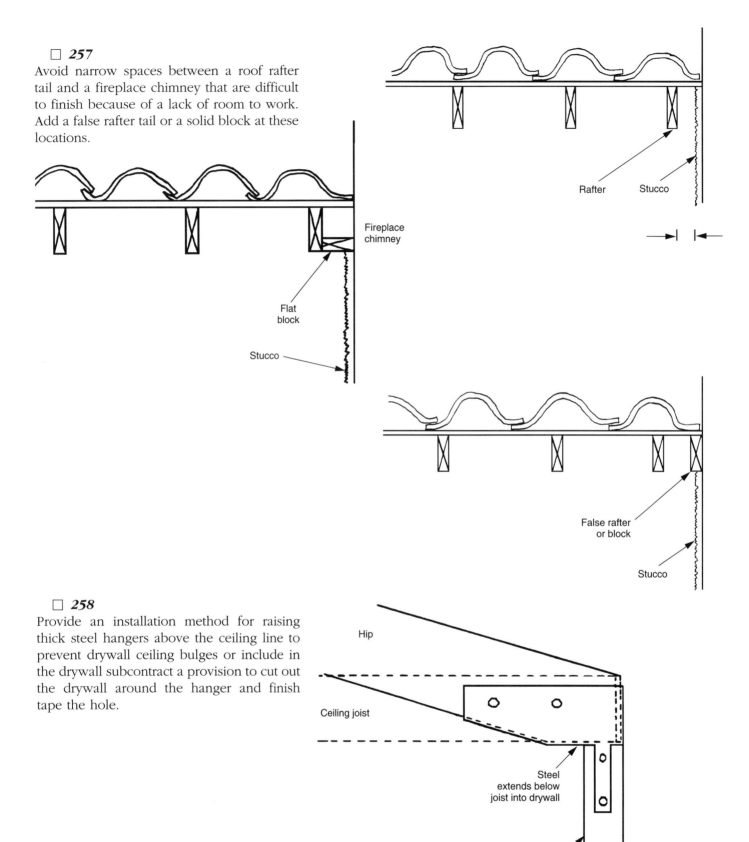

116 **Chapter 11**

☐ **259**
Remove exposed nail shiners in roof overhang areas immediately after the roof sheathing nailing is complete. It is easier to push up the nails from underneath and remove them from above than to cut them off later from underneath, using metal cutting pliers and a ladder, because the roofing already will have been installed.

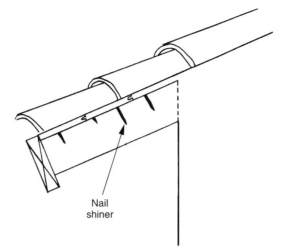

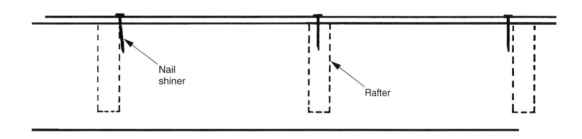

END VIEW

☐ **260**
Fill in the open spaces above the roof overhang outlookers with solid wood blocking to prevent pigeons and other birds from using this open space as a perch. Otherwise there will be bird droppings on the driveways or walkways below.

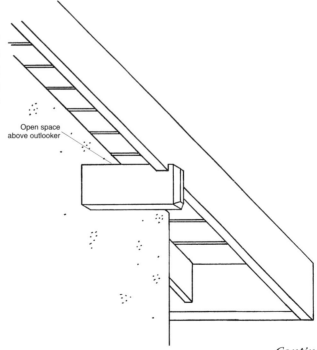

Continued on next page

Roof Framing *117*

Continued from previous page

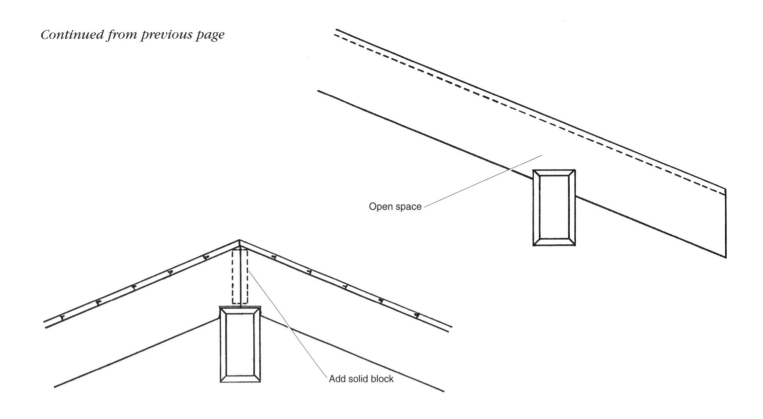

☐ *261*
When roof overhang eaves have decorative stuccoed details, check that there is enough clearance in the plans for these built-out eaves to be coordinated with the windows and entry arches below.

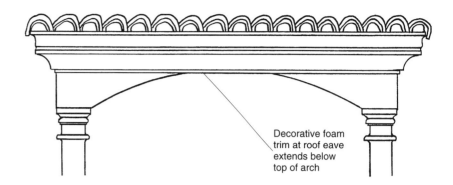

☐ **262**
Check that fascia joints are cut tight during roof framing.

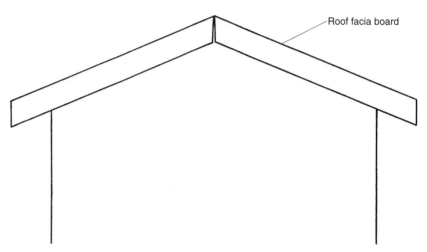

☐ **263**
Check that the ridge height for a hip roof does not place the roof within the line-of-sight view from a bathroom window, for example.

Roof Framing

☐ *264*
Whenever possible, avoid designing a flat roof section that is exposed and visible from a second floor hallway, for example.

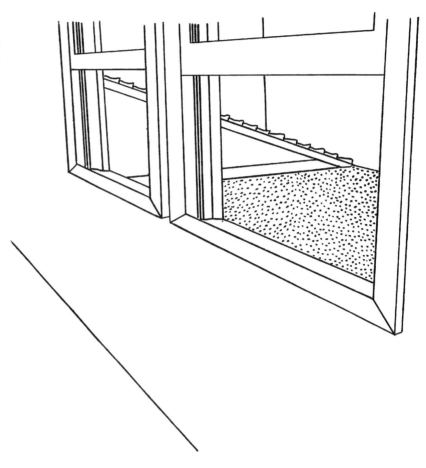

CHAPTER 12

Moment Frames

☐ **265**
When threaded bolts are cut from continuous metal rods and welded to steel moment frames or columns, specify that the fabricator tap the threads at the end of the bolts or install the nuts on the ends of the bolts; otherwise, specify that the steel fabricator use Nelson studs. This allows the framing carpenter to easily engage the nuts onto the bolt ends when installing the wood nailers and can save a lot of time and frustration on a large project.

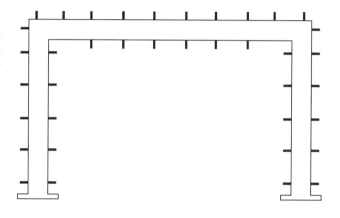

☐ **266**
Design the moment frames around sliding glass door openings with extra width at each side to create a mudsill that is long enough to contain at least two anchor bolts for strength and rigidity.

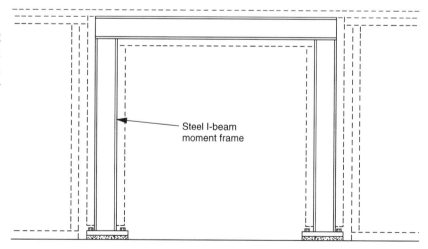

Continued on next page

Continued from previous page

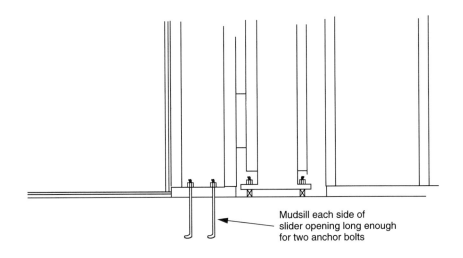

Mudsill each side of slider opening long enough for two anchor bolts

☐ *267*
Consider the extra framing that is required to fur out a standard 2×4 wall centered above a steel I-beam that is wider than 4 inches at a stairway or an open second floor hallway, for example. The beam appears as a single line on the plans but is actually wider than the wall. The location of the I-beam may be determined early in the construction by anchor bolts embedded in the concrete for a supporting steel column and thus cannot be shifted over to accommodate the hallway wall.

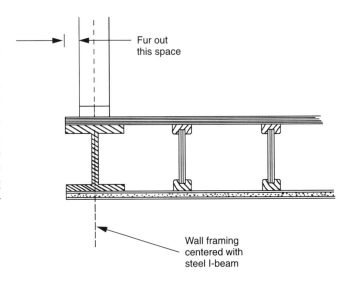

Fur out this space

Wall framing centered with steel I-beam

☐ *268*
A one-piece, full-length steel column with a saddle and a bucket attached at each end will not fit between two wood beams that are already framed in place. The steel column must be installed on top of the bottom beam before the upper beam is set, or the post must be delivered to the job site in pieces and then assembled and field-welded in place.

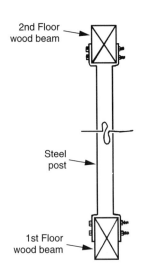

2nd Floor wood beam

Steel post

1st Floor wood beam

122 **Chapter 12**

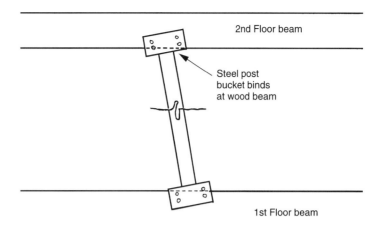

☐ **269**
Where threaded bolts might interfere with the layout of large wood beams, provide predrilled holes in the steel I-beam flanges to give the framer the option to choose bolt locations that are out of the way.

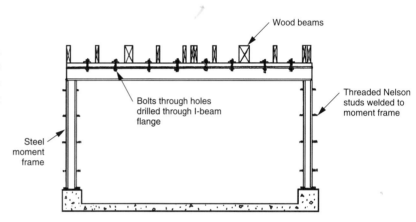

☐ **270**
Schedule and coordinate the work so that steel moment frames and/or columns are delivered to the job site and installed before the start of the rough framing so that the surrounding framing will not be in the way.

Moment Frames 123

☐ **271**
Check whether structural columns as designed actually fit within the wall framing. This eliminates the need to add an unwanted chase or soffit to box out around a column that doesn't fit within a wall, which will then look out of place with the surrounding walls.

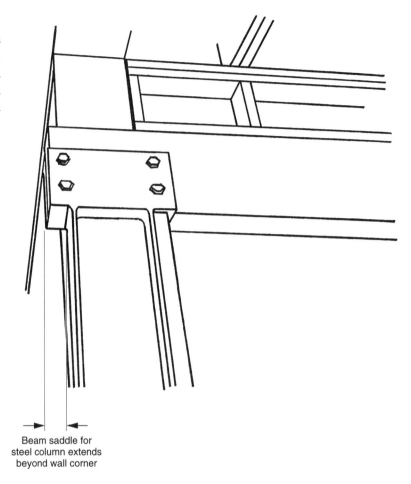

Beam saddle for steel column extends beyond wall corner

CHAPTER 13
Exterior Framing

☐ **272**
Get the elevation of exterior pop-out banding high enough to provide clearance for stucco weep screed.

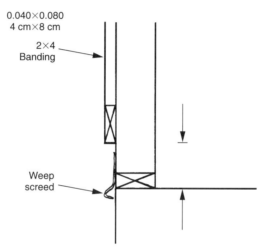

☐ **273**
Have a design detail and a construction method to install exposed exterior wood corbels securely in place.

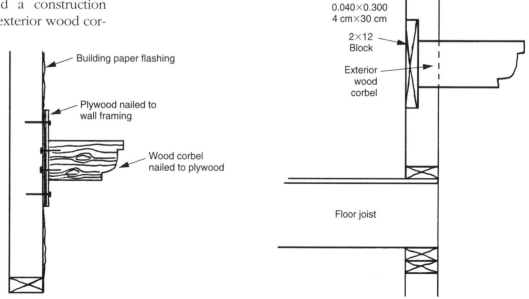

☐ **274**
Design balcony desks so that the edge of the deck and the handrail are set back from the wall corner. Everything will die into the wall without extending beyond the corner.

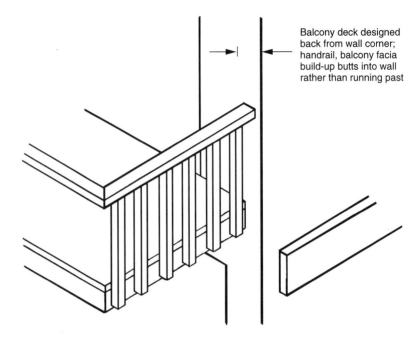

Balcony deck designed back from wall corner; handrail, balcony facia build-up butts into wall rather than running past

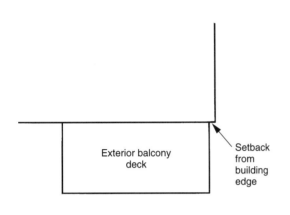

PLAN VIEW UNIT

☐ **275**
When wood boards are used for balcony deck floors, consider using clear boards without knots. This eliminates the possibility that a knot will come loose and someone will trip over it and be injured.

☐ **276**
Fur out behind pop-out banding at shear panel plywood to keep the banding straight.

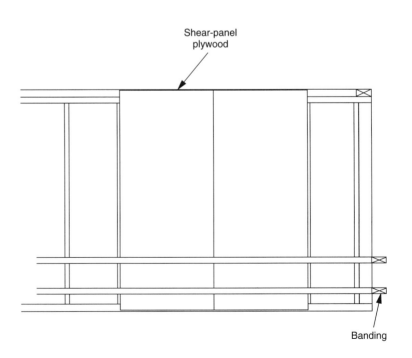

126 **Chapter 13**

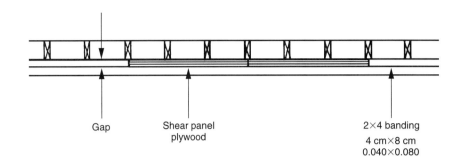

☐ **277**
Back cut the edges of plantons at stucco so that joint cracks are hidden behind the plantons.

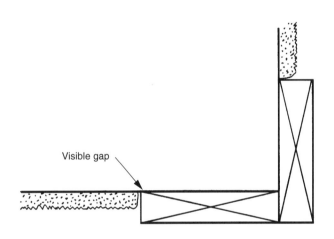

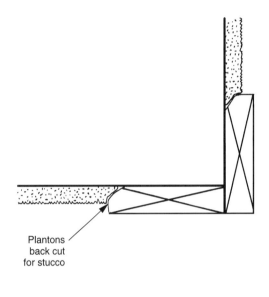

☐ **278**
Check that the tops of columns exposed to view from a second floor window are covered with finish materials, not just the tops of the rough framed column.

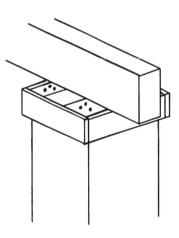

☐ **279**
When wood siding boards are finished with transparent or semitransparent stains, shorter pieces above and below windows should be cut from the same piece of wood siding to match the vertical continuity of the full-height siding boards outside the window areas.

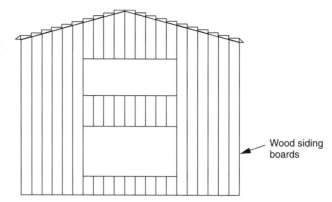

☐ **280**
Use rough-surfaced, hot-dipped galvanized nails for exterior wood siding to help prevent unsightly rust stains from running down the siding. Another option is to use stainless steel nails.

☐ **281**
Select and separate out of the lumber stacks good material for large cross-sectional lumber to be used in exposed applications.

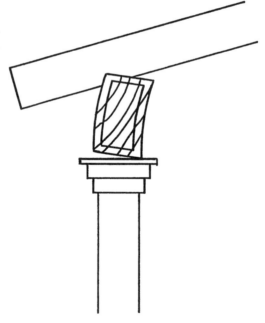

☐ *282*
To avoid unsightly cracking of decorative wood pieces such as corbels and angle braces, use high-quality lumber with tight grain patterns when finished with transparent stain and 2×4 lumber laminated together with exterior-grade glue when finished with paint or opaque stain.

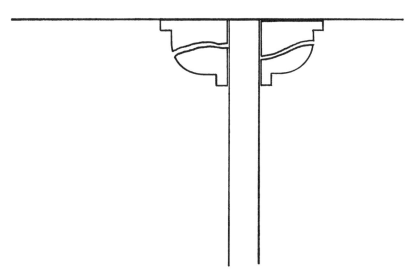

☐ *283*
Sight the garage door header stucco mold wood trim for straightness during installation. It should not be installed to match the crown of the garage door header, as the header may not be straight.

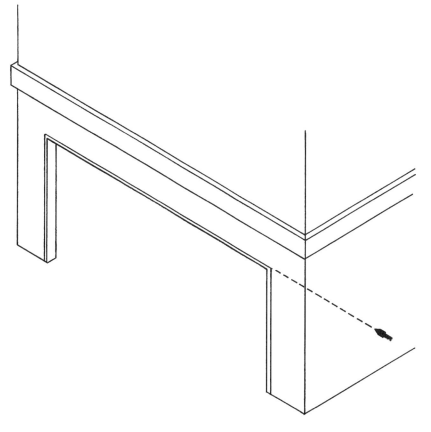

CHAPTER 14

Miscellaneous Framing

☐ **284**
Add plywood to the tops of interior pot shelves to provide support underneath the drywall.

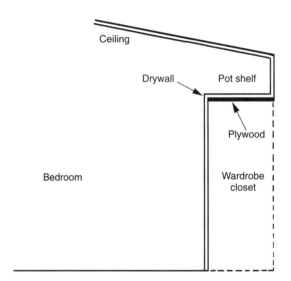

☐ **285**
Extend the plywood tops on water heater platforms beyond the framing to the thickness of the drywall to protect the top edges of the drywall.

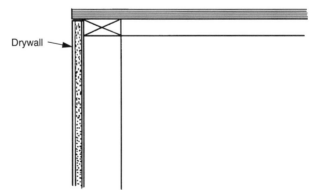

☐ **286**
To create a recessed space for the water heater between two side walls, cut the plywood top short on each side to the thickness of the drywall so that the plyood top will fit later, after the drywall has been installed.

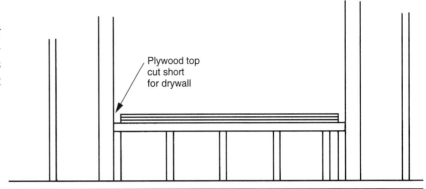

☐ **287**
Get the top of the framing for a pony wall level so that the wood trim will cover the small gap when a prefabricated countertop is installed level on top of the pony wall.

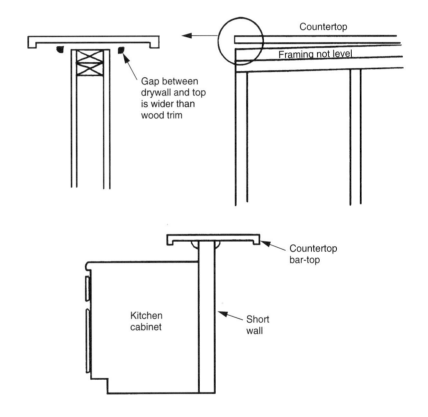

Miscellaneous Framing   131

☐ **288**
Wood side jambs at garage doors should be 2 inches wider than the garage door headers to prevent gaps at the sides of the garage doors.

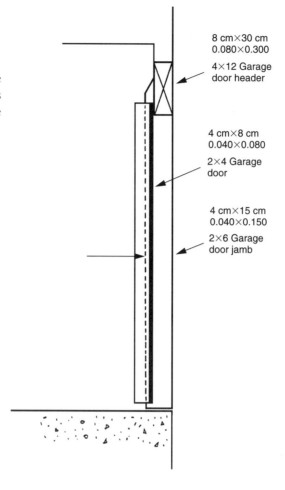

☐ **289**
Check that injected foam polyseal is not applied in such large beads that it adversely affects the drywall installation around windowsills and thus window reveals.

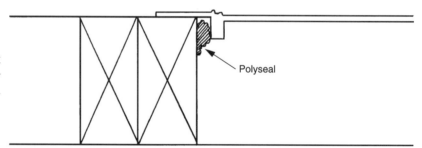

☐ **290**
Check that no excess asphalt tar is dripped or spilled on the plywood subfloor as a result of hot mopping bathtubs and showers. Drips of tar will interfere with the finish flooring in the bathroom; tar adheres to plywood and is difficult to remove after it has hardened.

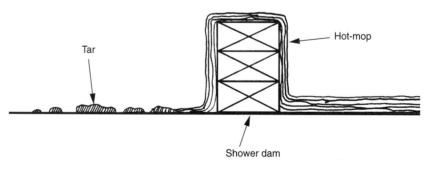

☐ *291*
Check for any required chipping off of excess concrete at exterior and interior walls before stuccoing and drywalling begin.

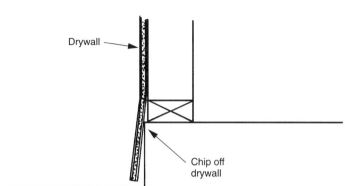
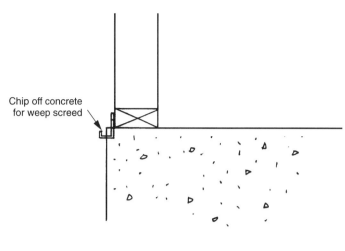

☐ *292*
Nail a scrap piece of plywood over the cutout hole in the top of an FAU furnace platform in a garage. This prevents tradespersons from throwing their lunch trash in the return-air plenum underneath the furnace by way of the open cutout hole.

☐ *293*
Have the framing subcontractor separate out and stack the usable lumber from the lumber scraps to be thrown away before each rough cleanup phase. This prevents the framing subcontractor from being able to complain that the cleanup subcontractor is carelessly throwing out good lumber along with the cutoff scraps and trash.

Miscellaneous Framing

CHAPTER 15

Framing for Bathtubs

☐ **294**
Design the shelves around the borders of bathtubs to have full-size tiles to simplify the construction.

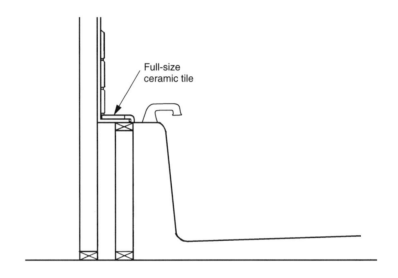

☐ **295**
Check that the bathtub spout is long enough to extend beyond a horizontal shelf and reach the basin of the tub.

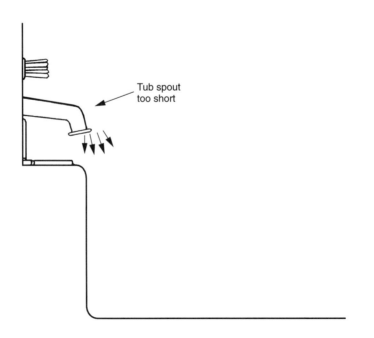

☐ **296**
Analyze the design of fiberglass bathtubs and showers and consider the need to add plaster or foam bedding underneath the bottom of the tubs and showers to provide support and prevent hairline stress cracks.

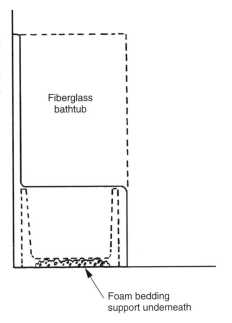

☐ **297**
Recess the shower dam so that all the tilework ends up set back from the wall corner. This allows the baseboard to make a clean 90-degree return into the tile.

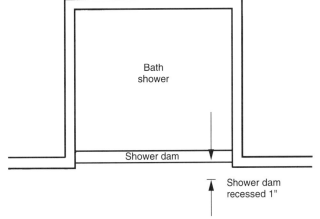

Framing for Bathtubs **135**

☐ **298**
Fill in the gaps between the wall framing studs and the fiberglass shower enclosure sides with wood furring. This provides a solid feel to the shower enclosure sides.

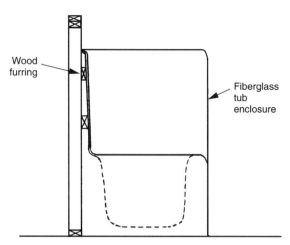

☐ **299**
Fur out the transition above a fiberglass shower enclosure at the top nailing flange so that the drywall can be nailed over this joint without kicking outward at the flange.

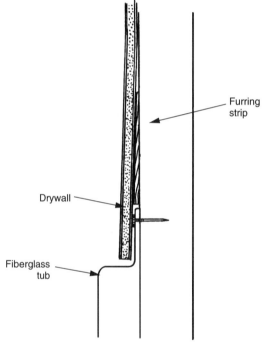

☐ **300**
Consider scheduling the installation of the motors for Spa bathtubs late in construction to prevent them from being stolen.

CHAPTER 16

Backing

☐ *301*
Adequately nail ceiling backing to the top of wall framing double top plates when ceiling joists run parallel to the wall so that drywall nailing does not cause the backing to come loose.

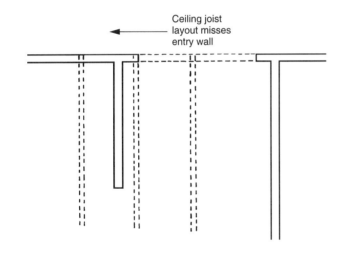

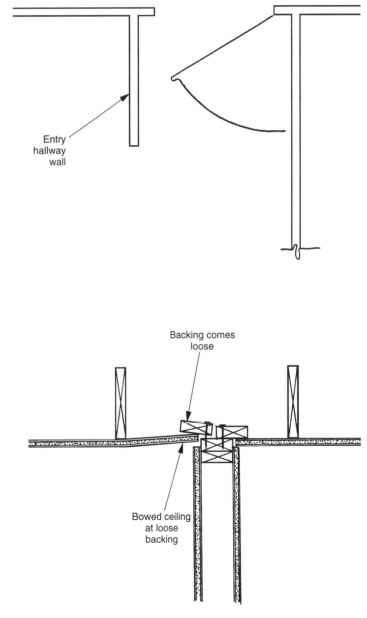

☐ *302*
Provide backing in the wall at the elevation of the wood handrail cap, not just at the wrought-iron portion of the handrail. The wood cap then can be nailed solidly to the wall.

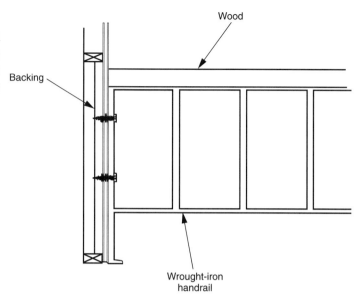

☐ *303*
Provide backing in the wall for wardrobe closet bumper jambs when the wall is flat and continuous without a king stud and trimmer return.

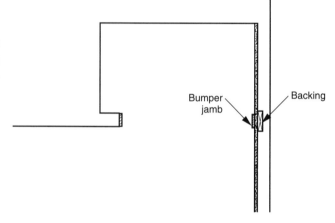

☐ *304*
Provide backing for the nailing of baseboard at the upper level of a split-level floor.

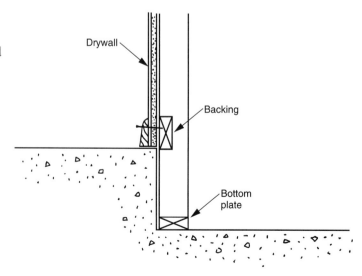

☐ *305*
Provide wood backing in the walls for the vertical legs of wainscot wood trim that will be installed later, during the finish carpentry phase.

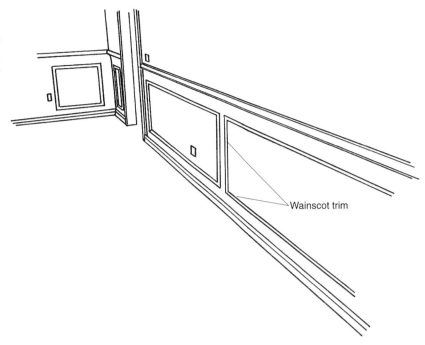
Wainscot trim

☐ *306*
Provide solid blocking between ceiling joists for floor-to-ceiling-height wardrobe closet doors with metal tracks attached to the ceiling. The predrilled holes in the track will not otherwise match the ceiling joist layout.

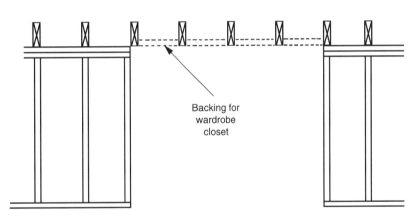
Backing for wardrobe closet

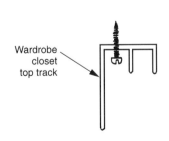
Wardrobe closet top track

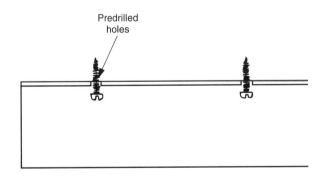
Predrilled holes

Backing *139*

☐ *307*
Provide backing on all four sides for garage foundation vents.

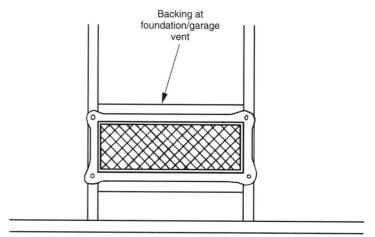

☐ *308*
Provide backing on all four sides around utility service boxes.

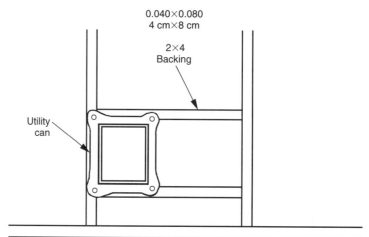

☐ *309*
Provide backing that is wider than the edge of a 2×4 wall framing stud for the ends of metal grab bars.

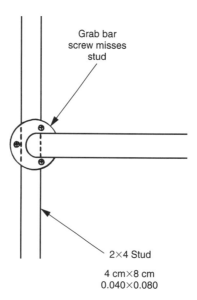

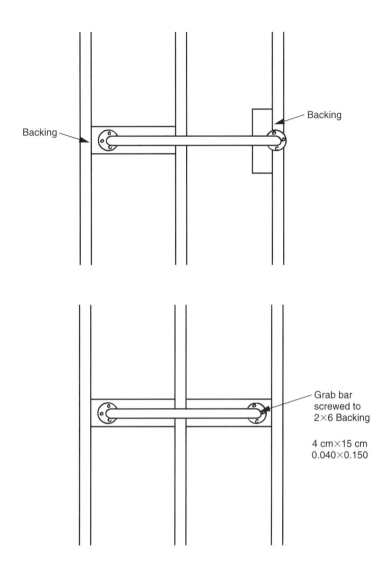

☐ **310**
When the exact location of future fences and gates is known or can be determined early in construction, add wood backing to the wall framing at those locations for easier attachment later.

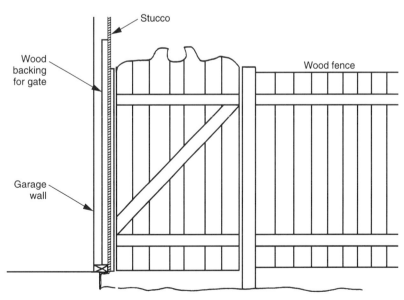

☐ *311*

Place backing in the wall for the magnetic catch that holds a corridor fire door open at the location of the front edge of the door or frame the doorway so that the door opens parallel to the wall. Otherwise the doorknob may hit the drywall before the two pieces of the magnetic catch can engage.

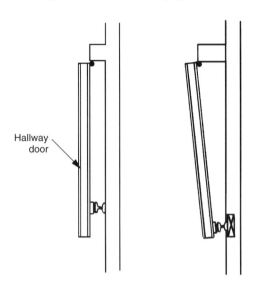

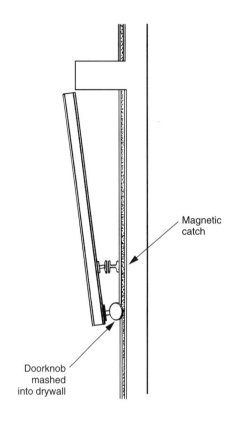

☐ *312*

For condominium and apartment corridor hallway handrail, mark the locations of wall framing studs for backing by predrilling the bracket holes before painting the walls. The small holes in the drywall will remain and be visible after the walls have been painted. This schedules the search for handrail backing at a time when the studs are easiest to find—after the drywall is textured but before the walls are painted. This method eliminates the cost of patching drywall holes caused by having to search for the wall framing studs after the walls have been painted.

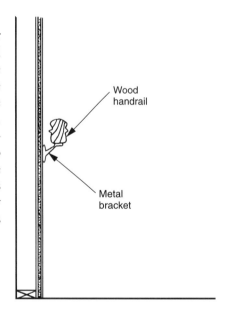

142 **Chapter 16**

☐ *313*
Provide adequate backing for the attachment of a fireplace's solid wood mantel.

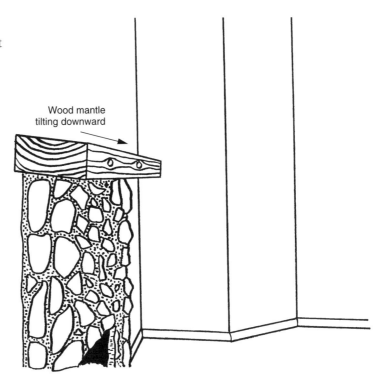

CHAPTER 17

Straightedge

□ *314*
Mirror walls

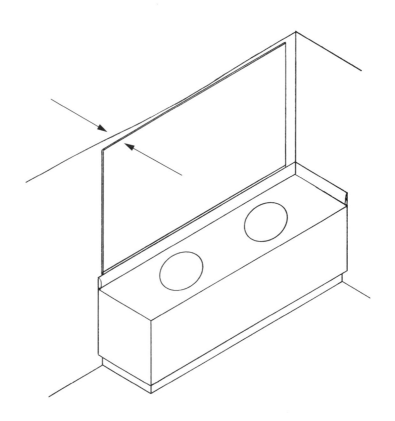

144   Chapter 17

☐ *315*
Tile walls

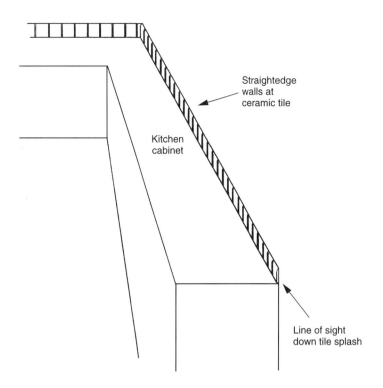

☐ *316*
Bar light walls

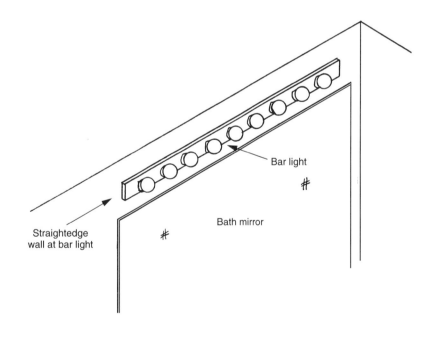

Straightedge *145*

☐ *317*
Lavatory top walls

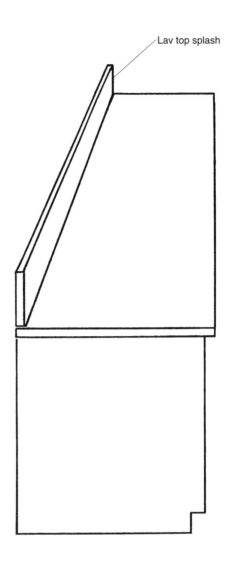

☐ *318*
Wall corners plumb for mirrors

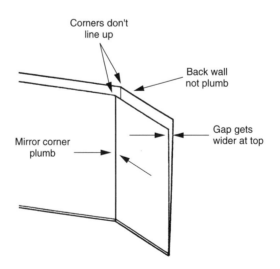

☐ **319**
Hard surface flooring walls

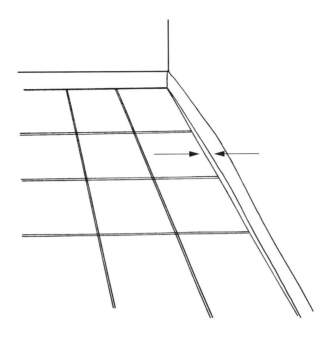

☐ **320**
Hallways

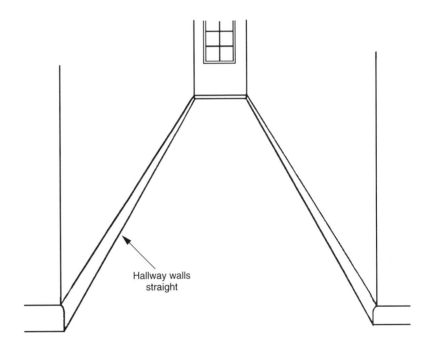

Hallway walls straight

☐ *321*
Entry handrail

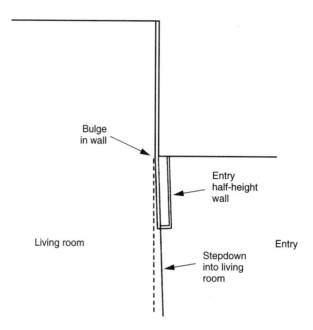

☐ *322*
Check for and remove bowed wall framing studs next to sliding glass door frames that will push the drywall out beyond the edge of the sliding glass door frame.

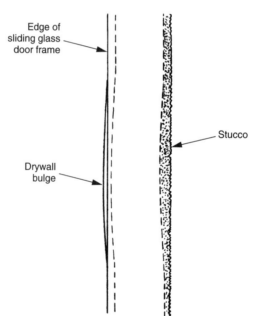

☐ *323*
Check for and remove crooked wall framing studs and posts before the start of the rough electrical work. It is easier to remove badly bowed or twisted material before the situation is complicated by having electrical wiring in place.

☐ *324*
Stair skirtboard walls

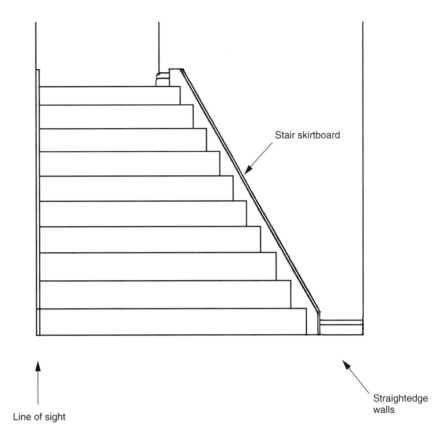

☐ *325*
Ceiling spring lines

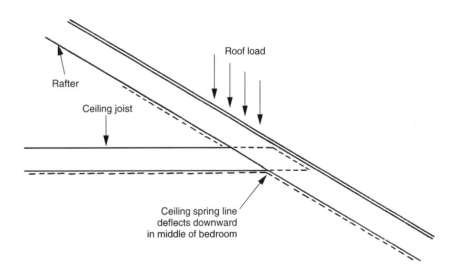

Straightedge *149*

CHAPTER 18

Rough Plumbing, Electrical, and HVAC

☐ *326*
Toilets require a minimum of 30 inches of clear space with 15 inches on each side. Check that toilets will have this minimum space after shear panel, drywall, bathtubs, and cabinets have been installed.

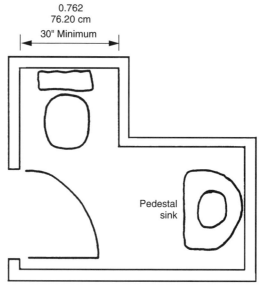

Powder bath

☐ *327*
Consider the showerhead heights in relation to the heights of the shower enclosures and ceramic tile. Do the showerheads go above the tops of the shower enclosures and ceramic tile and penetrate through the drywall, or do they go below the top edge and penetrate out through the fiberglass shower enclosure or the ceramic tile?

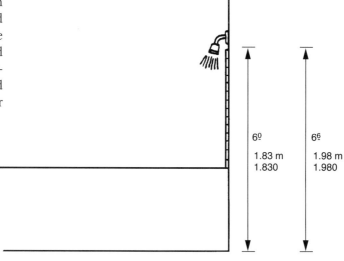

☐ *328*
Analyze the placement of solar panels on the roof in relation to the floor plan layout of the bedrooms below. In weather below 42 degrees the noise created by a circulating pump designed to keep the liquid in motion to prevent freezing in the pipes may be annoying over a bedroom at night.

☐ *329*
When the builder purchases bathtubs and shower enclosures directly, consider in the subcontracts who will move the bathtubs and/or shower enclosures up to the second or third floor.

☐ *330*
Install toilet water supply valves high enough above the floor so that the round escutcheon cover plates clear the floor baseboard.

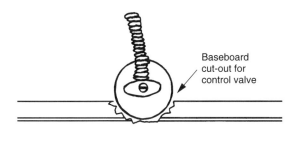

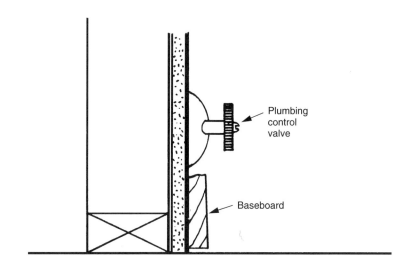

☐ *331*
Check that the elevation of the plumbing cleanouts is high enough that the cover plates will end up above the floor baseboard.

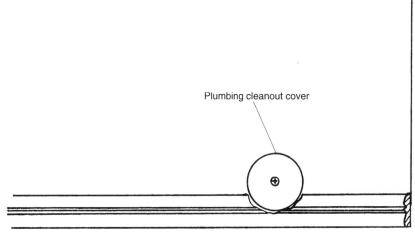

Rough Plumbing, Electrical, and HVAC

☐ *332*
Check that the dryer vent and the gas pipe for the clothes dryer in the laundry room are high enough so that they are above the floor baseboard.

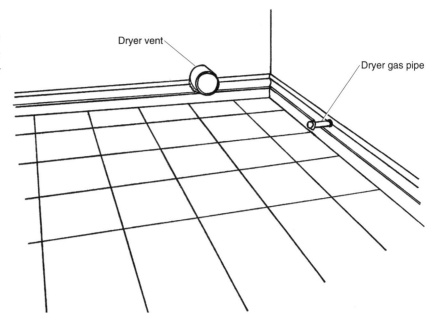

☐ *333*
For a plumbing cleanout concealed in a lower cabinet, check that the elevation of the cleanout coordinates with the cabinet bottom and/or midheight shelves. The cleanout should come through the wall above the shelf.

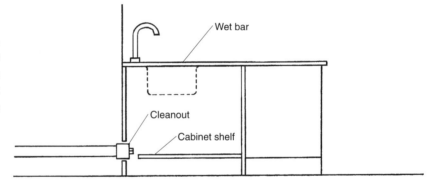

☐ *334*
Use short gas pipe nipples behind dryers in laundry closets to allow dryers to fit tightly against the walls. This provides more clearance at the front of the laundry closet at the bifold doors.

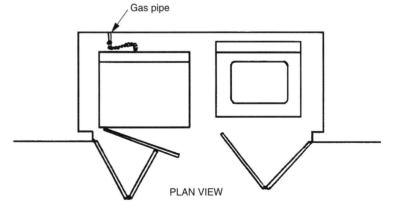

☐ *335*
Get plumbing stub-outs and outlets placed on the walls behind ranges at the correct locations and elevations to allow the range to fit tightly against the walls.

☐ *336*
Consider the placement of the fireplace log lighter during the rough plumbing phase so that it falls outside the decorative tile, marble, or brick that will be applied to the fireplace face later.

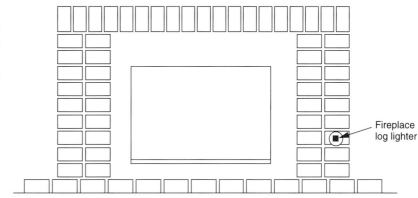

☐ *337*
Check that there will be enough clearance around an exterior hose-bib handle after brick or stone veneer is later added to the face of the wall.

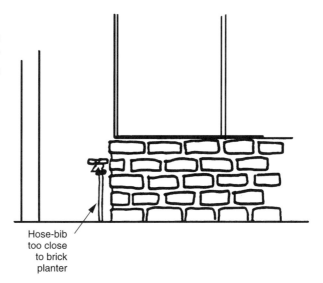

☐ *338*
Check that there are no nails in the bottoms of bathtubs and shower pans that will create rust stains when the tubs are filled with water for the plumbing inspection.

☐ *339*
Check during the rough plumbing phase that plumbing pipes do not run across the centerline of bathroom ceilings and later interfere with flush lights and bath fans.

☐ *340*
Because fiberglass shower enclosures sometimes are installed early in the rough framing phase, check that the exterior painting of roof overhang fascia and planton wood trim does not result in paint overspray drifting through the rough framed window openings and onto the shower enclosures.

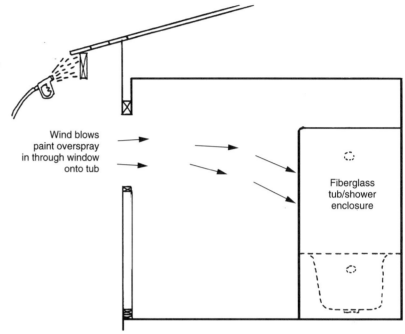

☐ *341*
Analyze the requirements for fire sprinkler monitoring and have the underground conduit in place at the sales models building to tie the sales models into the rest of the project at the end of construction.

☐ *342*
Don't place fire sprinkler heads in the center of rooms; they might interfere with ceiling fans installed later.

☐ *343*
For condominiums and apartments with private patios in the front yards, provide hose bibs at the front and rear yards of the units, not just one hose bib at the garage location. This will allow homeowners to wash off their patios.

☐ *344*
Check that bathtubs and shower enclosures are installed plumb and level whether or not the subfloor is flat and level. A shower enclosure next to an adjacent doorway will provide an immediate visual comparison of the relative plumbness of the shower enclosure and the door casing.

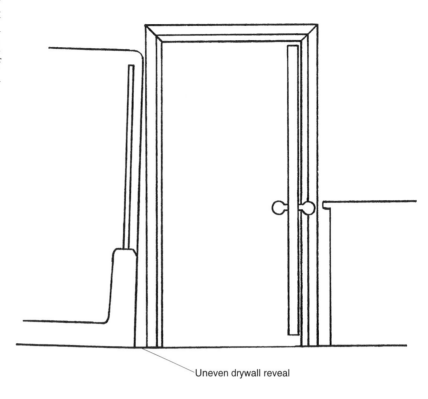
Uneven drywall reveal

☐ *345*
For bathrooms with ceramic tile and bullnose tile cap for baseboard, check that a plumbing cleanout is placed high enough above the floor to clear the tile baseboard. If a round metal escutcheon cover plate will be installed over the cleanout, the cleanout must be placed several inches above the tile baseboard.

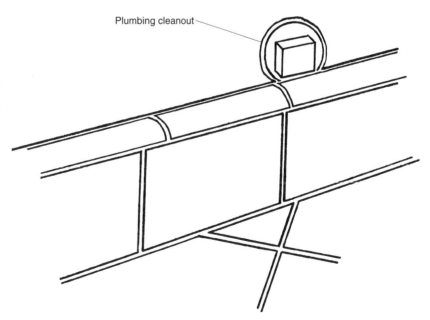
Plumbing cleanout

Rough Plumbing, Electrical, and HVAC

☐ *346*
Check that the bathroom layout does not place an access door to a bathtub behind or next to a toilet or bidet, especially an access door to a spa tub motor. It will be difficult to reach the plumbing valves or the spa motor inside the wall after the toilet has been installed.

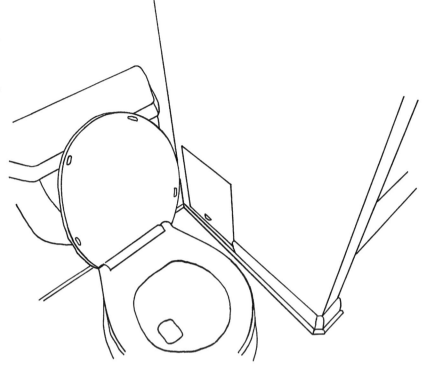

☐ *347*
For hose bibs, gas meters, and fire sprinkler flow switches located at a side yard concrete walkway, for example, consider placing these utilities inside a recessed cavity space framed into the exterior side of the garage wall. This prevents the utilities from projecting into the width of the walkway.

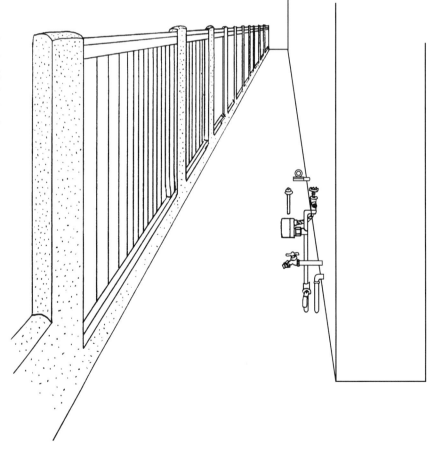

Utilities projecting into sideyard walkway.

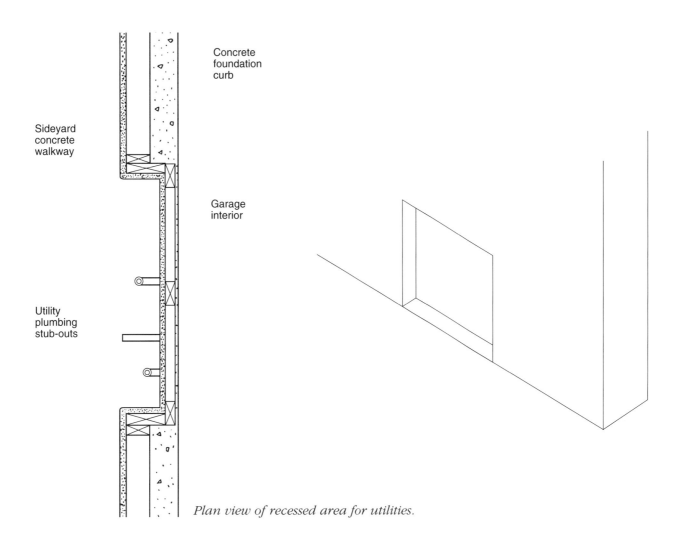

Plan view of recessed area for utilities.

☐ **348**
When you are searching for the ends of sewer laterals at the street curb for a house's sewer service connections, if the trench is mistakenly dug too deep, check that the dirt is properly backfilled and compacted in the overexcavated area. This will prevent future subsidence at the street curb parkway, for example.

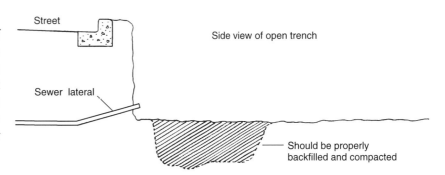

Trench dug too deep during search for end of sewer lateral.

☐ *349*
For condominium and apartment buildings with a bank of multiple gas meters in a row, label each gas pipe stub-out with the correct lot number or address, using an embossed metal tag wired to the gas pipe stub-out. This will help the gas company's meter installer find the correct units when installing the meters at the end of construction.

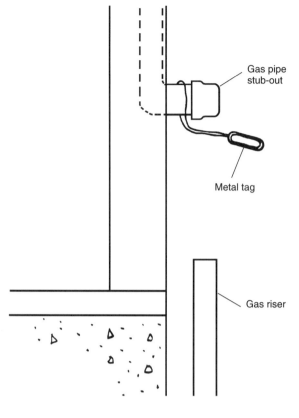

☐ *350*
Check that the HVAC return-air register will fit in the space provided for it.

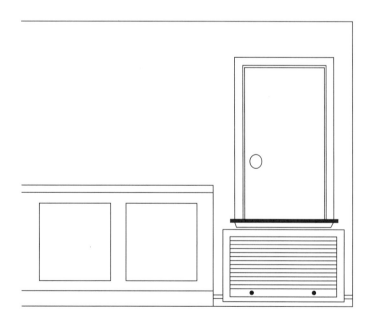

FRONT VIEW

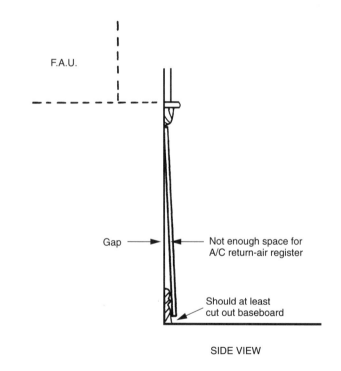

SIDE VIEW

158 **Chapter 18**

☐ *351*
Make sure that the distance the dryer vent must travel to reach an exterior wall, plus the number of 90-degree turns, does not exceed the allowable length for the dryer vent diameter pipe used. Check the local building codes.

☐ *352*
Make sure the A/C condenser units placed in side yards do not encroach into the minimum side yard setback clearances.

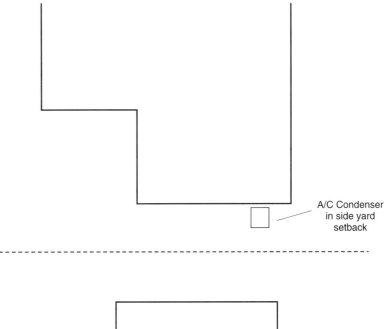
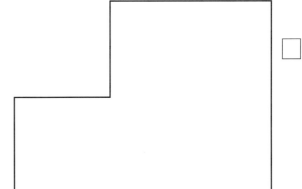

Check that A/C condenser layout on plans does not encroach on sideyard setbacks.

Rough Plumbing, Electrical, and HVAC

☐ **353**
Install HVAC return-air registers high enough above the floor to miss the baseboard.

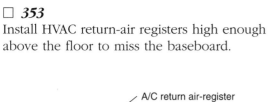
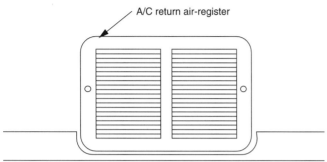
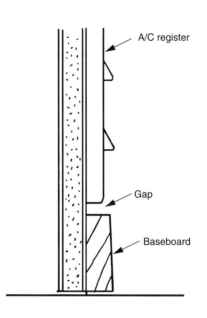

☐ **354**
Don't place the HVAC return-air register too close to a wall corner at a doorway. Otherwise, the register door may hit the door casing before being fully opened.

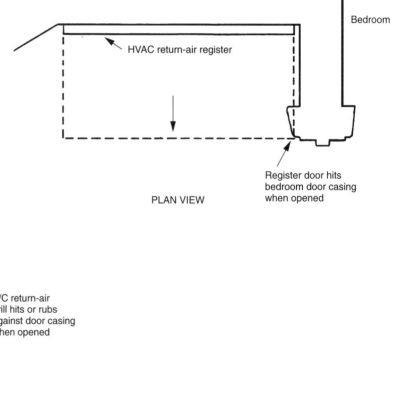
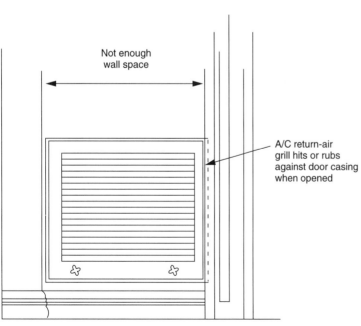

☐ **355**
Even when an HVAC rough duct fits underneath a sloping ceiling, check that the larger-size register cover also will fit.

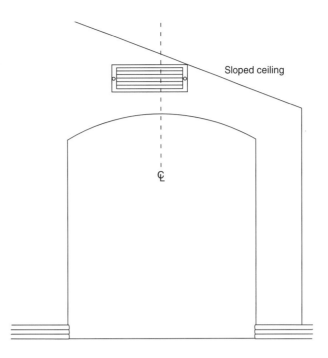

☐ **356**
Place dryer vents far enough away from wall corners and the floor to provide clearance around the dryer vents for fingers and hands to slip the flexible vent hose from the dryer over the round sheet metal duct.

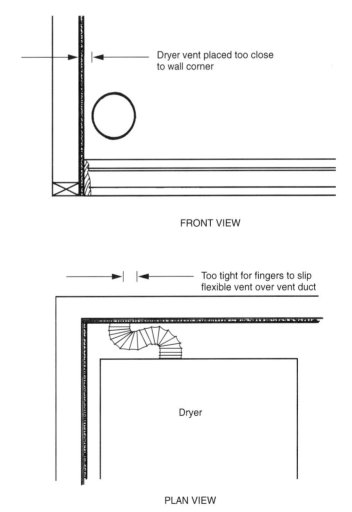

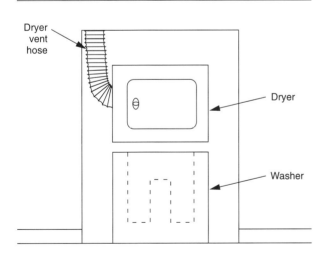

Rough Plumbing, Electrical, and HVAC

☐ **357**
When several sheet metal vents are in a row on an exterior wall for second floor laundry rooms and bathroom vents, for example, get all the vents straight, plumb, and level. When two or more vents are placed close together, this creates a group that provides a visual comparison in regard to each part's relative straightness.

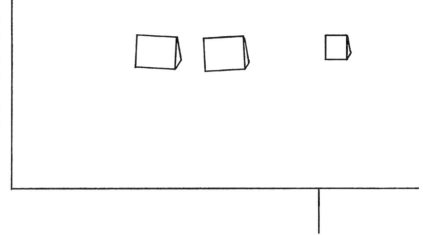

☐ **358**
Check that an HVAC return-air register at a stairway wall does not end up directly behind the stair handrail, giving the appearance of a design and construction mistake.

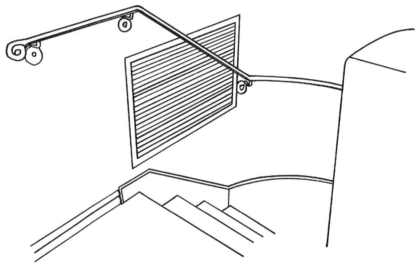

☐ **359**
Check that the locations drawn on the architectural plans for A/C condensers coordinate with the finish grades shown on the precise grading plans.

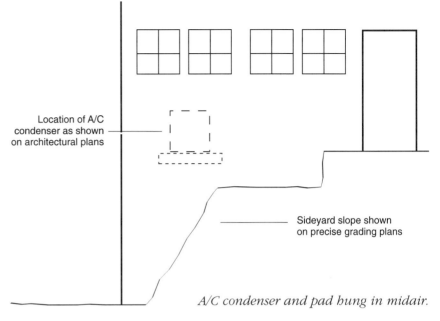

A/C condenser and pad hung in midair.

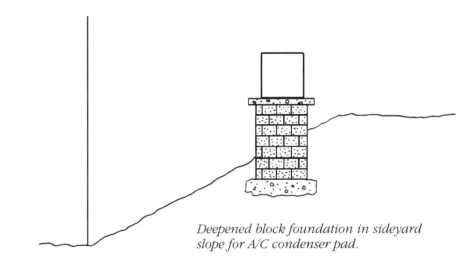

Deepened block foundation in sideyard slope for A/C condenser pad.

☐ **360**
For an A/C furnace placed in the roof attic on top of the ceiling joists, check that flush lights are not shown on the plans in the same location. You then will have to place the furnace on a raised platform framed on top of the ceiling joists to provide the required clearances above the flush lights.

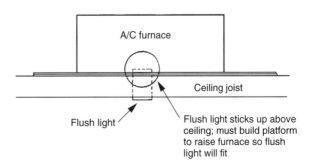

Rough Plumbing, Electrical, and HVAC

☐ *361*
When electrical meters and circuit breakers are shown on the plans to be located inside a swimming pool equipment room, check whether this is allowed by code. Pool equipment rooms are sometimes considered "wet areas," and the electric company will not allow meters and breakers in those areas.

☐ *362*
Check that the layout of large switch gear panels in electrical meter rooms for condominium and apartment projects will not interfere with the swing of the electric meter room's door.

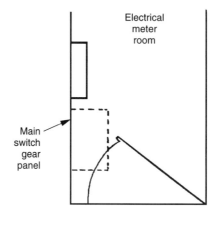

PLAN VIEW

☐ *363*
Get the utility company's specifications for the heights and clearances around its various meters and panels and check whether there is enough space in the electrical meter room or closet for everything to fit.

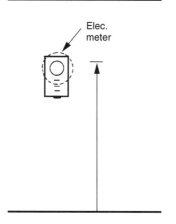

☐ *364*
For duplexes, town houses, and condominiums that have electrical breaker panels located at the entry sides of the garages because of a lack of usable wall space elsewhere, use small breaker panels to improve the aesthetics and consider an architectural design that hides the breaker panels from the front elevation view.

☐ *365*
On condominium and apartment projects, check whether the police department requires that address numbers be illuminated by a nearby light fixture.

☐ *366*
On condominium and apartment projects, find out whether a single-line drawing is required by the building department for the rough electrical inspection.

☐ *367*
Check that hanging light fixtures in hallways, for example, will not be within the swing of upper cabinets' doors.

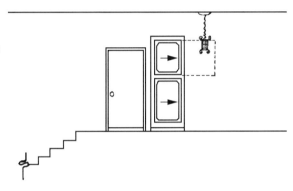

☐ *368*
Don't place flush lights on an entry ceiling 18 feet above the floor, for example. This is too high for homeowners to reach to change the light bulbs.

☐ *369*
Check that globe-type light fixtures will not interfere with the swinging of upper cabinet doors.

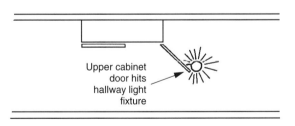

Rough Plumbing, Electrical, and HVAC   165

☐ *370*
Verify the correct height at which the flex conduit should penetrate through the drywall for undercabinet fluorescent light fixtures.

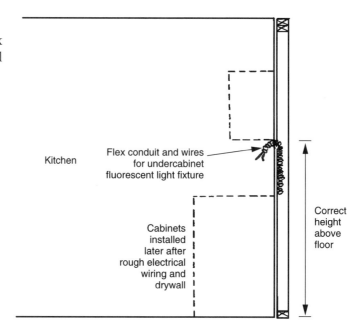

☐ *371*
Take into account the reduction of the room width after cabinets are added when you are determining the centerline location for a hanging light fixture in a breakfast nook, for example.

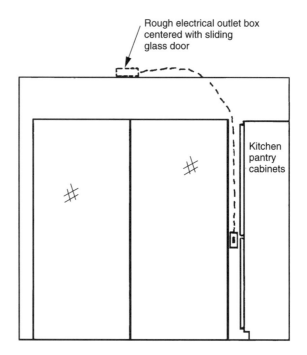

☐ **372**
Coordinate the location of a walk-in closet light fixture with an attic access drywall panel.

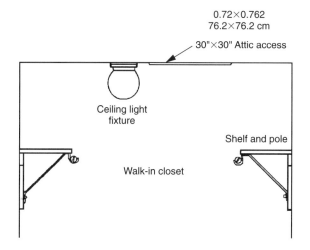
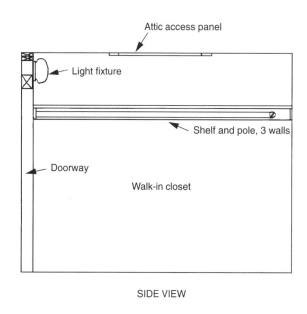

☐ **373**
Make sure doors do not open into wall sconce light fixtures.

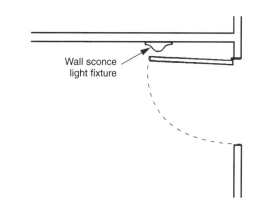

☐ **374**
In conditions where an exit door swings outward, check that a light fixture mounted on an exterior wall is placed outside the swing of the door.

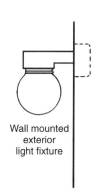
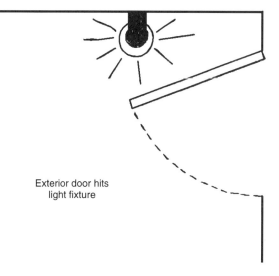

☐ **375**
Turn the receptacle outlet horizontal in bathroom splash tile when there is no wall space available because of the mirror and you choose not to cut out a hole through the mirror itself. You need to determine the correct distance to the centerline of the splash tile.

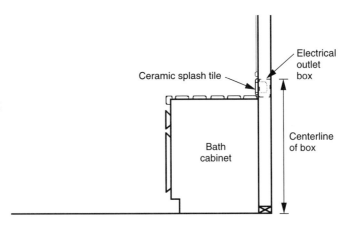

☐ **376**
For a kitchen's electric cooktops, place the electrical receptacle in the upper shelf space of the lower cabinet to avoid having to drill a hole through the shelf.

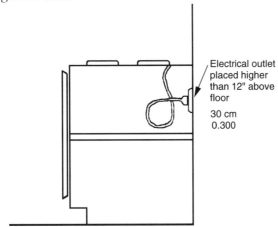

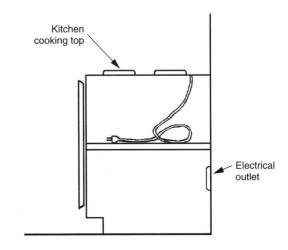

☐ **377**
Don't place the electrical outlet and water pipe on opposite sides of the same stud; the cover plates will overlap. This condition sometimes occurs inside the cabinet underneath the kitchen sink.

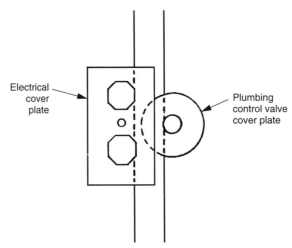

168 Chapter 18

☐ **378**
Don't place the electrical receptacle on a stud that is directly behind the future location of the range hood's vent duct. If there is not enough clearance behind the duct, you won't be able to plug in the range hood vent's motor extension cord. Install the receptacle offset from the centerline of the upper cabinet.

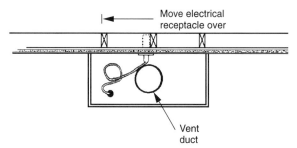

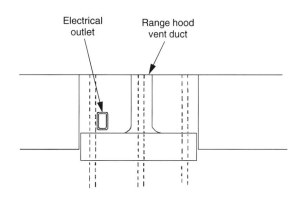

☐ **379**
Don't place the cold water ground directly behind the water heater. There will be no working space to check the bonding clamp grounding screw connection to the water pipe if the water heater is in the way.

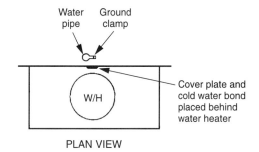

☐ **380**
Check that the location where the electrical flex conduit penetrates through the wall surface for a cabinet end plug will actually fall inside the cabinet.

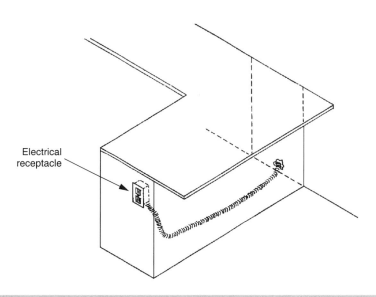

Rough Plumbing, Electrical, and HVAC   169

☐ *381*

For an exterior balcony deck with decorative stuccoed foam trim around sliding glass doors, check that the waterproof electrical receptacle is placed far enough away from the sliding glass door to allow clearance for the foam trim.

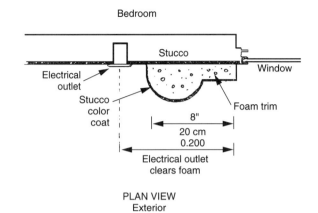

☐ *382*

When a combination mirror and cosmetic box is surface mounted to a bathroom wall, install the electrical receptacle away from the wall corner so that the receptacle is not buried between the cosmetic box and the wall.

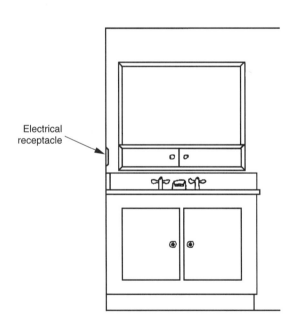

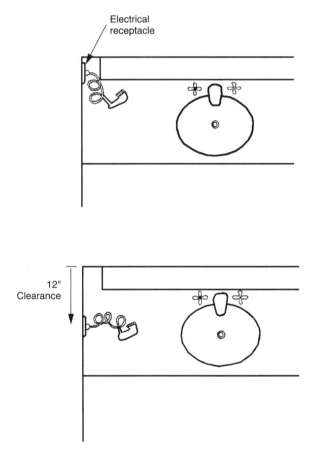

☐ *383*
For interior doorways with oversized, detailed wood casing trim, check that the electrical switches and outlets are placed far enough away from the roughly framed openings for the casing and cover plates to fit.

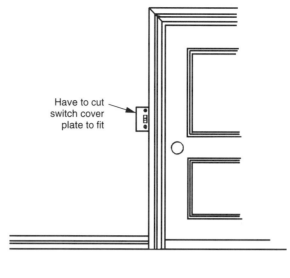

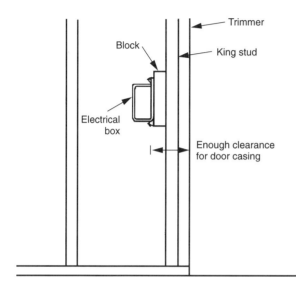

☐ *384*
Lay out electrical outlets on exterior walls with wood siding board and batt so that the outlets do not fall on a vertical batt. It is then more difficult to coordinate the outlet box depth and the weatherproof cover plate.

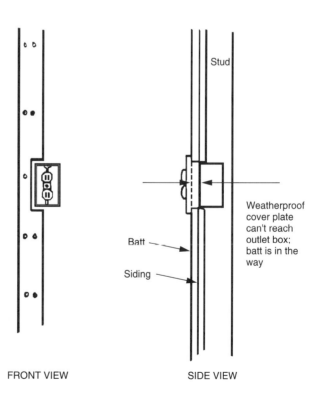

Rough Plumbing, Electrical, and HVAC 171

☐ **385**

For walls with two or three wall coverings such as shear panel plywood, drywall, and wood siding, for example, check that the electrical outlet boxes are installed at the correct depth so that they end up flush with the finish surface of these combined materials.

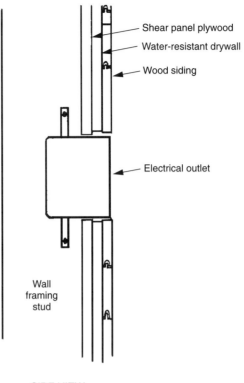

SIDE VIEW

☐ **386**

For two- or 3-gang electrical outlet boxes, check that the cutout in the drywall does not bend the plastic box up or down; as an alternative, consider using metal boxes for outlets that are two-gang or larger. Outlet boxes that are bent or twisted as a result of a sloppy drywall cutout make cover plate installation more difficult and sometimes result in cover plates that are crooked and out of level.

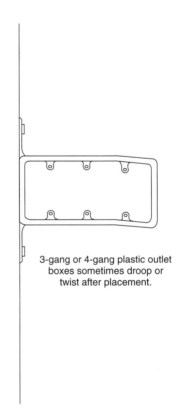

☐ *387*
Consider the layout of telephone and television outlets in bedrooms. The telephone jack should be placed on the wall where the bed headboard probably will be located, and the television outlet should be on the opposite wall, where the foot of the bed probably will be located.

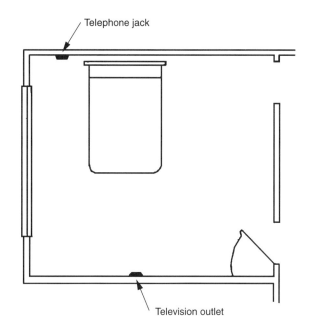

☐ *388*
Place electrical outlets above the kitchen splash tile and on the drywall surface. This makes tile installation easier because you will not have to cut out tile around outlets.

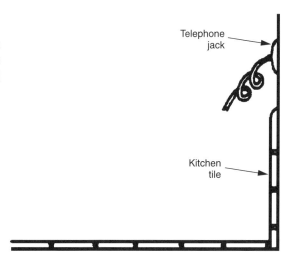

☐ *389*
Do not place the intercom/radio main panel on a dining room wall, taking up valuable wall space at the probable location of a furniture hutch cabinet.

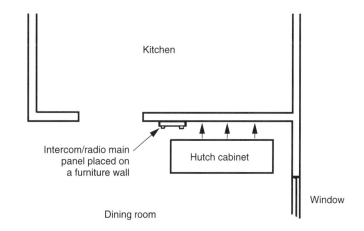

Rough Plumbing, Electrical, and HVAC

☐ *390*

For homeowner options such as ceiling fans, install a future electrical receptacle slightly above the ceiling joist line, spray-paint its location on the subfloor below, and then drywall over the box.

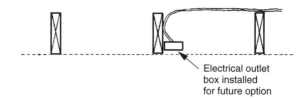

☐ *391*

Use an extension plastic or metal tube for the entry doorbell wire and install it to end up flush with the exterior finish wall surface. This prevents the doorbell wire from being mistakenly buried in the stucco, for example.

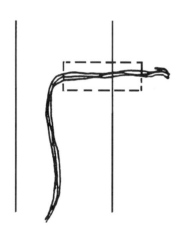

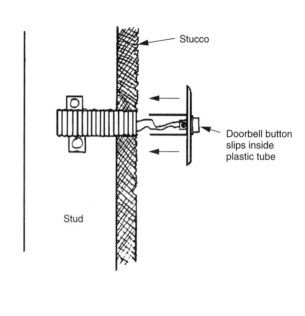

☐ *392*

Check that the rough wiring for the entry doorbell is high enough and is offset from the location where the handrail will butt into the exterior entry wall.

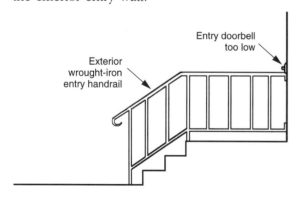

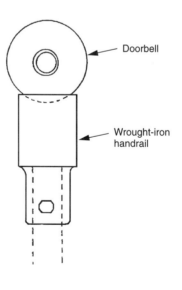

174 **Chapter 18**

☐ *393*
Entry doorbells should be audible in all parts of the house, especially in the laundry room, master bathroom, and kitchen.

☐ *394*
For two- or three-story condominium and apartment projects, loop the electrical wiring 6 feet or higher in some of the wall framing bays for easier drywall stocking. You then will not have to disassemble the windows to provide access.

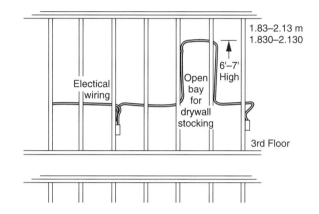

☐ *395*
Don't allow electrical wiring to squash HVAC soft ducts.

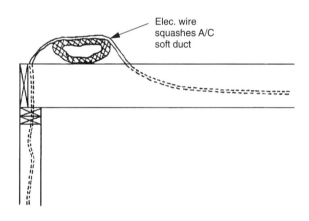

☐ *396*
Don't allow electrical wiring to squash insulation.

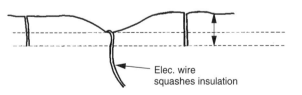

Rough Plumbing, Electrical, and HVAC

☐ *397*

Don't "make up" the wiring for the A/C condenser disconnect. Instead, pull out the wires a few inches from the box before stucco plastering so that the box is not buried by the stucco. Stucco will get onto the ends of the wires anyway, so strip the wire ends when the disconnect is installed.

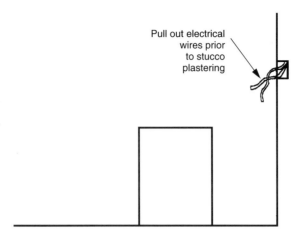

☐ *398*

When a keyless light fixture is installed in the roof attic space to provide lighting for the servicing of a forced-air unit, for example, make a provision in the electrical subcontract for the electrician also to install a light bulb.

☐ *399*

Place the GFI receptacle with the reset button in an accessible location—on one of the bathroom walls or in a nearby closet—for a spa tub motor. That way, if the receptacle with the reset button is mistakenly placed inside the enclosure surrounding the tub, if the GFI reset button needs to be pushed in, the homeowner will not have to remove the access panel to reach the receptacle.

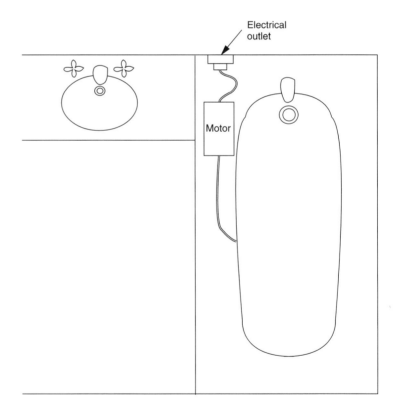

☐ **400**
Check who supplies television cable for new construction—the local cable television company or the electrical subcontractor. Sometimes the television cable company will supply its own cable to ensure quality, and the cable will be installed by the electrician. If this is the case, order the cable with enough lead time for it to be delivered to the project before the start of the rough electrical work.

☐ **401**
Center an electrical outlet with the future bathroom cabinet and mirror, not with the centerline of wall space.

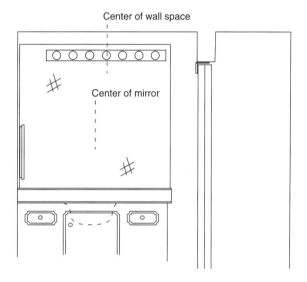

☐ **402**
Make sure that bar lights placed above bathroom mirrors will not be too close to the ceiling. Heat from light bulbs can burn the paint, causing unsightly brown spots.

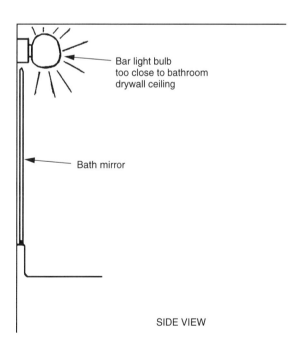

Rough Plumbing, Electrical, and HVAC *177*

☐ **403**
Install a ceiling A/C return-air register above a wall sconce light fixture so that the register door does not open into the light fixture.

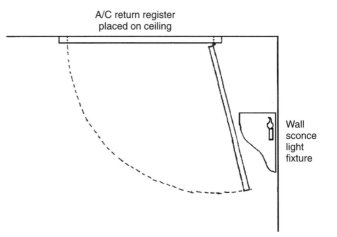

☐ **404**
Lay out wall framing studs to provide an open framing bay at the centerline of a future wall sconce light fixture. This allows a full-depth outlet box to be used, not a flat pancake-type box. With a pancake box it is sometimes difficult to pull the wall sconce fixture tight against the wall because the threaded hickey hits the back of the shallow pancake box first.

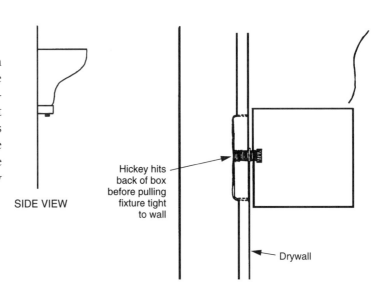

☐ **405**
Lay out and install the electrical outlet boxes at the correct elevation so that the light fixtures fall at the center of exterior wood siding boards, not at the middle of a splice joint.

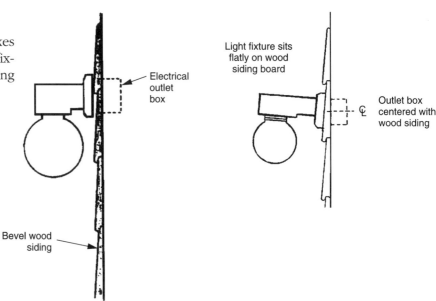

178 **Chapter 18**

☐ **406**

For light fixtures that are larger than the width of a bevel wood siding board, fill in the gaps between the fixture and the siding with pieces of siding turned upside down. Two pieces of bevel siding together create a flat surface for the light fixture.

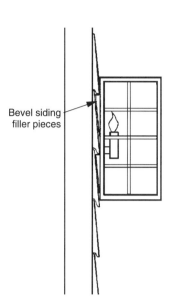

☐ **407**

Analyze and coordinate the layout of light fixtures with exterior board and batt wood siding when the outlet boxes are installed during the rough electrical phase. If the light fixtures are smaller than the space between the vertical wood batts, the rough electrical outlet boxes can be placed so that the fixtures end up centered between the batts.

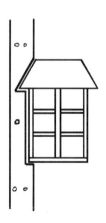 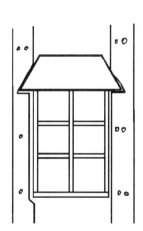 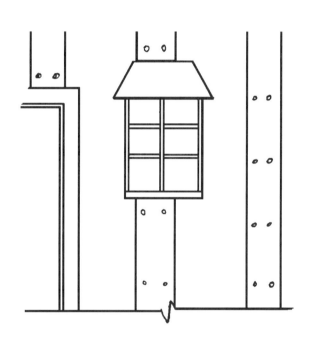

☐ **408**

Get the floors clean and free of loose debris before the start of the rough electrical work. This allows the electricians to write layout notes on the floors and lets them look upward toward the ceilings while walking around without having to worry about tripping over something.

Rough Plumbing, Electrical, and HVAC *179*

☐ *409*
Use a chalk line to align the electrical outlets above countertop splashes instead of measuring off the floor; the floor may not be straight.

☐ *410*
Make sure the electrician measures and installs the rough outlet box for a light switch at a stairway so that the light switch ends up above the stair handrail. If a structural post or another condition at the wall corner requires that the outlet box be shifted over horizontally, the outlet box also should go up vertically to clear the handrail.

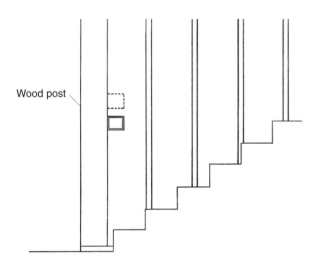
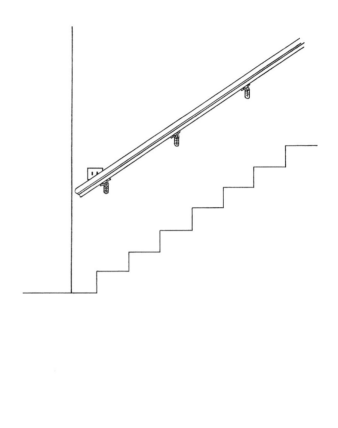

☐ *411*
Place the electrical outlet for the dishwasher appliance pigtail cord in the lower cabinet space underneath the kitchen sink, not directly behind the dishwasher. This allows a serviceperson to check whether the electrical outlet works without having to remove the dishwasher to reach the outlet.

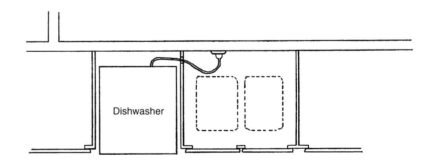

180 **Chapter 18**

☐ *412*

Check that the layout of kitchen countertop electrical outlets during the rough electrical phase does not place an outlet partially or totally behind the range. During this phase no cabinets are in place to show the future location of the range.

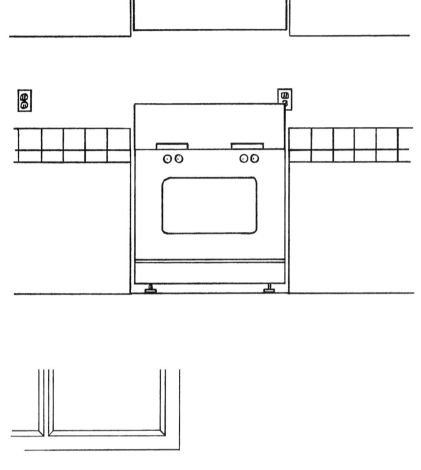

☐ *413*

Consider providing both 110-volt and 220-volt electrical outlets in the laundry room for the benefit of homeowners who have electrical dryers. This saves the new homeowners the expense of adding a 220-volt outlet after moving in or having to purchase a new gas dryer.

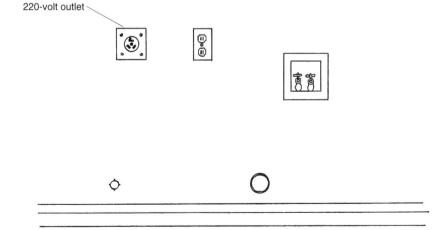

☐ **414**
For medicine cabinets with wood doors and face frames, check that there is enough clearance between the bottom of the medicine cabinet and the top of the countertop splash for electrical outlets to fit and then get the rough outlet boxes and finish cover plates installed in line and straight so that the wall reveals will be uniform and even.

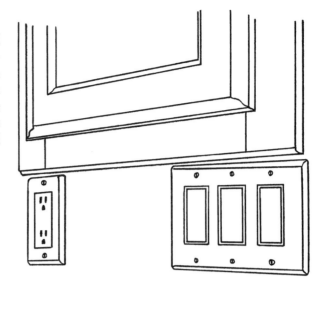

☐ **415**
For flush lights installed at the tops of curved arches, design and build the arched areas with soft curves or a slightly flat area at the top. Flush light covers are straight and will not sit flat against curved surfaces, resulting in an unsightly gap that has to be filled with caulking and touched up with paint.

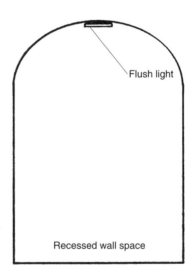

☐ **416**
During the rough electrical phase, check that electrical outlet boxes are installed far enough away from laundry valves to clear the laundry valve box covers that will be installed later, during the finish plumbing phase.

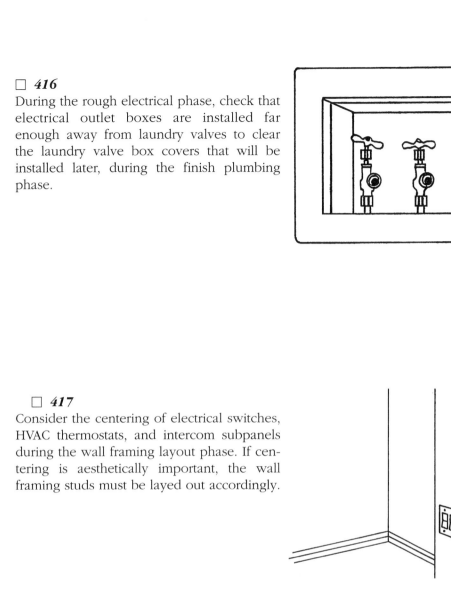

☐ **417**
Consider the centering of electrical switches, HVAC thermostats, and intercom subpanels during the wall framing layout phase. If centering is aesthetically important, the wall framing studs must be layed out accordingly.

☐ *418*
Check during the rough electrical phase that the main intercom panel in the kitchen will clear the countertops after the finish panel cover has been installed.

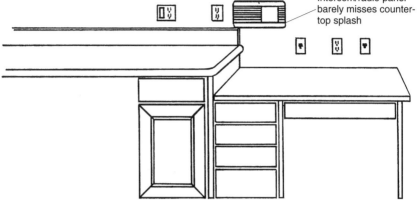

☐ *419*
Make sure there will be enough wall space revealed between the bottom of the intercom main panel in the kitchen and the top of the tile countertop, for example, after the intercom finish cover is installed. A leftover gap of 1/8 inch or less looks bad and is difficult to finish.

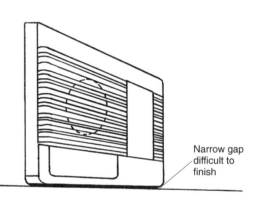

184   **Chapter 18**

☐ *420*

Whenever possible, install electrical outlets several inches above the countertop splash to hide or soften any slight elevational differences between the outlets in relation to the countertop splash. If the outlets are all 1/4 to 1/2 inch above the splash, any difference in elevation between the outlets is immediately noticeable in relation to the splash.

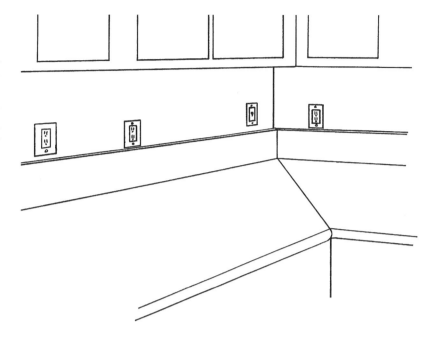

☐ *421*

For flush lights on the underside of exterior balcony decks or cantilevered rooms, lay out or head out the ceiling joists so that you will be able to center the flush lights in relation to the surrounding architectural members that will be installed later, such as wood corbels and siding.

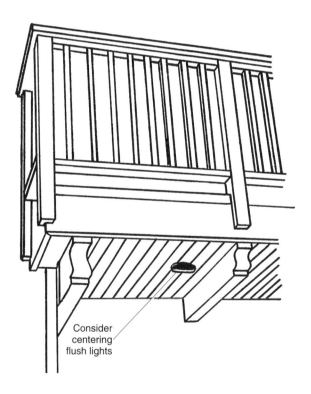

☐ **422**
When ceramic tile splashes have bull-nose tile at the top or when the size of the splash tile is upgraded to 6 inches instead of 4 inches, make sure during the rough electrical phase that the outlet boxes are being installed at the correct height so that the cover plates will clear the tile.

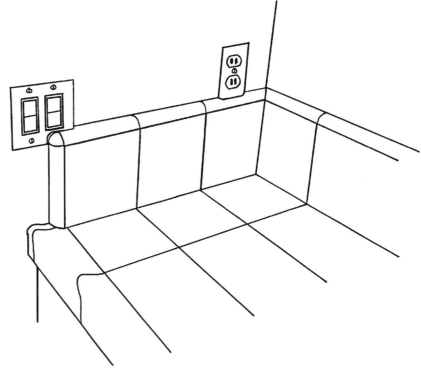

☐ **423**
When canvas awnings are installed above exterior entries or elsewhere on exterior walls, check during the rough electrical phase that the outlet boxes are layed out so that the exterior light fixtures will fall outside of and clear the awnings.

☐ **424**
Consider ahead of time the location and height above the floor of an electrical outlet for a central vacuum tank located in a garage that may be separated from the vacuum tank by a storage shelf built late in the construction or by the homeowners after they have moved in.

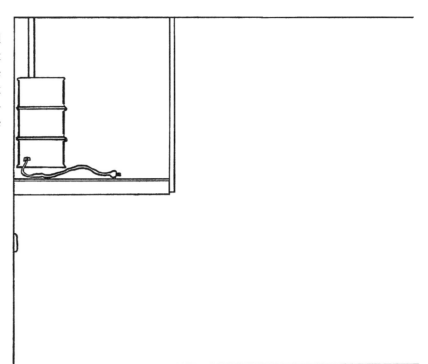

☐ **425**
For electrical outlets placed on the back walls inside cabinets for built-in appliances, coordinate the location and height of the outlets so that they do not end up at the same location as cabinet shelves.

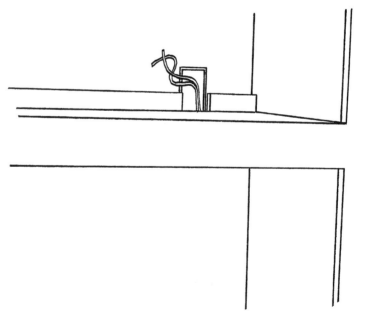

☐ **426**

For large, two-story houses with two separate A/C condensers, one for the first floor and the other for the second floor, vertically stack the electrical disconnect outlets in line to achieve the 30-inch working space clearance required by electrical codes. This is the most efficient way to use wall space because the 30-inch clearance is shared jointly by the electrical disconnect outlets.

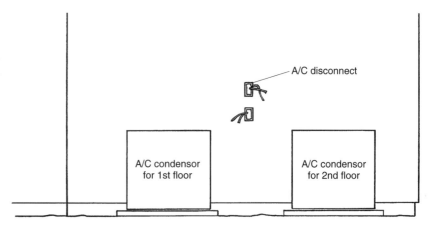

☐ **427**

For a condominium or apartment project, check that the finish floor elevation for an electrical meter room is coordinated with the exterior finish grade shown on the precise grading plans. If the electrical meter room must be lowered several feet to match the surrounding finish grade to avoid the need to add concrete stair steps, the heights and locations of the penetrations through the back wall of the electrical meter room for the various utilities also must be considered and changed. The electrical meter room may be mistakenly shown on the building plans to have the same finish floor elevation as the interior of the structure, when it should be lower to match the exterior grades.

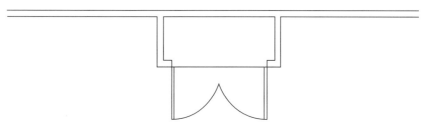

Plan view of electrical meter room.

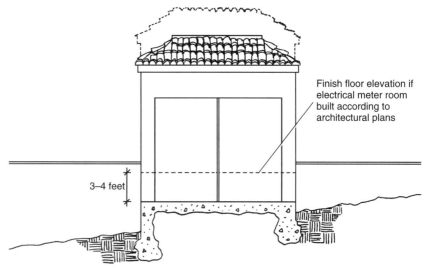

Precise grading plans used to determine the correct finish floor elevation for electrical meter room.

☐ *428*

For entertainment centers recessed into wall areas that will require interconnecting wires and cables between the television and the sound system, install empty electrical outlet boxes in each entertainment space, connected by flex or rigid conduit, so that the homeowners have a means of pulling wires and cables from one space to the next.

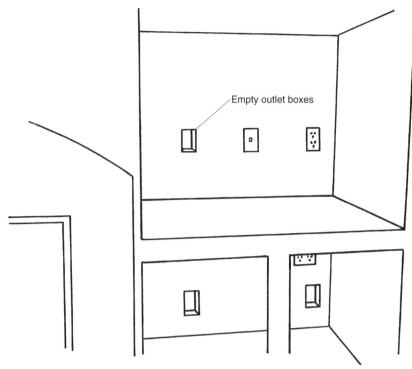

CHAPTER 19

Drywall

☐ **429**
Clean up exterior yard areas before stocking drywall into the units. This will provide easier access for the forklifts that spread the drywall.

☐ **430**
Stock drywall on the second and third floors in stacks that are equal in height and are spread out uniformly to distribute the weight evenly.

☐ **431**
When windows must be disassembled temporarily to spread drywall into the second and third floors, the drywall stockers must reinstall the window parts so that they match their original condition exactly.

☐ **432**
When wall framing studs must be removed temporarily to spread drywall into the second and third floors, the drywall stockers must reinstall the studs in the correct location and with the correct nailing.

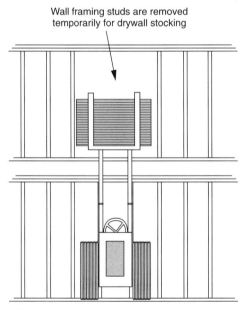

Wall framing studs are removed temporarily for drywall stocking

Drywall stockers replace studs that were removed

☐ **433**
Avoid stacking drywall too close to a wall that will get shear panel plywood. You won't be able to nail off the bottom of the plywood with a nail gun or by hand if drywall stack is in the way. This usually occurs at walls that get both sides covered with plywood, with the second side being covered after the electrical wiring and plumbing inside the wall have been inspected and after the drywall has been stocked.

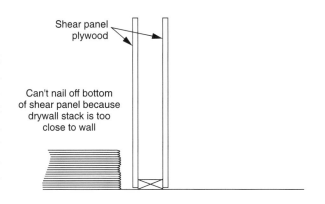

☐ **434**
When drywall must be stocked in a few high stacks on a second or third floor because of the tight layout of the floor plan, provide temporary bracing underneath the floor area with the high stack of drywall. Use a 4×4 post or two 2×4's nailed together and wedged between the floor below and the underside of the floor joists directly beneath the high stack of drywall.

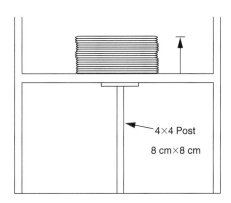

☐ **435**
For condominium and apartment projects, install preliminary drywall in the cavity between the shower enclosure's sides and the party wall framing before the installation of the shower enclosures to maintain the continuity of the sound and fire separation between the units.

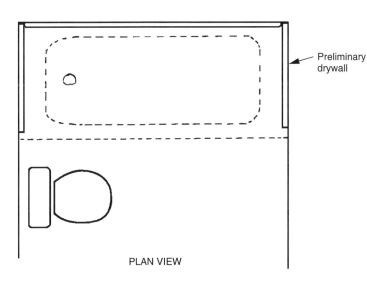
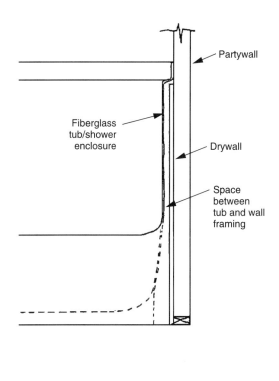

Drywall **191**

☐ **436**
On condominium and apartment projects, when the drywall must extend up to the bottom of the floor sheathing at each floor level for the fire rating, install preliminary drywall over the floor joists above the elevation of the wall framing before the start of the rough mechanicals. It is easier to install the drywall over the joists without electrical wiring and plumbing pipes in the way.

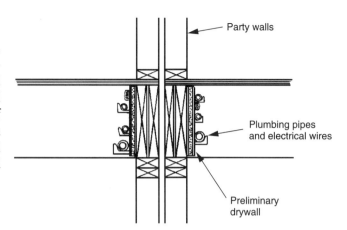

☐ **437**
Consider the strategy of installing preliminary drywall on the garage party wall, for example, before building a raised water heater platform. A small piece of preliminary drywall sandwiched between the water heater platform and the party wall framing provides the required fire separation and eliminates the need to put the drywall on top of the platform.

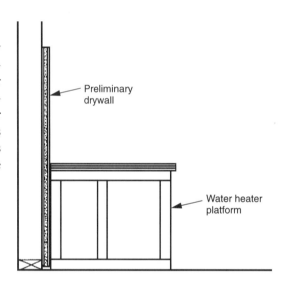

☐ **438**
Consider the depth of the window reveal at the bottom sill for windows with stool and apron wood trim in terms of installing or not installing drywall on the bottom sill. You don't want the stool and apron to end up above the edge of the window frame.

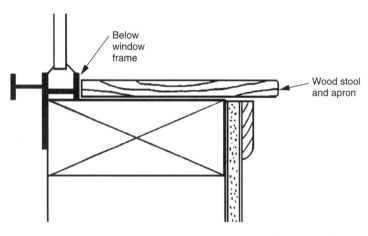

Continued on next page

192 **Chapter 19**

Continued from previous page

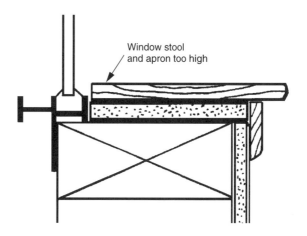

☐ **439**
Have the drywallers trim off any excess drywall that projects into the interior door openings so that it is flush with the trimmers and headers. This saves the finish carpenter from having to chop off this excess drywall in the doorway openings when the carpenter installs prefit doors.

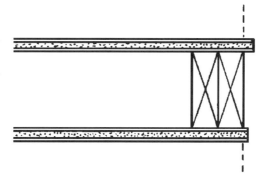

☐ **440**
Check that the cutouts in the drywall for electrical outlet boxes are accurate and close to the outside edges of the boxes. This provides a firm drywall surface for the receptacle tabs or "ears" to rest against.

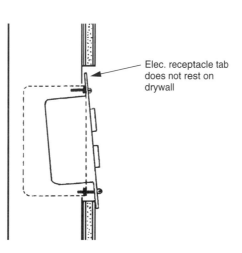

☐ **441**
Break drywall joints over the middle of a doorway. This prevents hairline cracks from forming above the door latch side of the doorway when the doors are slammed shut.

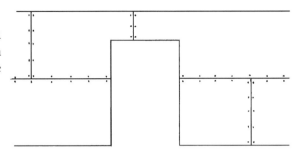

☐ **442**
In locations where leftover scraps of drywall are used to fill in small wall areas, check that the drywall extends all the way down to the floor. This provides a flat wall surface for the baseboard.

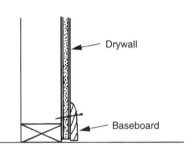

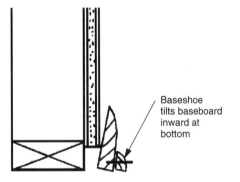

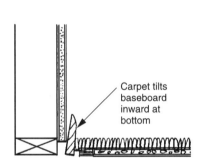

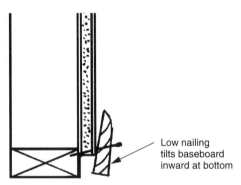

☐ **443**
Use screws rather than nails to attach drywall to the sides of pocket door frames. Screws can be shorter than nails for the same thickness of drywall and therefore will not penetrate through the wood slats and into the pocket door cavity, scratching the pocket door.

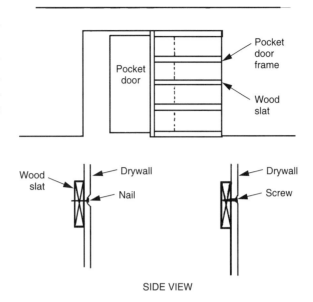

☐ **444**
Check that the piece of drywall cut to fit between the wall framing and the stair stringer extends below the inside corner of each stair step. It is difficult to finish these gaps with taping mud, and the carpeting may not be thick enough to cover them.

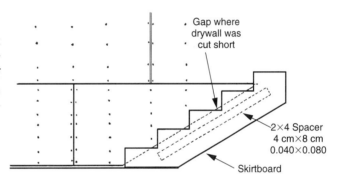

☐ **445**
For corner fireplaces, consider using fire-resistant fiberboard instead of drywall to allow more freedom in designing clearances to adjacent wall surfaces, pop-outs, and mantels.

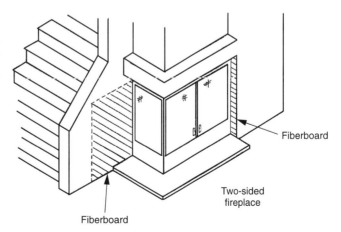

Drywall 195

☐ 446
For garages that are drywalled, hold back the drywall on the inside face of the garage door header so that the garage door can close flush against the header. Finish the bottom edge of the drywall with metal mill-core trim.

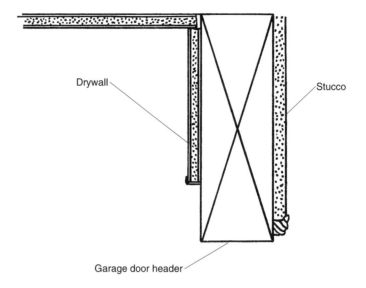

☐ 447
For a group of windows in a row or for interior pot shelves that are grouped in a row, align the corner bead for all the openings in a straight line.

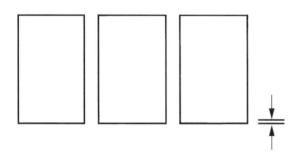

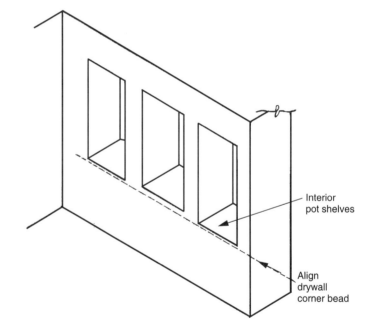

☐ **448**
For half-height pony walls, use a level to install the corner bead.

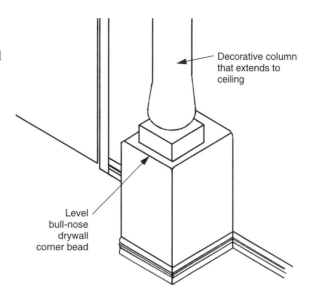

☐ **449**
Cut out the drywall around rough framing metal plate straps and hardware stirrups and fill in those holes during the finish taping phase. This will prevent drywall bulges and bowed wall areas.

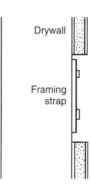

☐ **450**
Check for drywall bulges and bowed wall areas before the start of drywall taping. When they are discovered before taping, the problems causing the bowed walls can be fixed, and the resultant taping repairs can be worked into the normal drywall taping without becoming extras.

☐ **451**
When a second floor balcony deck is located above a garage that is drywalled, schedule and apply the first coats of waterproofing to the balcony deck before insulating and drywalling the garage ceiling and before scaffolding is built on top of the deck for lathing and stucco. This will protect the garage ceiling drywall from water damage.

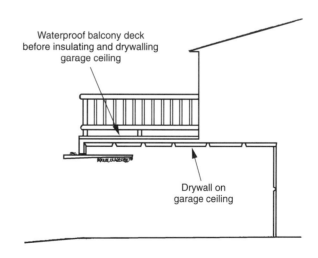

☐ **452**
Solid wood beams tend to shrink and twist. Apply extra nailing to the corner bead at drywalled wood beams to reduce or eliminate unsightly cracks.

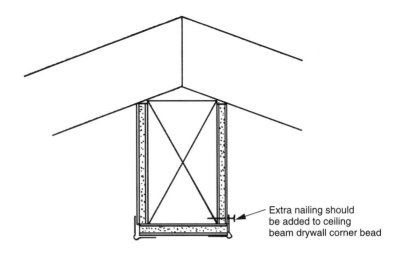

☐ **453**
For a stair wrought-iron handrail that is installed after the walls are covered with drywall, include within the drywall subcontract the patching of the holes that are cut out in the drywall for the handrail attachment metal flanges and lag bolts; otherwise this work will become an extra that is payable to the drywall subcontractor.

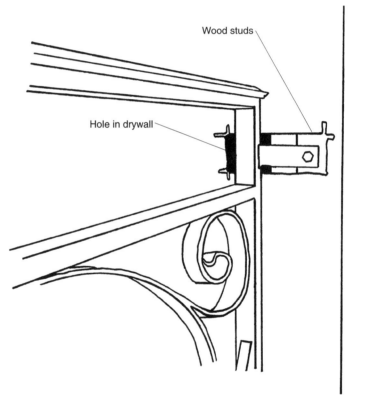

☐ **454**

Consider using bull-nose drywall corner bead at the windowsills for round windows. It is easier to soften and hide slight imperfections in the drywall and taping within the curvature of the bull-nose corner bead than it is in the sharply defined edges of standard corner bead.

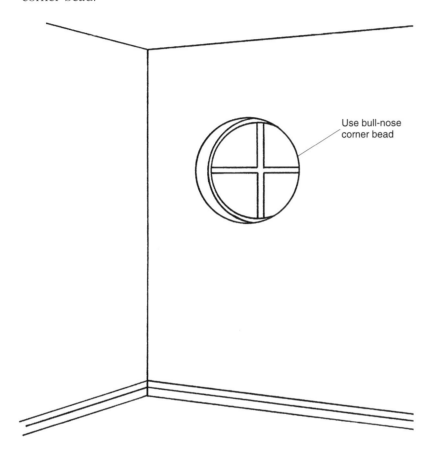

☐ **455**

When attic access drywall panels are held in place by metal T-bar track squares, include within the drywall subcontract the recutting of the drywall access panels to fit after the T-bar tracks are installed later in the construction process.

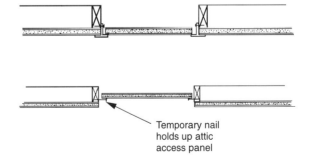

Drywall 199

☐ **456**
Consider the option of screwing drywall instead of nailing it so that interior drywall hanging and exterior stucco plastering can proceed simultaneously, accelerating the construction schedule.

☐ **457**
When exterior stucco plastering and interior drywall work are occurring simultaneously, coordinate the masking of the exterior sides of the windows by the plasterer and the scrapping of leftover drywall pieces by the drywaller. If necessary, designate one window on the second and third floor levels, for example, where the masking can be removed and drywall scraps can be thrown outside.

☐ **458**
Place pallets containing boxes of drywall taping mud far enough inside garages to be out of the way during the later installation of garage doors. Check that they are not casually placed at the front edge of the garage slab for the convenience of the driver of the drywall delivery forklift.

☐ **459**
Door casing wood trim will not always cover the drywall nail dimples at door openings completely, especially with prefit door units that are installed favoring one side of the opening. Get the taping mud application around the door openings to be the same quality as the filling of nail dimples in the middle of a wall.

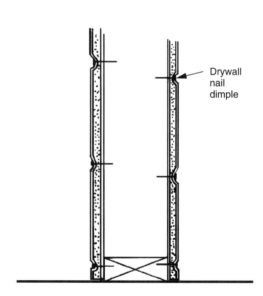

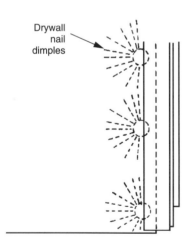

☐ **460**
Decide ahead of time whether to texture drywalled windowsills. For production tract housing it is easier to texture the windowsills because there are no transition lines to get straight.

200 **Chapter 19**

☐ **461**
Scrape and clean drywall taping mud off metal window frame edges to provide a straight edge for the painting of the windowsill drywall.

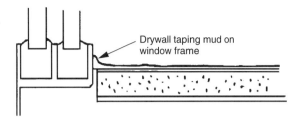

☐ **462**
Scrape and clean excess drywall taping mud off the top of fiberglass tub and shower enclosures at the joint with the drywall to provide a clean and straight edge for the painting.

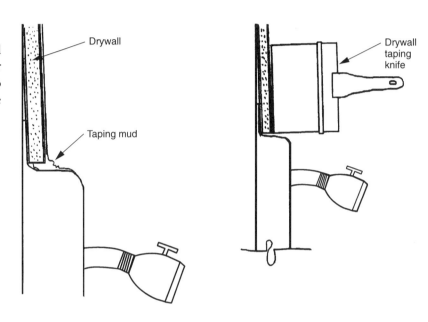

☐ **463**
Have tapers remove excess mud that accumulates at the ends of taping seams at interior doorway openings so that the finish carpenters do not have to hammer off this drywall mud at the time of interior door installation.

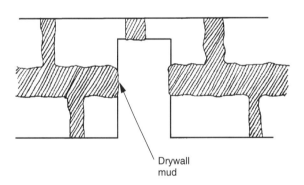

☐ **464**
For interior lath and plaster, wash off the plaster that gets onto the edges of doorjambs. When hardened, plaster adheres to the wood jambs and is difficult to remove without damaging the wood; it looks unsightly if it is left on the jambs and painted over.

Drywall *201*

☐ **465**
Consider walls or wall areas that will get wallpaper and do not apply drywall texture to those areas. The texture may be noticeable underneath the wallpaper.

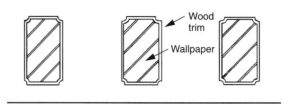

☐ **466**
When bathtubs and/or showers are lathed and scratch coated with mortar before drywall texturing, mask off the bathtub and shower areas. Drywall texture overspray on the edges of the mortar forms a barrier that prevents adhesion to the subsequent brown coat of mortar.

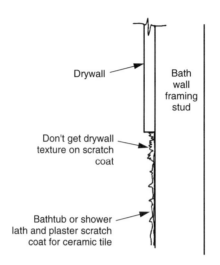

☐ **467**
Have the drywall tapers remove any excess taping mud that gets into electrical outlet boxes.

☐ **468**
When a glue-on ceramic tile application is used around kitchen pass-through countertops and splashes, don't apply taping mud to the inside and outside corners where the tile will be installed later. Untaped corners allow the tile setter to obtain square corners by using thinly set mortar during the tile installation without having to contend with or tear out excess taping mud.

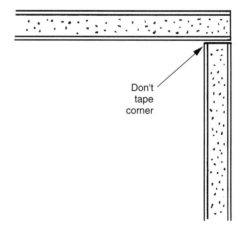

☐ **469**
Have the drywall texture cleanup crew scrape and clean off the excess taping mud around sliding glass door frames to provide a clean and straight edge for painting.

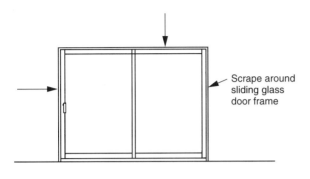
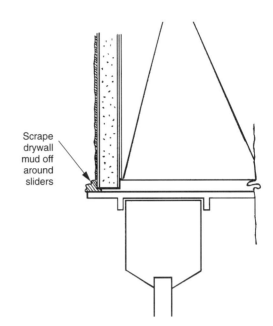

☐ **470**
For two- or three-story condominium or apartment units the drywall texturing should start in the upper floors and move downward. This prevents the texture spray gun hose from brushing up against the stairway walls, damaging the newly sprayed and wet texture, which then must be repaired during the later drywall prepainting phase. When one works from the top downward, the hose is always in front of the texture application in the stairways rather than trailing behind.

☐ **471**
Check that drywall texturing at stairway walls is uniformly sprayed around all inside and outside corners of the stair steps.

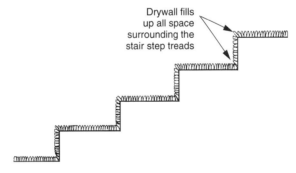

☐ **472**
Have the floors clean and free of loose debris before the start of drywall texturing so that the texture spray gun operator can move around without having to worry about tripping over something on the floor. This results in a more uniform application of texture.

Drywall 203

☐ **473**

Do not allow the drywall texture cleanup crew to use excess water on second floor and third floor plywood subfloors for the cleanup and removal of texture. This can cause the plywood to swell, loosening nails that cause floor squeaks later.

☐ **474**

Check that the quality of the taping matches the type of texture being used and account for the extra time in the construction schedule required for better taping in preparation for a light texture application.

☐ **475**

Before the drywall prepaint repairs, lightly circle with a carpenter's pencil the scratches, dents, and dings on the drywall surfaces that need repair. This establishes the quality and extent of the drywall prepaint repairs instead of leaving it up to the drywall subcontractors.

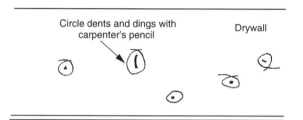

☐ **476**

To gain a head start on the drywall taping, start the taping around the electrical outlets and ceiling flush lights before the drywall nailing inspection, as these areas will not cover up or affect the inspection. Get preapproval from the building inspector.

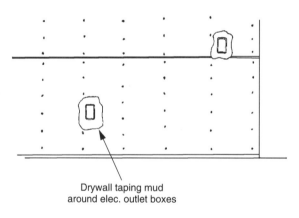

Drywall taping mud around elec. outlet boxes

☐ **477**

Have the texture cleanup crew clean out the channels between the stairway steps and the wall surfaces of all drywall scraps, taping and texturing debris. This keeps the finish carpenter from having to clean out those channels later in preparation for the installation of stair skirtboard.

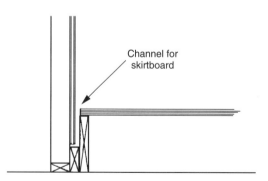

204 **Chapter 19**

☐ *478*

When an attic space requires two layers of drywall for a two-hour fire rating and the ceilings are 9 or 10 feet high with a ladder truss above the wall framing, shift the ladder truss over ⁵⁄₈ of an inch during the framing phase to accept the second layer of drywall. This allows the entire height of the wall to be flush without a ⁵⁄₈-inch bump at the 8-foot-high point for the second layer of drywall.

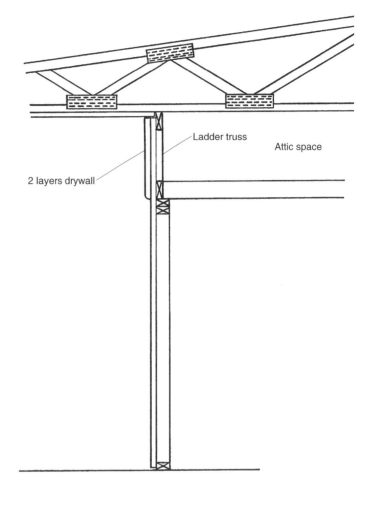

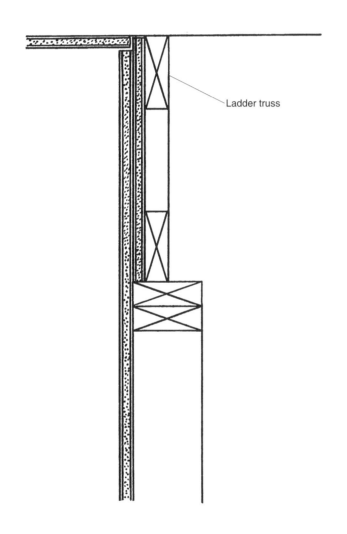

☐ *479*
When walls and ceilings get a knockdown drywall texture finish, the electrical cover plates will not sit flat against the wall on all four sides because of the drywall texture. Consider ahead of time whether to include in the painting subcontract provisions for caulking the gaps around the electrical cover plates.

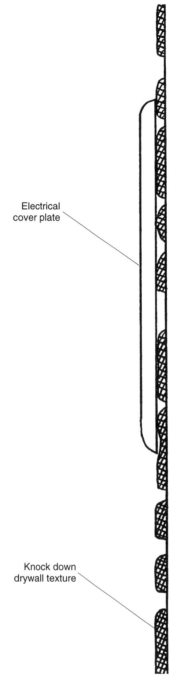

☐ *480*

For interior doors with bull-nose drywall corner bead instead of wood casing trim, provide a wide enough reveal on the door-jamb edge so that the doorknob striker bolt does not hit or rub against the corner bead when the door closes.

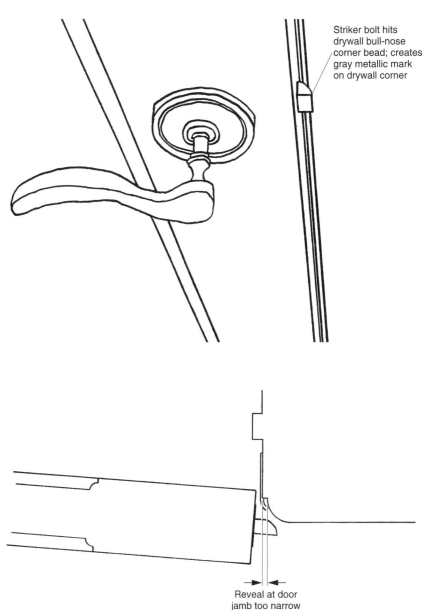

CHAPTER 20

Cabinets

☐ **481**
Check that the cabinets have face frames wide enough to allow the opening and closing of drawers without hitting the door casing.

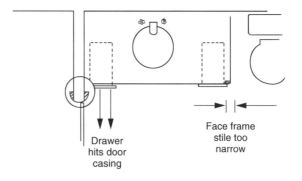

☐ **482**
Island cabinets should have two full-length sides that extend down to the floor, without toe kicks, to provide rigidity to the cabinet. Placing toe kicks on four sides creates too many connections for the cabinets to achieve a firm and solid feel.

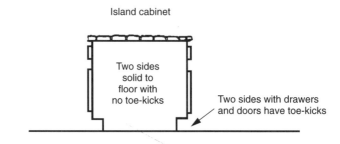

☐ **483**
Provide pass-through countertop overhangs with enough legroom so that people do not hit their knees before pulling up close to the countertop.

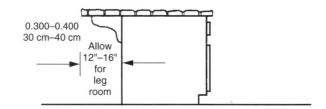

☐ *484*
For floor-to-ceiling kitchen pantry cabinets with removable and adjustable-height shelves, install one fixed shelf at mid-height to provide support to the side panel or install wood cleats on the back wall.

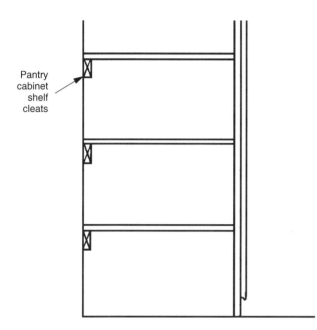

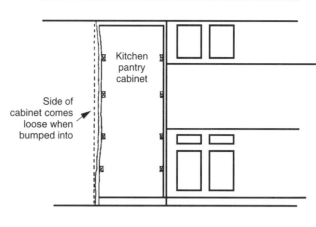

FRONT VIEW

☐ *485*
For production tract housing, kitchen pantry cabinets with fixed, nailed-in-place shelves are easier to manage than are cabinets with loose, adjustable shelves. Fixed shelves cannot be lost, stolen, or damaged.

☐ *486*
For wide pantry cabinets with adjustable-height shelves, provide a method to support the front of the shelves at the center face frame stile, for example.

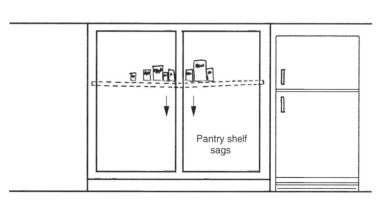

FRONT VIEW

Continued on next page

Cabinets 209

Continued from previous page

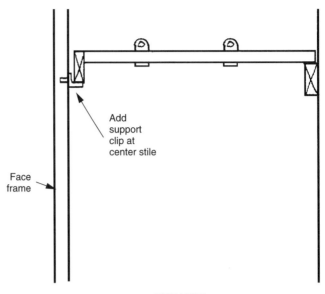

SIDE VIEW

☐ **487**
Check that the upper cabinet valence's wood trim for the undercabinet fluorescent light fixtures is deep enough. The valence hides the light fixture.

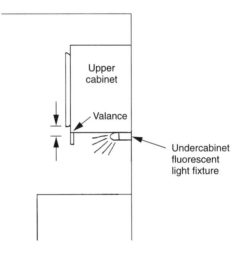

☐ **488**
In areas where new home buyers may choose hard surface flooring such as wood, ceramic tile, or marble, for example, raise the cabinets off the floor an additional amount by increasing the depth of the toe kick and then install the hard surface flooring in the appliance areas as well.

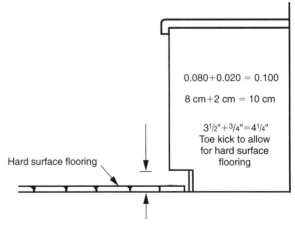

210   **Chapter 20**

☐ **489**
When cabinet knobs are used, check that the corner stiles are wide enough so that drawers do not open into the cabinet knobs of the adjacent drawers and doors.

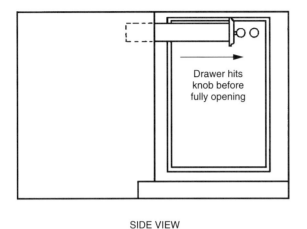

SIDE VIEW

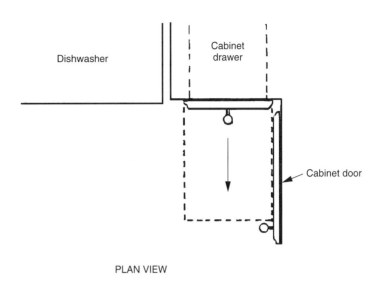

PLAN VIEW

☐ **490**
Lay out the cabinets and appliances on the kitchen subfloors of the sales models, using colored chalk lines, to check whether everything will fit with the correct reveals and margins.

☐ **491**
Check the kitchen floor plan and make sure the refrigerators are not designed to be placed too close to dishwashers so that the refrigerator is within the downward swing of the dishwasher door.

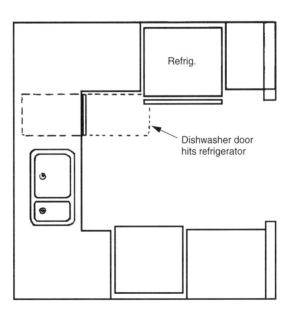

☐ **492**
Check that there is enough dimensional width in drop ceiling soffits for both flush lights and cabinet crown molding trim to fit.

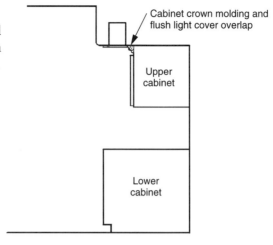

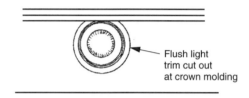

☐ **493**
Make sure the upper cabinet crown molding will not project beyond the adjacent wall corners.

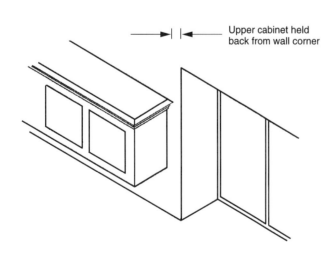

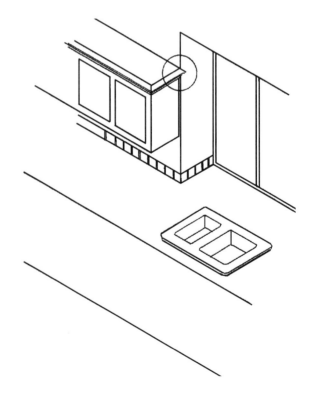

☐ **494**

If the interiors of the houses are not ready to receive a delivery of cabinets because the drywall texturing or wall painting is behind schedule, reschedule the cabinet delivery for a later date. Do not have the cabinets delivered into the garages or into another house, to be spread later by the job site superintendents and laborers.

☐ **495**

Do not perform a quality sign-off walk-through with the cabinet company's field supervisor. This will turn into a "run-through" without enough time spent looking closely for cabinet defects that need repairs.

☐ **496**

For cabinets that have wood countertops instead of ceramic tile, devise a method to protect the countertops during the course of construction.

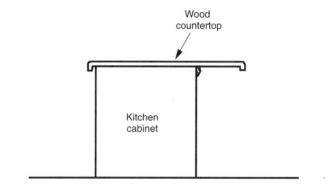

☐ **497**

For cabinets with backside panels, have the cabinet installers make neat cuts around electrical outlets so that standard-size cover plates can be used.

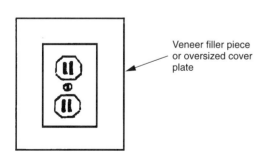

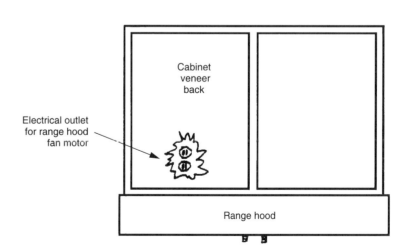

Cabinets 213

☐ **498**
For cabinets with backside panels, have the cabinet installers make neat, square cuts around plumbing pipes without saw kerfs running past the corners of the cutout. Cabinet cutouts for plumbing pipes should be made with the same care used for other exposed finish materials.

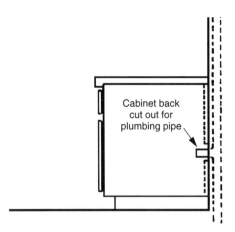

☐ **499**
Cut cabinet scribe molding wood trim at angled walls with the same angle to match the walls.

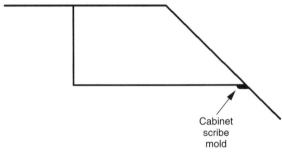

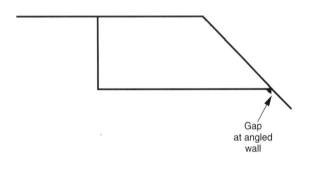

☐ **500**
Install scribe mold wood trim around all the sides and back walls of each linen cabinet shelf to cover the gaps; this looks better.

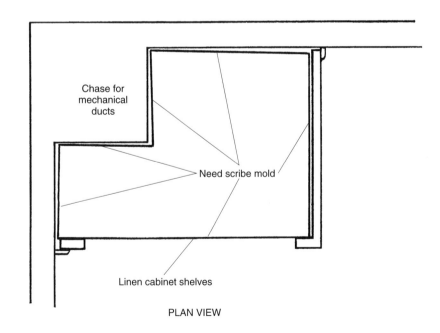

☐ **501**

When coved vinyl flooring is installed, have the cabinet scribe mold only cut and tacked in place so that it can be removed easily and recut to a shorter length to fit after the coved vinyl flooring is in.

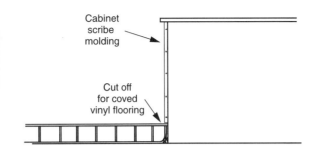

☐ **502**

With cabinets that extend from floor to ceiling, such as pantry cabinets in kitchens, make sure the gap at the floor is not wider than the standard base shoe. If it is, have the cabinet company provide and install matching wood trim that is wider than the base shoe or obtain wider material and prepaint or stain it to match; it will be installed by the finish carpenter with the base shoe.

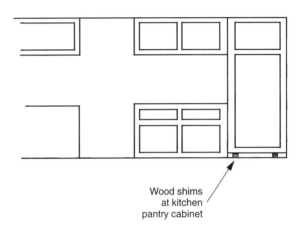

☐ **503**

Install a wood cleat on the rear wall at the range hood opening in the kitchen's upper cabinets that is equal in depth to the face frame at the front of the cabinets. This allows the range hood to be installed level.

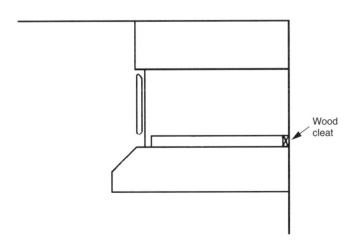
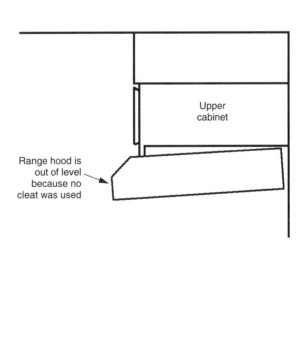

Cabinets 215

☐ **504**
Include as part of the cabinet installation the provision that the cabinet setter will cut out the holes in the bottom shelves of upper cabinets for the range hood's electrical extension cords.

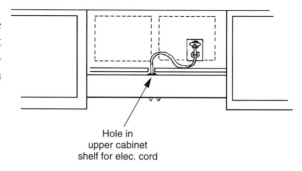

Hole in upper cabinet shelf for elec. cord

☐ **505**
Check that the openings in upper cabinets are wide enough for the microwave ovens to fit during the cabinet installation phase of construction. The openings should be $30^{1}/_{8}$ inches or $30^{1}/_{4}$ inches wide for a 30-inch-wide microwave oven, for example. The time to discover a problem with the openings is while the cabinet setters are on the job site.

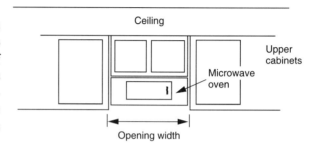

☐ **506**
When the cabinet setters are responsible for cutting out or drilling the holes through the bottom shelves of the upper cabinets for the electrical extension cords to microwave ovens, obtain a sample template ahead of time from the manufacturer (purchase one microwave and then make copies of the cardboard template) so that templates are on the job site showing the exact location of these holes at the time of cabinet installation.

☐ **507**
Include in the cabinet subcontract the installation of felt, rubber, or plastic bumper pads at the corners of cabinet doors or give the builder those bumper pads so that they can be installed later by the builder, after the cabinets have been painted or stained and lacquered.

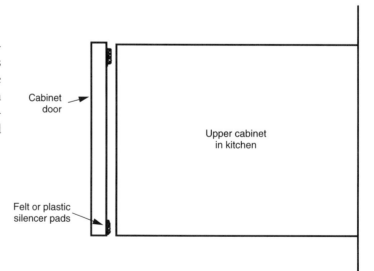

216 **Chapter 20**

☐ *508*
During cabinet installation, obtain from the cabinet subcontractor one breadboard and use it as a test piece to check that the breadboard slot openings are not too tight; then have these tight openings repaired by the cabinet setters. Breadboards usually are not included with the cabinet delivery or installed with the cabinets because they can be stolen. This practice of delivering breadboards later prevents the discovery of tight breadboard slots until late in construction.

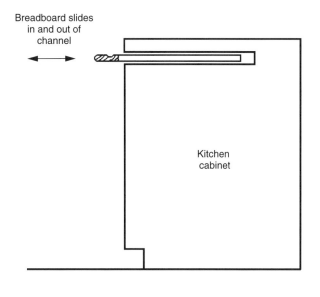

☐ *509*
Decide whether to schedule the installation of the plumbing angle stops before or after the installation of the cabinets.

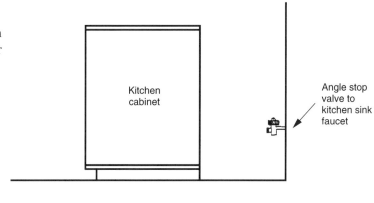

☐ *510*
Obtain a sample template for the hole cutout in bathroom cabinet plywood rough tops for the sinks and make several copies. Also determine the correct setback dimension for the hole from the front and rear edges of the cabinet rough top. The templates and the setback information are needed by the cabinet setters who cut out the holes for the bathroom sinks, yet this activity occurs long before the bathroom sinks are delivered to the job site for the finish plumbing phase.

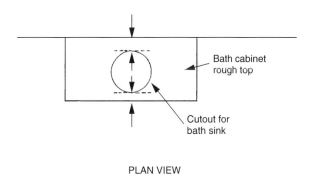

Cabinets *217*

☐ *511*
When the bathroom cabinet splash height establishes the elevation of the tile for an adjacent bathtub, for example, have the cabinet setters install the cabinets by starting with the bathrooms first. This allows the tilework to begin as soon as the bathroom cabinets are set and to proceed while cabinets are being installed elsewhere in the house.

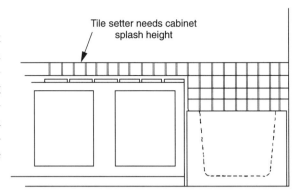

☐ *512*
As part of cabinet installation, have the cabinet setters of prefinished cabinets clean off smudges, pencil marks, and greasy fingerprints from the cabinets. This prevents this activity from carrying over unfairly to the final cleanup subcontractor.

☐ *513*
Install and tighten cabinet knob screws by hand, using a screwdriver rather than a motorized drill or screw gun. This prevents the screw slot from being chewed up by a motorized screwdriver bit turning inside the slot during the last revolutions, resulting in sharp metal burrs on the ends of the screws.

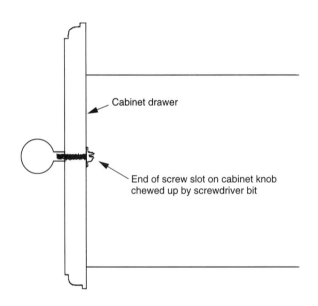

☐ *514*
Have the cabinet installers cut off the wood wedge shims that are used to raise, align, and level the cabinets flush with the front surfaces of the cabinets. If they are not cut off, the wood shims will be in the way of the vinyl flooring and base shoe installation later.

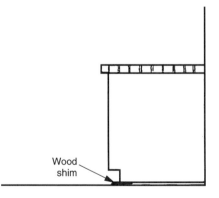

☐ **515**
Have upper cabinet wood valence trim pieces span entire lengths without splice joints, which are usually very noticeable and unsightly. The cabinet subcontractor should be able to provide full-length pieces for each length of cabinet.

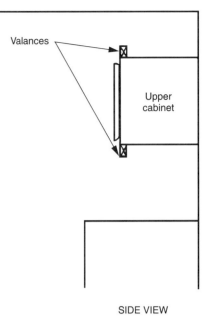

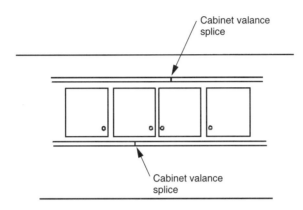

☐ **516**
For a bar-top countertop door, provide some means, such as a chain stop and/or a wall-mounted rubber bumper pad, to prevent the door from hitting the wall and denting the drywall.

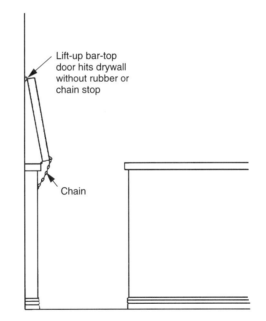

☐ **517**
When the builder offers home buyers the option of selecting kitchen cabinet styles and colors at the time of purchase, add to the cleanup subcontract or the cabinet subcontract a special cleanup clause that will follow the installation of the cabinets inside otherwise completed production units. This is a unique situation because the cabinets are installed outside the normal construction sequence and therefore require a special cleanup.

☐ **518**

For prefinished cabinets, check that the cabinet manufacturer supplies prefinished base shoe trim that matches the cabinets.

☐ **519**

Check that the crown molding on an upper cabinet does not extend too close to a windowsill, requiring the window to have standard drywall corner bead instead of bull-nose corner bead when the rest of the windowsills have bull-nose corner bead.

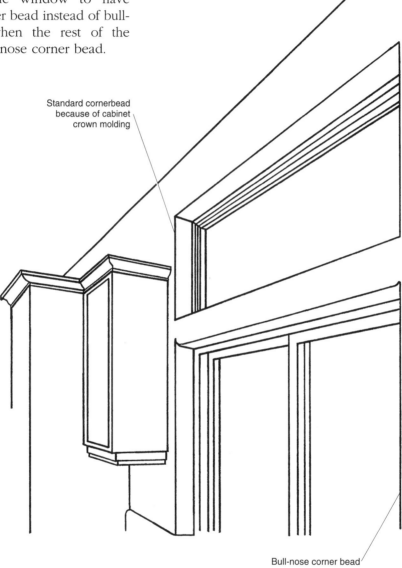

☐ **520**
Provide a wide enough cabinet face frame at a split-level step that has wood flooring with nosing trim at the front edge of the step. If the nosing projects 1 inch beyond the step, for example, the cabinet door may hit the nosing before fully opening.

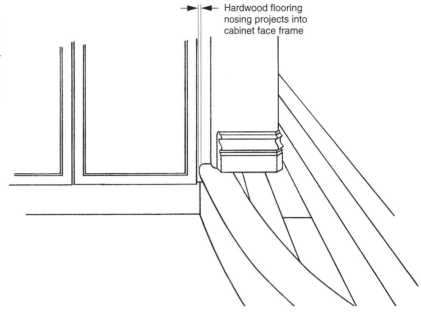

Hardwood flooring nosing projects into cabinet face frame

☐ **521**
Make sure the crown molding on an upper cabinet does not extend over into the casing trim of an adjacent doorway. On the plans, the upper cabinets are longer than the lower cabinets because of the crown molding.

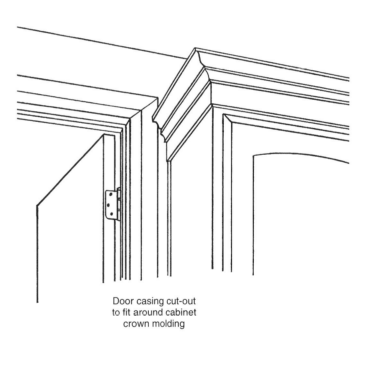

Door casing cut-out to fit around cabinet crown molding

☐ **522**
For laundry room cabinets that widen to accommodate a laundry sink, check that the swing of the door clears the corner of the cabinet.

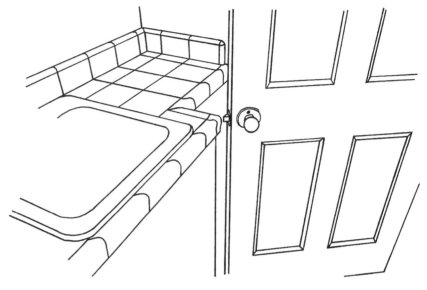

☐ **523**
For cabinets without exposed face frames, design and install the upper cabinet doors above a microwave oven with about $1/4$ to $1/2$ inch of clearance, exposing some of the shelf edge between the cabinet doors and the microwave. This allows the upper cabinet doors to sag slightly downward yet still open and close freely without touching the top of the microwave.

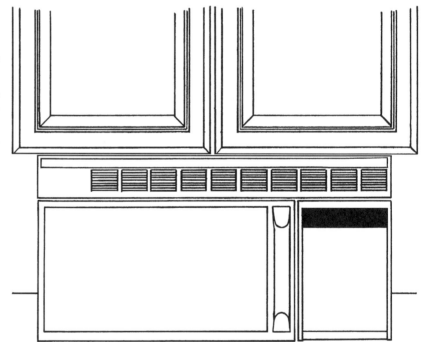

☐ **524**

When kitchen cabinets have a 45-degree section at a corner, consider the clearance between the countertop material and the cabinet doors and drawers. If the front edge of the countertop (granite, for example) must be cut off at the corner to miss the adjacent cabinet door, consider this ahead of time. The granite countertop should have this corner done in the factory with a glossy finish to match the rest of the countertop, or the cabinet face frame should be widened to provide more clearance from the countertop corner.

☐ **525**

For cabinets without face frames, check that there will be enough clearance between the bull-nose tile at the front edge of the cabinets and the top row of doors and drawers. If there is not, add more depth to the cabinet face frame's top rail or increase slightly the thickness of the ceramic tile floated mortar base.

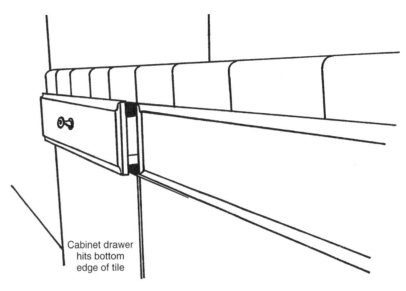

Cabinet drawer hits bottom edge of tile

☐ **526**

For kitchen upper cabinets that have crown molding instead of dropped soffits, consider ahead of time what to do with the range hood or microwave sheet metal vent. You don't want to add a special cabinet section to enclose the vent that looks out of place and unsightly, especially in kitchens with 9- or 10-foot-high ceilings.

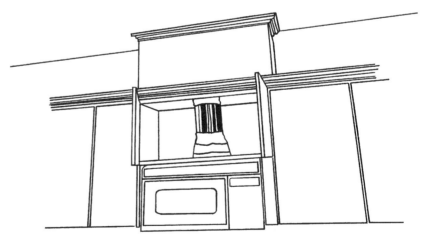

☐ **527**
For kitchen upper cabinets with crown molding and with cabinet tops that are exposed to view from an adjacent stairway or hallway, add dust shelves to the tops of the cabinets that come up flush with the top edges of the crown molding to add a finished look to the cabinet tops, such as fine furniture hutches.

Fill in top of cabinet with finish material

☐ **528**
Design and install cabinets in pantry closets to avoid closed-off dead spaces in the corners. Have a narrow vertical stile at the front corner with open space behind it that can be reached from the right or left side.

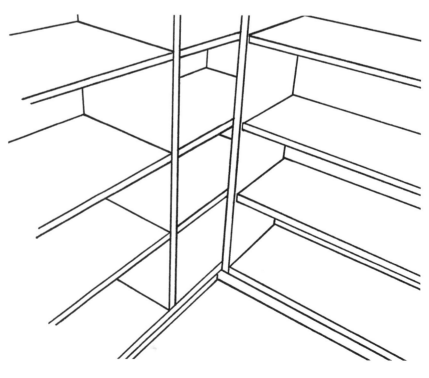

224 **Chapter 20**

☐ **529**
For drywall-covered dropped soffit beams in kitchen ceilings, for example, check that the crown molding wood trim at the top of the upper cabinets will not extend up into the dropped soffit beams, requiring that the crown molding be notched out to fit around the soffits.

☐ **530**
Avoid upper cabinet designs that feature wide cabinet doors that may sag downward from their own weight, resulting in uneven gaps at the middles and tops of the doors.

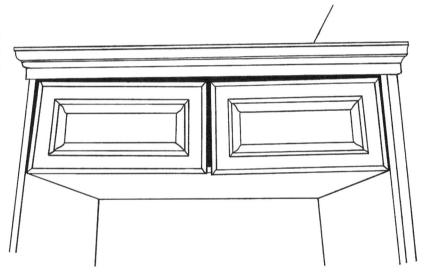

☐ **531**
For kitchen countertops that step down in elevation, check that the difference between the two countertops is greater than the thickness of the countertop finish material (marble in this case), plus a leftover amount for a reveal of the side of the upper level cabinet. You don't want the marble countertop at the lower level to slip underneath the marble countertop at the upper level with a gap of only $1/4$ or $1/8$ of an inch between the countertops.

□ *532*
For upper cabinets with crown molding trim in rooms with sloping ceilings, check that there is enough wall height at the backs of the cabinets for the crown molding to fit. Cutting the crown molding to fit against the sloping ceiling at the ceiling-to-wall corner creates the appearance of a design and construction mistake.

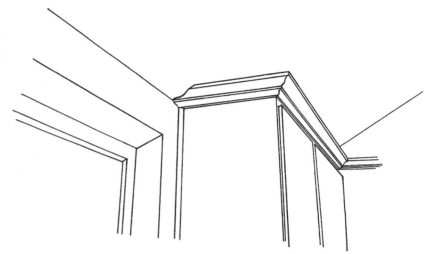

CHAPTER 21

Finish Carpentry

☐ **533**
Hang interior and exterior doors with a uniform gap between ¹/₈ and ³/₁₆ of an inch between the door edge and the jamb so that after paint is added, this gap will be about ¹/₈ of an inch and the doors will not bind or stick after settling or swelling slightly.

☐ **534**
For interior prefit doors in kitchens and bathrooms that may get vinyl flooring, cut off the bottom of one side of the jamb and casing rather than raising the opposite side when you are adjusting the door unit to achieve a proper fit. Vinyl flooring is not thick enough to cover the gap at a jamb that has been raised off the floor.

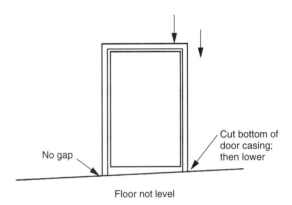

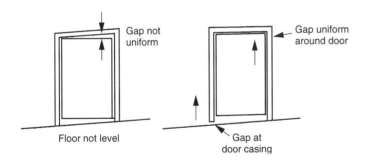

☐ **535**
When a few windows are accidentally broken during construction, devise a method to keep interior doors from being continually slammed open and shut by the wind before the time when they are painted and hardware installed. Wind can blow doors through the jamb stops, off their hinges, and onto the floor. Wedge the interior doors in the open position by using scrap pieces of baseboard or casing trim.

☐ **536**
Make sure doors close evenly against the jamb stops before painting so that jamb stops that do not fit the doors tightly can be moved over and adjusted without ruining the paint, caulking and/or exposing raw unpainted wood that will require touching up.

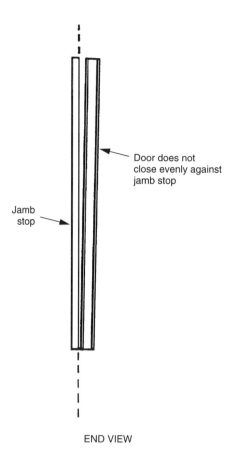

END VIEW

☐ **537**
Install pocket doors so that they close evenly against the striker side door jamb but are high enough above the floor to clear the carpeting.

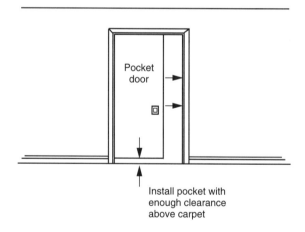

☐ **538**
When pocket doors rub against one side or both sides of the interior pocket channel, wedge pieces of wood baseboard trim between both sides of the pocket door and the pocket door frame for a 24-hour period. This will permanently bow out the sides of the frame, providing additional clearance around the door.

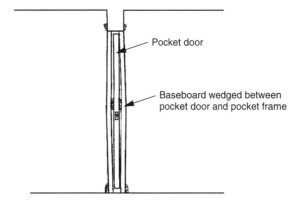

☐ **539**
Consider the option of having the finish carpenter install the pocket doorjamb on the striker side at the time when the pocket door and casing are installed during the finish carpentry phase. This allows the finish carpenter to shim the jamb straight and plumb to fit the door rather than adjusting the door to match a jamb that is out of plumb.

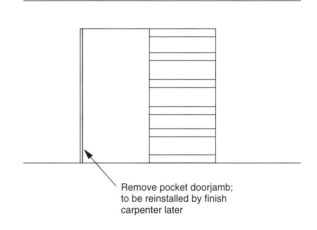

☐ **540**
Install the pocket doorjamb stop at the top header with an additional margin or gap so that the pocket door can be wiggled out of the top metal track for removal if necessary. If the jamb stop margin is too tight, the pocket door cannot be removed without first removing the wood trim of the jamb stop.

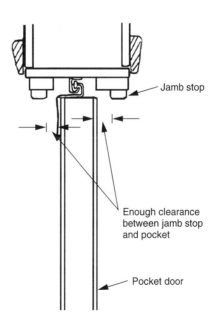

Finish Carpentry

☐ *541*

Consider the option of adding more nails to the striker side of interior doorjambs as standard practice to prevent the jambs, casing, and doors from coming loose as a result of the wind slamming doors open and shut before the doors are painted and the doorknob hardware is installed.

☐ *542*

For arched windows that get wood casing trim, order custom-made casing to match the radius of an arched window.

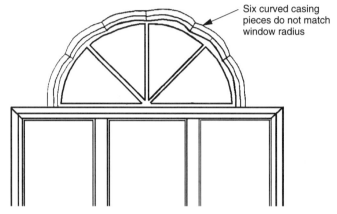

Six curved casing pieces do not match window radius

☐ *543*

Instruct the finish carpenter to remove and replace any piece of wood trim that cracks or splits upon installation. This saves the painter the trouble of caulking, spackling, and painting a piece of wood that will be rejected and replaced later.

☐ *544*

Instruct the finish carpenter to sandpaper smooth all exposed cut ends of wood trim as part of the installation.

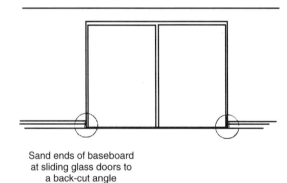

Sand ends of baseboard at sliding glass doors to a back-cut angle

☐ **545**
Order rough-sawn lumber and trim in the correct sizes so that pieces do not have to be ripped to narrower widths, resulting in smooth cut surfaces mixed in with rough-sawn surfaces.

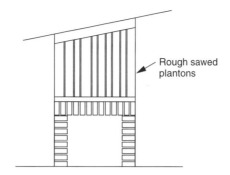

☐ **546**
When home buyers have the option of selecting mirror wardrobe doors and those selections have been made for some of the units before the start of the finish carpentry phase, inform the finish carpentry subcontractor ahead of time which closet openings have had mirror doors selected to avoid wasting time installing and painting standard closet doors in those openings.

☐ **547**
Adjust wardrobe closet bipass doors to match each vertical side wall or bumper jamb and high enough above the floor to clear the carpeting.

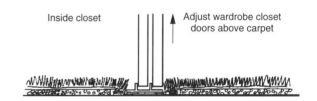

☐ **548**
Cut the wardrobe door header track to extend beyond the inside edges of the bumper jamb. The bumper jamb then will hide any gap that would show if the track were not cut perfectly tight against the jamb.

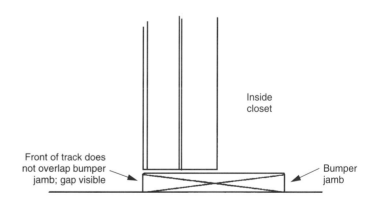

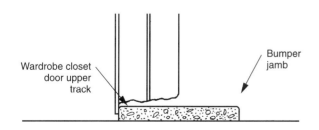

Finish Carpentry

☐ **549**
Shop prime paint the fronts and backs of false wood shutters to seal the wood before the shutters are installed over a stucco brown coat, for example; they will later be painted on the exterior exposed side along with the surrounding wood trim and roof fascia.

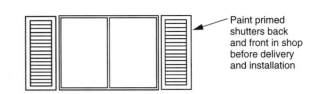

☐ **550**
Do not install shelf metal supports for a wardrobe closet directly against drywall. The shelf support should be installed over a wood hook-strip cleat. Drywall has little compressive strength, and weight placed on the clothes pole will mash in the drywall gypsum and break the paper surface.

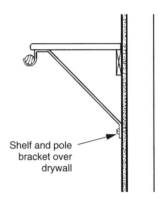

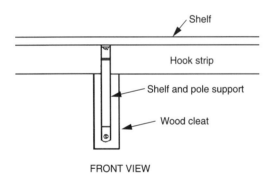

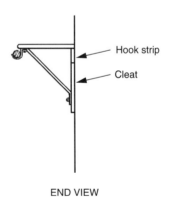

232 **Chapter 21**

☐ **551**
Install bumper jambs tight against the header of the door opening of the wardrobe closet so that if the metal header track is cut slightly short at either end, there will not be a gap or hole behind the track.

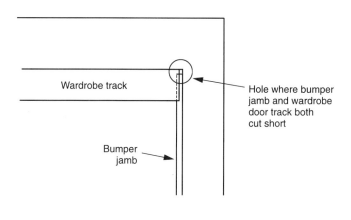
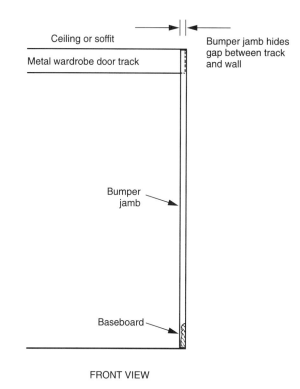

☐ **552**
Check that the nailing through particleboard results in nails that are set flush with or below the surface of the particleboard. The 85- or 90-psi of air pressure in a nail gun that drives nails correctly into soft pine door casing trim may not get nails all the way through particleboard closet shelving, for example, with the heads countersunk.

☐ **553**
Check that the saw kerfs are not overcut in stairway skirtboard that could be exposed by thin or tight grades of carpet.

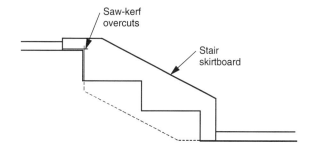

☐ **554**
Check that the ears at the top of the stairway skirtboard are equal on both sides.

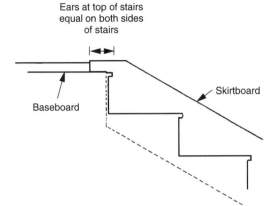

☐ **555**
Check that stairway skirtboard exactly matches the slope of the stair steps.

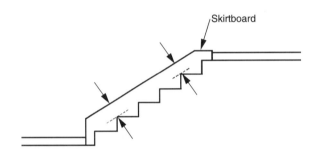

☐ **556**
Check that the finished factory edges of particleboard shelving are uniformly curved and smooth, especially when it is used for stairway skirtboard.

☐ **557**
Use two support brackets at a minimum, top and bottom, for short runs of stairway handrail. One bracket alone in the center of a short piece of wood handrail looks bad.

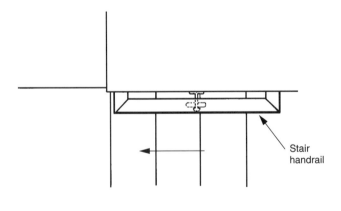

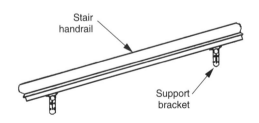

234 **Chapter 21**

☐ **558**

Instruct the finish carpenters installing stair handrail to predrill the holes for screwing into the wall backing to attach the handrail support brackets. If they are not predrilled, the last revolutions using a screw gun will strip out the ends of the screw heads. This makes it difficult to remove the screws later and leaves sharp metal burrs at the screw heads.

☐ **559**

Don't allow the baseboard installer to slip baseboard into narrow gaps between the door casing and the wall. Baseboard should always butt into another piece of baseboard.

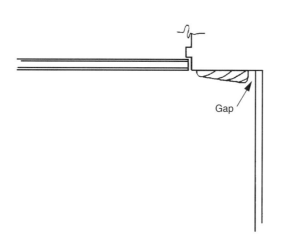

☐ **560**

Don't allow the baseboard installer to reduce the margin gap between the interior doors and jambs by forcing into the wall a bowed piece of baseboard that is cut slightly too long, between a wall corner and an interior door. This practice pushes over the door casing and thus the doorjamb, reducing the gap to the point where doors rub at the bottom after paint is added.

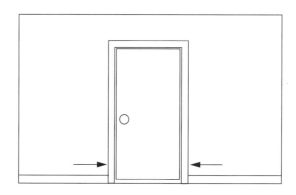

Finish Carpentry

☐ **561**
Use baseboard horizontally and vertically at stair steps in lieu of skirtboard at curved walls that have pie-shaped stair steps; 1×2-inch skirtboard will not bend around a curved wall.

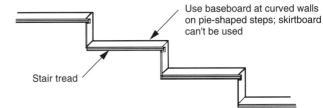

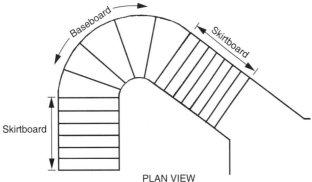

☐ **562**
Have the finish carpenter hand nail the baseboard. This pulls and holds the baseboard into the walls more firmly than do nails driven with air-compressed nail guns.

☐ **563**
Don't allow butt joint splices in baseboard. Baseboard splices should be mitered angle cuts so that if the wood shrinks slightly over time, there will be wood behind the front surface splice joint to hide the gap.

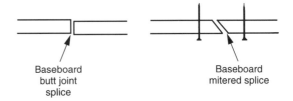

☐ **564**
For wide baseboard with detailing, raise the baseboard above the floor $1/2$ to $3/4$ of an inch to prevent the baseboard from being partially covered or buried by thick carpeting.

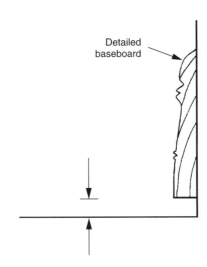

☐ **565**

For curved walls with tight radius bends, remember to order and obtain flexible rubber baseboard and base shoe in time for the finish carpentry phase.

☐ **566**

For walls that are covered with wood paneling, lay out the 4-foot intervals and paint vertical dark brown or black bands on the walls before installing the paneling. This provides a dark background behind the paneling joints if the factory edges of the panels are not perfectly straight. This technique also works well for random-length wood veneer kits. The entire wall surface should be painted to match the color of the wood pieces.

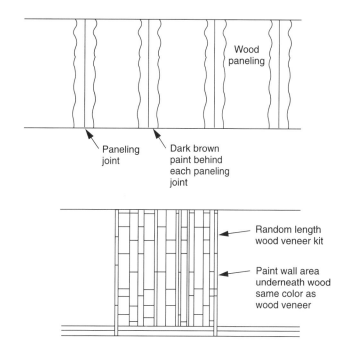

☐ **567**

For three-story condominium and apartment projects with elevators, get the elevators operational before the start of the finish carpentry phase. This saves the finish carpentry crew days of carrying materials up stairways to the units.

☐ **568**

If backcutting the ends of baseboard at sliding glass doors is not aesthetically acceptable, consider cutting the baseboard at a 45-degree miter with a 90-degree return into the floor to eliminate any exposed cut edges.

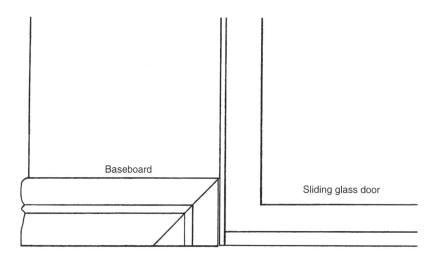

Finish Carpentry 237

☐ **569**
Consider ahead of time who will cut off the bottoms of doors that rub against or are stuck immovably within the thick nap of upgraded carpeting or hard surface flooring: the finish carpenter or the flooring installers.

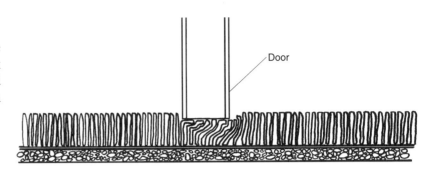

☐ **570**
Check that the top of an 8-foot-high entry door does not open into the bottom of an adjacent 8-foot-high ceiling.

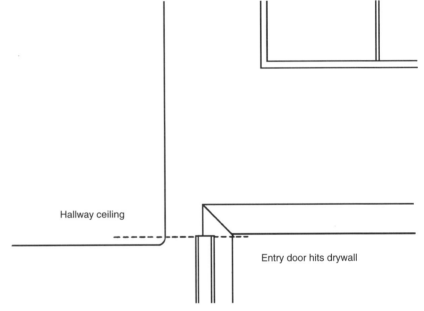

☐ **571**
Consider ahead of time a method to terminate the baseboard at an irregularly shaped precast fireplace hearth. Simply backcutting the end of the baseboard results in an awkward-looking joint.

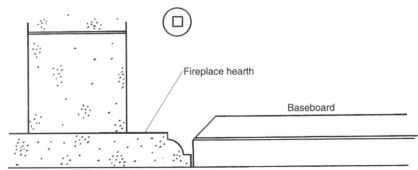

☐ **572**

For windows that get wood shutters, install the stool and apron wood trim on the windowsills several inches farther on each side so that they end up beyond the outside edges of the wood shutters.

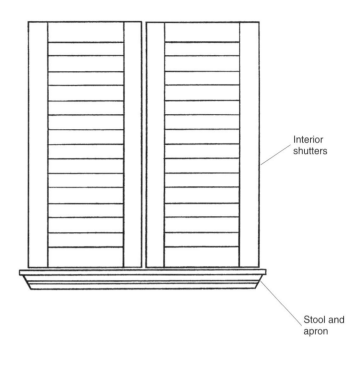

Interior shutters

Stool and apron

☐ **573**

For dropped ceiling soffits, check that there is enough depth for wood crown molding trim and HVAC registers to fit. Cutting out the crown molding trim to fit around the register covers looks like a design and construction mistake.

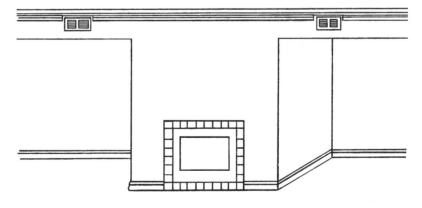

Finish Carpentry 239

☐ **574**
Analyze the various sizes and dimensions of wood cap and trim pieces around wrought-iron handrailing at the edge of a second floor hallway, for example. The wood trim covering the bottom of the wrought iron as it penetrates through the wood cap should not project past the edge of the wood cap.

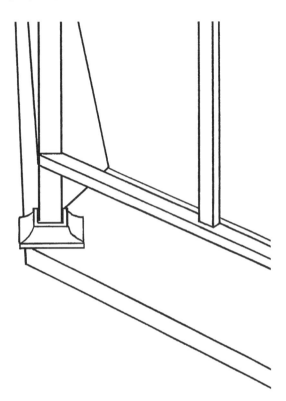
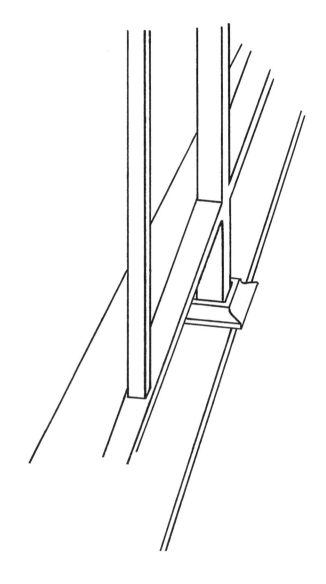

☐ **575**
When a built-out wall treatment of plywood and casing trim extends the surface of a wall until it is even with the adjacent door casing, consider ahead of time how to terminate the ends of the baseboard at the door casing. It can no longer butt into the door casing because the finish wall surface is now even with the front side of the casing.

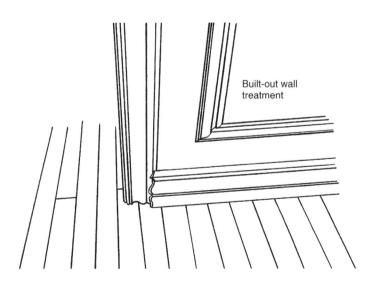

☐ **576**
Check that the dimension from a wall corner to the door opening is wide enough for the cabinet and tile countertop to fit without the need to scribe cut the door casing around the tile and the cabinet.

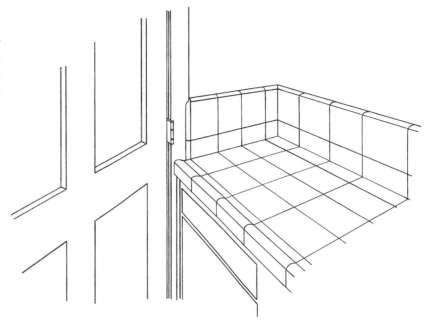

☐ **577**
When the other side of a hallway curved radius wall is a bedroom wardrobe closet with a shelf and pole, consider ahead of time how to finish the shelving, the support hook strip, and the round rosette for the clothes pole at the curved wall inside corner.

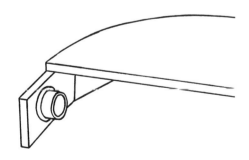

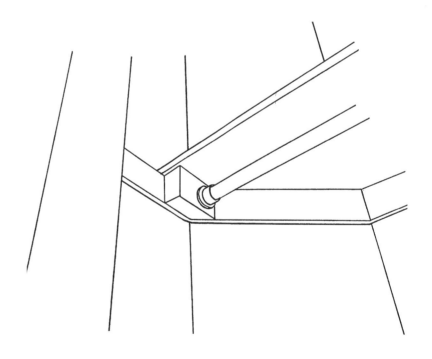

Finish Carpentry *241*

☐ **578**

When bifold doors must be removed by the carpet layers along with the other interior doors that are in the way of carpet installation, include within the finish carpentry subcontract a provision for the finish carpenters to come back and reinstall the bifold doors when carpeting is being installed in groups of several houses at a time. Carpet layers usually do not know how to reinstall and properly adjust bifold doors so that they have the correct margins and gaps.

☐ **579**

Analyze the spindle spacing for stair wood handrailing in terms of the horizontal run of the handrail section and choose a number of spindles that results in a balanced spacing of the spindles without the two spindles at each end being only 1 or 2 inches away from the wall and the newel post when the rest of the spindles are uniformly spaced 4 inches apart.

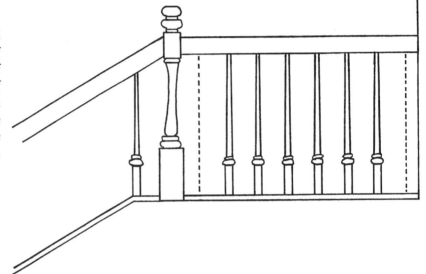

☐ **580**

When the bottom cap piece of a wood handrail is installed over a wood-framed and drywalled short-height wall and the wood cap and handrail members are finished with transparent stain and lacquer, install the bottom cap over a bead of white caulking to fill in any air gaps between the cap and the drywall. It is difficult to caulk the sides of the wood cap after it is installed without getting some of the caulking bead onto the bottom of the cap; the contrast in color between the stained wood cap and the painted drywall and caulking makes this a difficult joint to finish cleanly after the cap is in place.

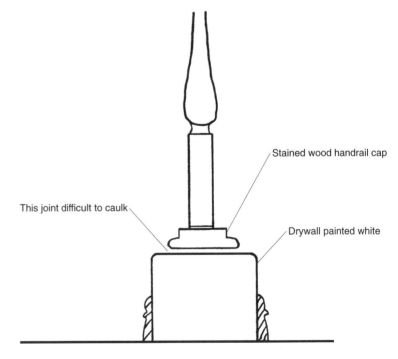

Continued on next page

Continued from previous page

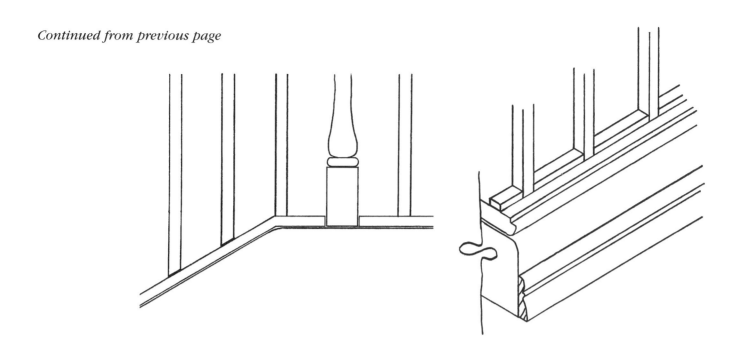

□ **581**
When wood handrail newel posts have wood trim caps at the top, decide ahead of time how to finish the cap at a wall corner when the cap extends past the corner. The ear, as shown in the drawing, looks awkward from the front elevation view.

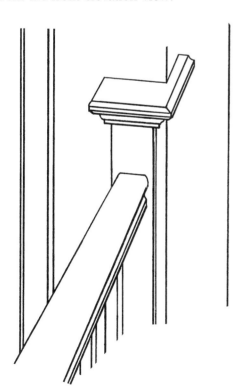

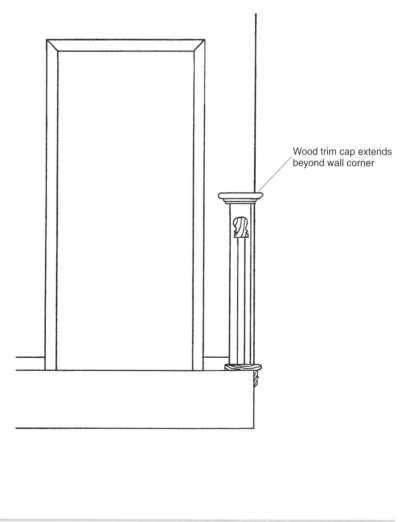

Wood trim cap extends beyond wall corner

Finish Carpentry

☐ **582**
When two dissimilarly detailed wood trim pieces join at an awkward angle, such as at the joint between detailed baseboard and the stair skirtboard, install a block of wood trim between the two pieces so that they butt into the flat block of wood trim rather than into each other.

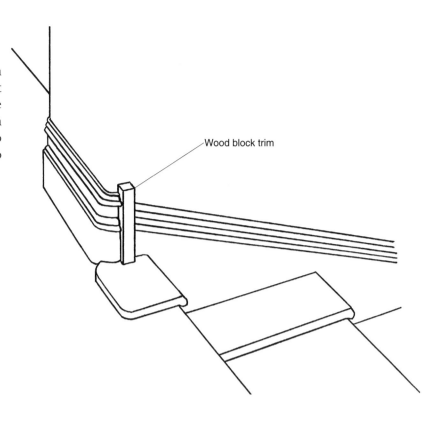

☐ **583**
For wide, detailed baseboard, one method to resolve the conflict between the baseboard and the escutcheon cover plate for the water pipe for the toilet is to install a wood trim piece around the water pipe coming out through the wall that provides a flat surface for the escutcheon cover plate. The baseboard then butts into the sides of the wood trim.

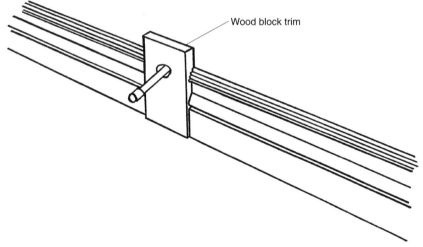

☐ **584**

For a hardwood border around the outside perimeter edges of a second floor hallway that is stained dark brown, for example, also stain and lacquer the edges of the wood border pieces that will be exposed if a thin carpet type is used. You don't want the raw, unfinished edges of the wood border pieces exposed to view as the carpet tucks downward into the wood border.

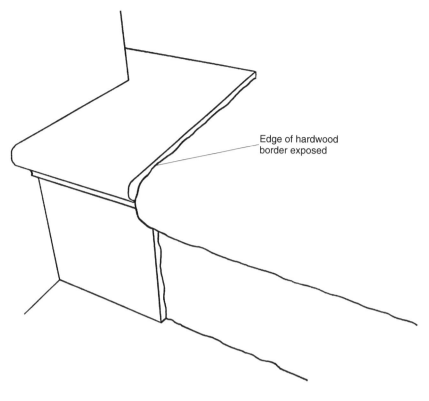

Edge of hardwood border exposed

☐ **585**

For stair pony walls with a hardwood skirtboard and a wood cap trim on top of the wall, check that the height of the pony wall will not result in a narrow leftover reveal of drywall between the top of the skirtboard and the bottom of the cap apron trim. You should have 2 or 3 inches of exposed drywall reveal at a minimum or have the space completely filled in with hardwood trim.

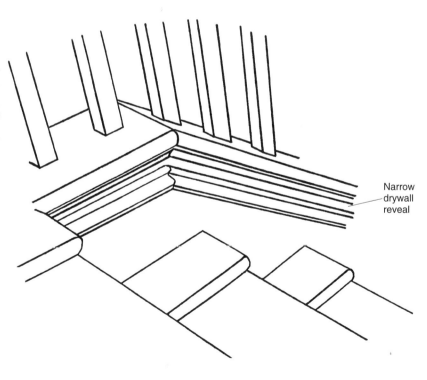

Narrow drywall reveal

Finish Carpentry 245

☐ **586**
For stair handrail newel post tops with cap trim, check that the wall section that the handrail butts into is not narrower than the cap trim. The ends of the cap trim that project beyond the wall corners are visible from the backside, and this has the appearance of a design and construction mistake.

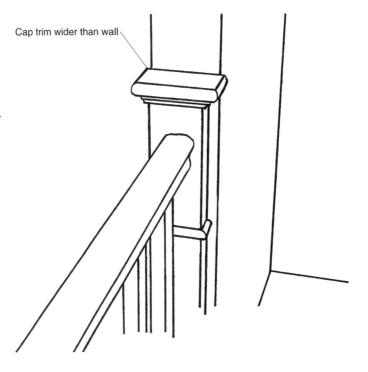

☐ **587**
Check that the alignment between a hallway ceiling above and a stairway landing below does not result in the wood handrailing ending up halfway underneath the corner of the hallway ceiling.

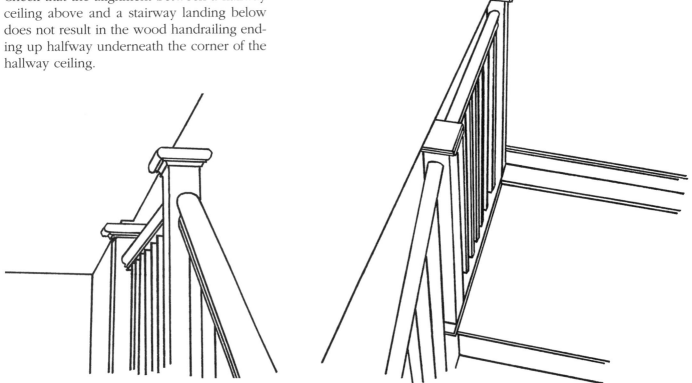

☐ **588**
For windows with wood casing trim, make sure there is enough clearance to an adjacent upper cabinet to leave some wall reveal behind the cabinet scribe mold trim and the window casing. A narrow gap of 1/2 of an inch or less is difficult to finish and looks like a design and construction mistake, especially when the tile underneath the upper cabinet extends over to the window casing.

☐ **589**
For a window in a bathroom that has wood casing trim, check that the casing will not extend over into the bathroom mirror, resulting in the need to cut the bathroom mirror to a length that is a few inches short of the end of the countertop splash.

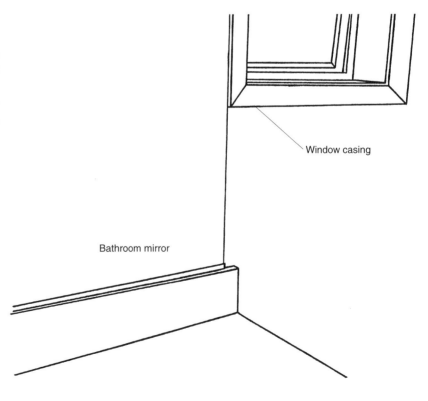

☐ **590**
Design the height of pot shelves to coordinate with the elevation of the stairway handrail.

CHAPTER 22
Ceramic Tile

☐ **591**
Lay out the tile to avoid small pieces at the ends of countertops and other locations.

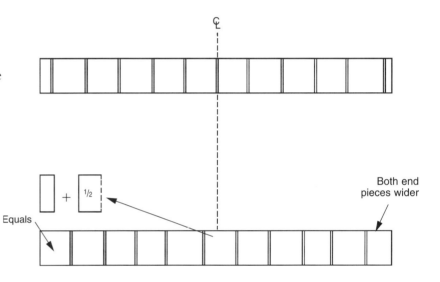

☐ **592**
Lay out the tile to get equal drywall reveals around windows.

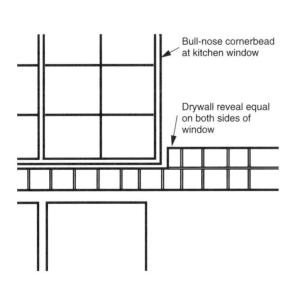

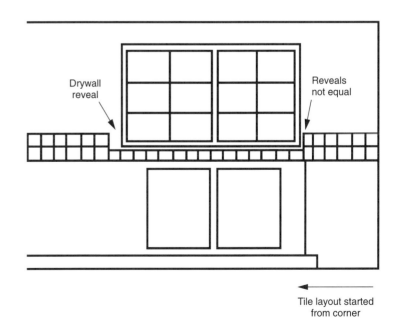

Ceramic Tile 249

☐ **593**
Install wood casing trim around windows before tiling and have the tile butt into the wood casing rather than having the thicker casing butt into the thinner tile.

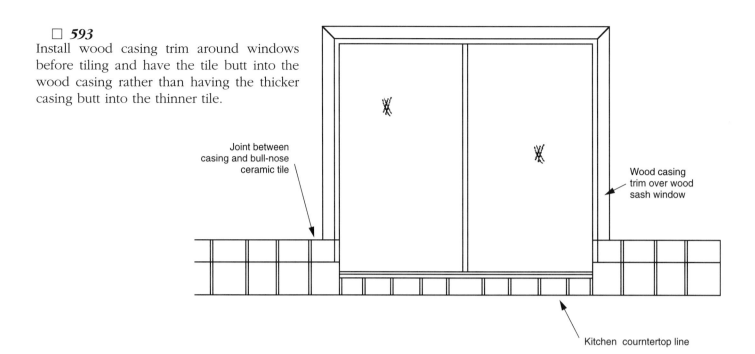
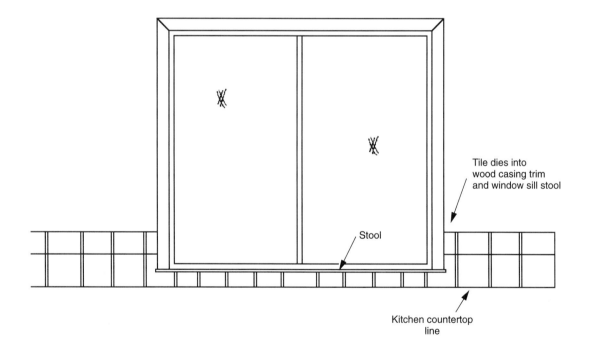

☐ **594**
When tile extends from the bathtub or shower to the ceiling, avoid a narrow top row of tile by cutting the bottom row to a narrower width. The amount cut off the bottom row of tiles is added to the top row.

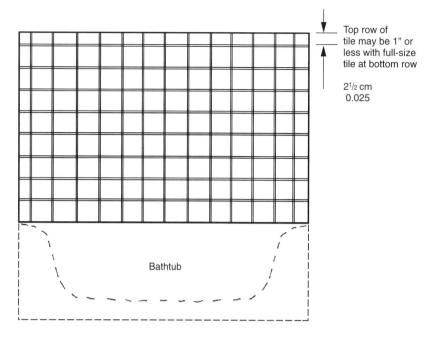

☐ **595**
Check that corner shower dams are square so that the tile setters do not have to float out the dams with mortar to achieve squareness, thus requiring an additional row of cut tiles to complete the top surface of the shower dam.

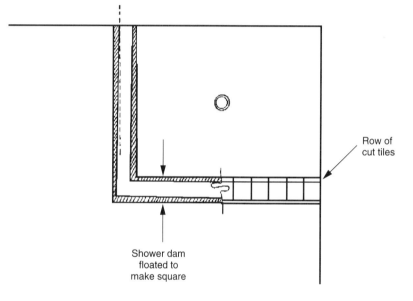

☐ **596**
Avoid installing tile directly against cabinet sides by using a glue on application. The tile may not adhere well to the cabinet surface.

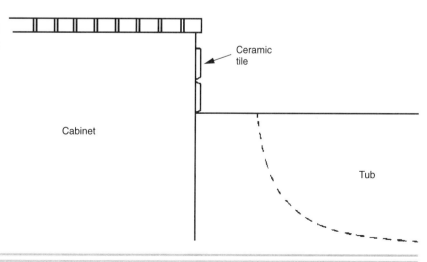

Ceramic Tile 251

☐ **597**
Lay out the tile in bathtub and shower locations so that the tile ends up at least one row higher than the top of the glass shower door enclosure that is installed later. This prevents an awkward-looking and difficult to caulk gap when the shower door's top rail ends up above the tile.

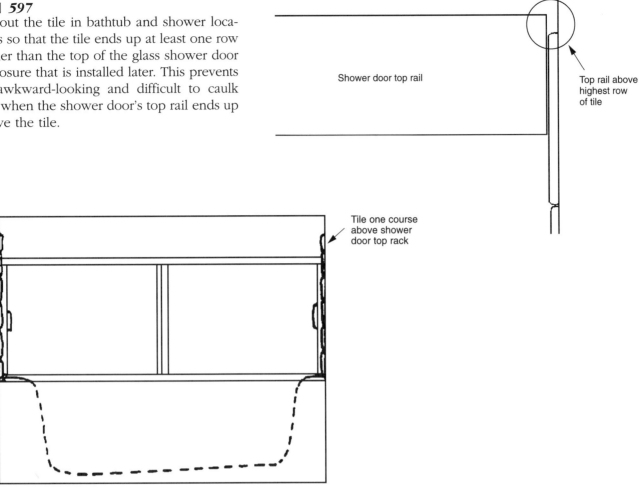

☐ **598**
Check that the combined thicknesses of the cabinet rough top wood and the tile application are not deeper than the bar sink attachment metal clip. If they are, reduce the floated mortar thickness or rout out the underside of the rough top wood before tile and sink installation.

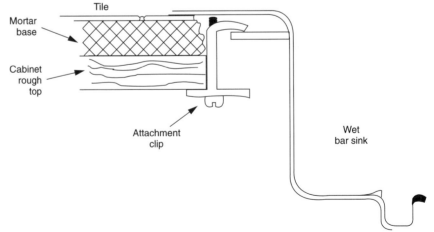

Continued on next page

Continued from previous page

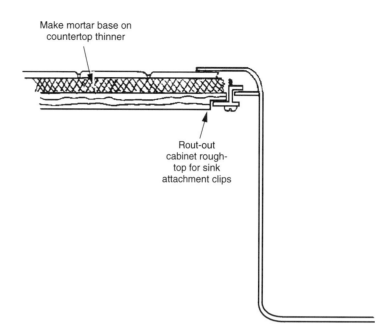

☐ **599**
Use colored caulking that matches the tile grout at locations that are prone to cracking: kitchen sinks, bathtubs, fireplace faces, and spa bathtubs.

☐ **600**
To prevent cracks in the tile grout at the countertop-to-splash corner, use wood at the cabinet rough top that has a low moisture content and has already shrunk and use a lot of nails and/or screws to attach the cabinets to the walls.

☐ **601**
Use extra care to protect colored bathtubs from being scratched during the ceramic tilework. Colored bathtubs, especially dark blue or black ones, show scratches more noticeably than do white bathtubs.

☐ **602**
With the tile subcontractor check bathtubs for scratches, chips, and holes before the start of the ceramic tilework to establish the before and after damage to the bathtub surfaces. This will determine at the end of construction what damage was caused by the tilework and help apportion some of the costs of repairing bathtub surfaces to the tile subcontractor.

☐ **603**
Check that the bathtubs are installed level so that the bottom tile grout joint is uniformly the same narrow width at the bathtub, not $1/8$ inch wide at one end and $1/2$ inch wide at the other end of the bathtub.

Ceramic Tile 253

☐ *604*
The installation of a spa bathtub should result in the tub being firmly secured in place. The vibration from the spa motor and the water jets may create grout cracks if the bathtub is not secure.

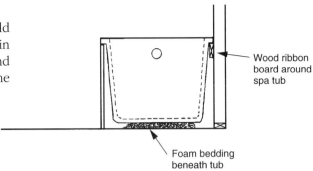

☐ *605*
Frame the wall opening oversize for a shower shampoo shelf and fill in the overage with tile mortar so that full-size tiles are used up to, through, and above the shampoo shelf.

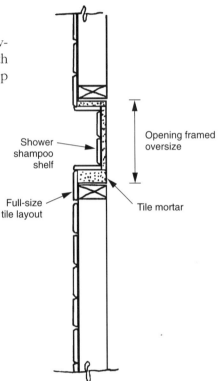

☐ *606*
Check that all the cylindrically shaped plastic valve covers are in place in showers before and during the ceramic tile installation. This provides clearance for the installation of the finish plumbing fixtures later.

☐ *607*
When the floor tile material at the entry door is supposed to match the tile used on the fireplace face and hearth, coordinate the work of the flooring subcontractor and the tile subcontractor so that they both obtain the tile from the same supplier and from the same batch of tile.

☐ *608*
In conditions where vinyl flooring will be installed around a fireplace, such as in a family room, finish the exposed edge of the fireplace hearth with tile or grout to conceal the underlayment material. Vinyl flooring is not thick enough to cover the edge of the hearth.

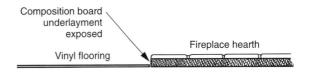

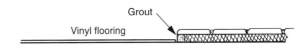

☐ *609*
When the exposed factory edges of tile are unattractive, one method to finish the exposed tile edges at a fireplace face, for example, is to backcut the tile edges at a 45-degree angle and then fill the edges with grout.

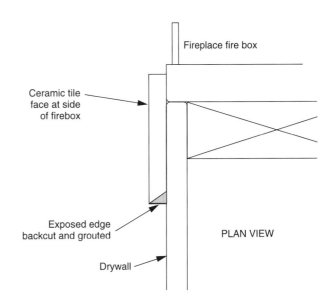

☐ *610*
Have the tile installers sand smooth the cut edges of ceramic tiles, especially at rows of tiles cut at a 45-degree angle for diagonal tile installation.

☐ *611*
For prefinished metal fireplaces, check the manufacturer's specifications for the setback dimension from the edge of the firebox opening to the edge of the decorative face tile for the required minimum clearance for the installation of the glass doors.

☐ *612*
Devise a method to protect dark tile in kitchens and bathrooms from being scratched during construction. Scratches in dark tile show up more clearly and result in the need to replace individual tiles or entire countertops.

☐ *613*
Don't install fireplace hearth tile directly over a plywood subfloor. There should be some type of underlayment to act as a buffer between the subfloor and the hearth tile.

☐ **614**
Have the tile setters cut off the excess black building paper at the front edges of cabinet countertops.

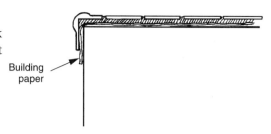
Building paper

☐ **615**
When bathtubs are placed next to walls with sliding pocket doors, check that the scratch mortar coat does not ooze out between the underlying paper joints and project into the pocket door cavity, scratching the pocket door as it slides back and forth inside the pocket channel.

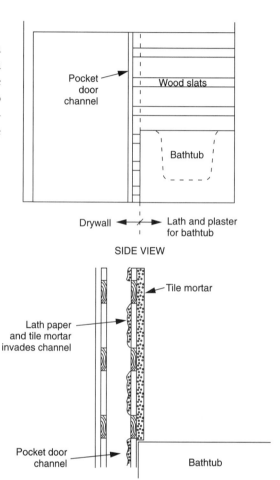

☐ **616**
Have the tile setters clean off any mortar splatter that gets onto the walls adjacent to the ceramic tilework as part of the tile installation. If it is left on the walls to harden, the grout splatter is difficult to remove without damaging the drywall texture, yet rough splatters of mortar cannot simply be left on the walls and painted over.

☐ **617**
On medium-size to large projects, have the tile subcontractor maintain a storage container bin on the job site rather than use an empty garage to store materials and equipment. This prevents a garage from being locked up and inaccessible and prevents boxes of tile from being stolen.

☐ **618**
Check prefabricated Pullman tops for sharp edges and have them filed and sanded smooth as part of the installation.

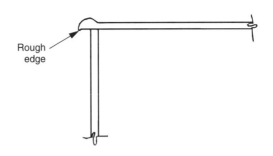

☐ **619**
Check that the ends of the Pullman tops are finished edges, not merely cut edges that have been filed and sanded smooth. Depending on the manufacturer, this may or may not look good.

☐ **620**
Check that the metal clips used to attach sinks to prefabricated Pullman tops are delivered to the job site along with the Pullman tops. The metal clips are used by the plumber to install the bathroom sinks. The clips should be handed to the builder or the job site superintendent, not just left in a plastic bag in one of the bathroom cabinet drawers.

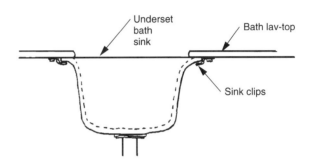

☐ **621**
When possible, avoid Pullman tops that come in two or three pieces for a corner cabinet. It is more difficult to get all the edges flush and even.

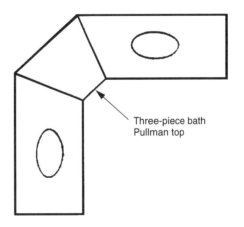

☐ **622**
For conditions where a countertop made of Corian®, marble, or granite has a front edge lip and is installed over a cabinet without face frames, check that there will be clearance between the edge of the countertop material and the tops of the cabinet's doors and drawers. If there is not, build up the cabinet rough top and add a finished trim piece at the cabinet front.

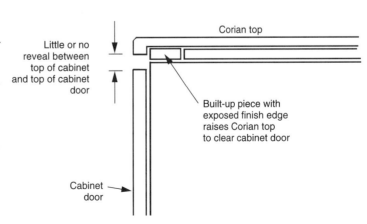

Ceramic Tile

☐ *623*
Check that the continuation of mortar floated tile on the wall behind a toilet from an adjacent bathtub does not affect the required dimension from the wall to the floor closet ring for the toilet to fit.

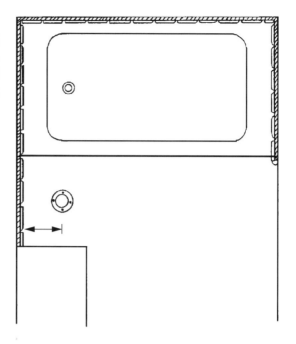

☐ *624*
When new home buyers are given the option of choosing kitchen and bathroom tile colors, for production inventory units that are uncompleted up to the tile phase until they are sold, consider in the tile subcontract the issue of the added cost of trip and setup charges out to the project for the tile subcontractor each time a unit is sold.

☐ *625*
Design short partition walls between cabinets and bathtubs so that the dimensions allow full-size tiles to be used without the need to cut any tiles.

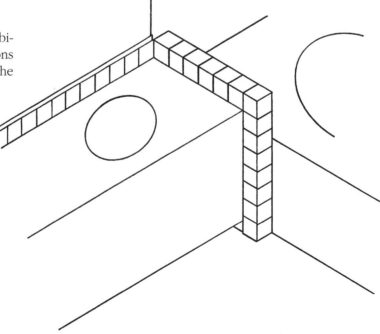

☐ **626**
When a row of tiles extends above the countertop splash for a bathroom cabinet, hold back the tile even with the edge of the countertop splash so that the bathroom mirror, which will be installed later, can extend all the way over to the end of the countertop without hitting the top row of tile.

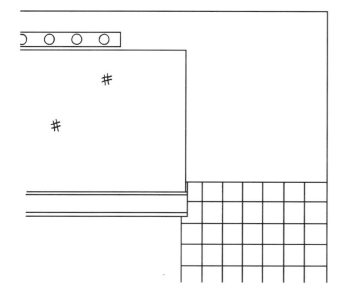
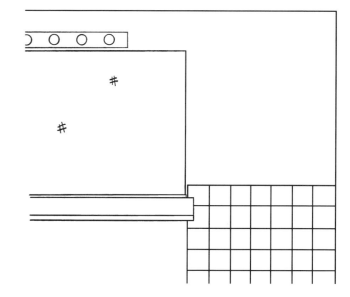

☐ **627**
Check that there is enough dimensional space within a bathroom to provide some empty wall reveal between the door casing and the bull-nose tile for an adjacent shower. If the gap is too narrow, this wall area is difficult to finish and has the appearance of a design and construction mistake.

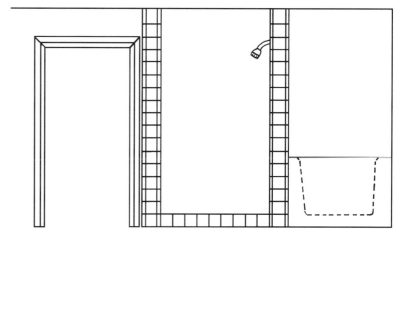

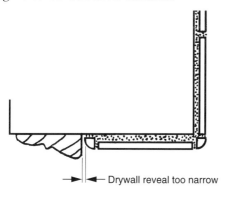

Drywall reveal too narrow

PLAN VIEW

Ceramic Tile 259

☐ **628**
When tile wraps around a windowsill, plan the rough framing around the window to provide an adequate window frame reveal for the tile to butt into. If the window frame reveal is too narrow, the thickness of the tile application will extend beyond the edge of the window frame and over into the window glass.

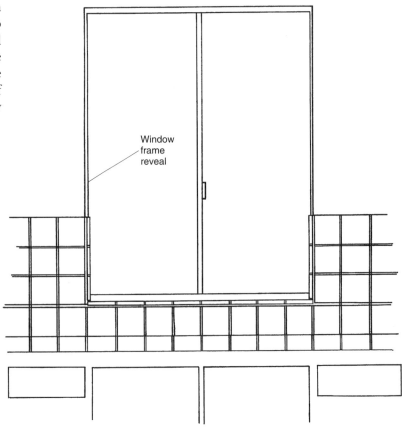

☐ **629**
When the tile installation will result in exposed factory edges, select a tile product that has aesthetically good-looking finished edges.

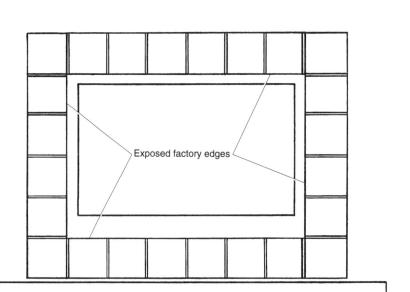

260   Chapter 22

☐ **630**
Coordinate the location of the toilet and the dimension from the back wall to the toilet closet ring on the floor in relation to the tile for an adjacent bathtub. If 6×8-inch tile is installed over a floated mortar base, for example, the tile may extend behind the toilet and prevent the toilet from sitting flat and straight against the wall.

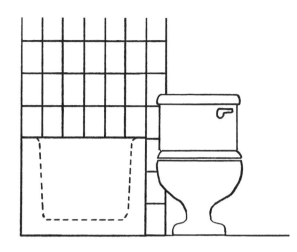

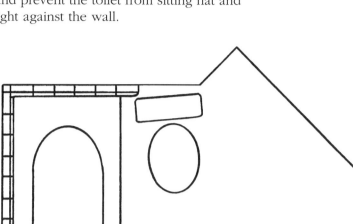

☐ **631**
Check that the top elevation of a countertop tile splash with a floated mortar base and bull-nose tile cap at the top is coordinated with the elevation of the curvature of bull-nose drywall corner bead at an adjacent windowsill. The curvature of the bull-nose drywall corner bead makes the usable wall space at the windowsill lower in elevation than the sharply defined edge of standard drywall corner bead.

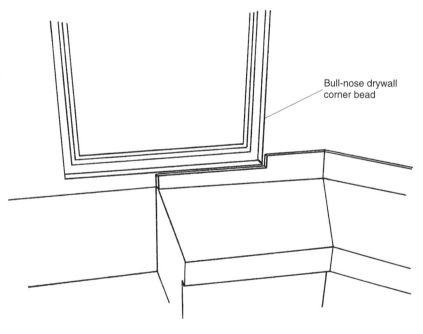

Ceramic Tile 261

☐ **632**
When a row of decorative tile with a raised relief pattern of fruit, leaves, and branches, for example, is installed within the countertop tile splash, install the electrical outlets during the rough electrical phase so that the outlets will not end up within this row of decorative tiles. Not only will the outlets look out of place, the electrical outlet cover plates will not sit flat against the raised relief pattern of the tile.

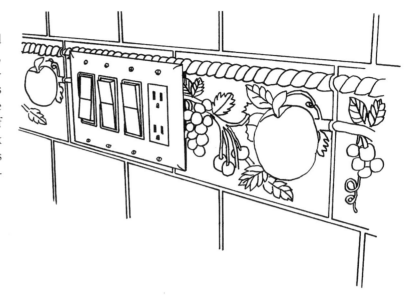

☐ **633**
For ceramic tile, granite, or marble countertops that have splashes with inlays of different colored tiles that create a decorative pattern, plan the locations of the electrical outlets during the rough electrical phase to coordinate with or be out of the countertop splash. You don't want a white two-gang switch cover plate to extend partially into and cover a diagonal square piece of red marble inlaid into a light brown marble splash. The electrical outlet cover plate will then stand out as the visually dominant architectural feature.

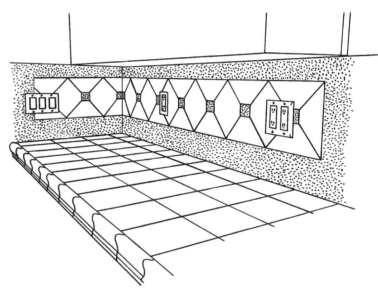

☐ **634**
For greenhouse or canopy windows located at the kitchen sink, match the elevation of the window with the height of the countertop tile so that the edges of the tile are not exposed at the exterior edge of the window.

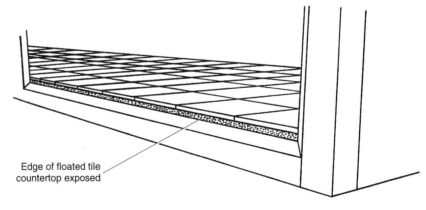

Edge of floated tile countertop exposed

262 **Chapter 22**

☐ **635**
Check that there is enough dimensional space in a bathroom for the bathroom door to miss the tiled shower pan, especially in an irregularly shaped bathroom with angled walls.

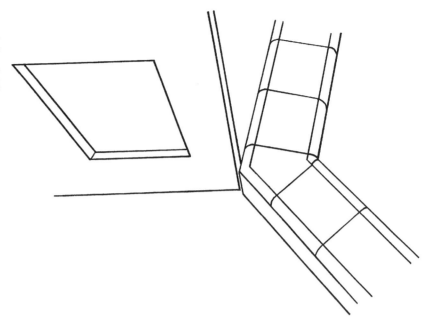

☐ **636**
Coordinate the elevation of windows with the height of bathroom cabinet countertops so that marble splashes, for example, do not have to be cut down to a narrower width at the bottom sill of an adjacent window, giving the appearance of a design and construction mistake.

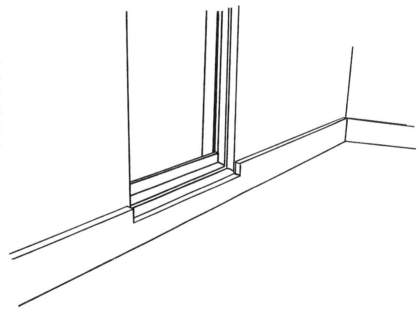

☐ **637**

End the tile or marble at the top point of the curvature of the bathtub edge or extend the tile or marble beyond the bathtub and add a vertical leg down the side of the tub. Having the tile or marble terminate even with or slightly beyond the edge of the bathtub leaves an awkward-looking unfinished area at the curved edge of the tub.

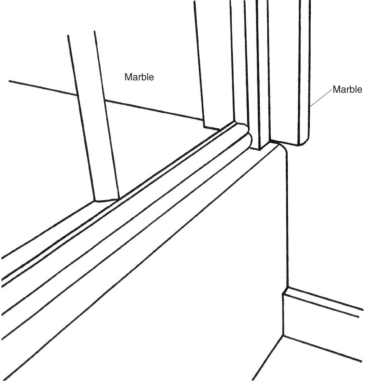

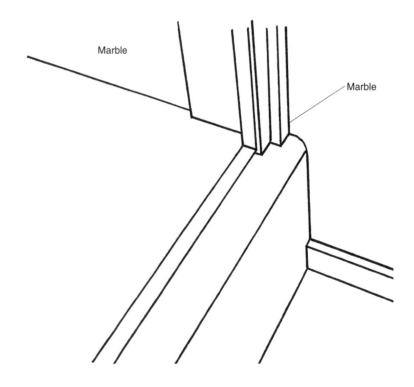

☐ *638*
Check the measurement for and installation of a bathroom cabinet marble countertop, for example, so that the countertop does not extend past the typical 1-inch projection beyond the cabinet. The countertop may project into and violate the minimum 15-inch clearance required on each side of the adjacent toilet in a small bathroom.

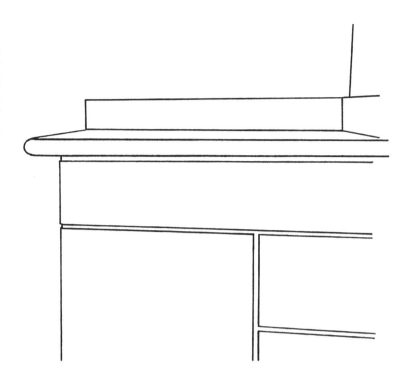

☐ *639*
Check that the location of an oval window with wood casing trim in a bathroom does not project into the tile for an adjacent shower.

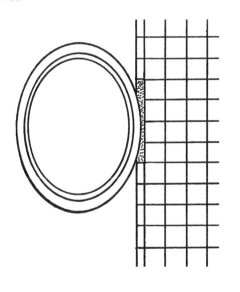

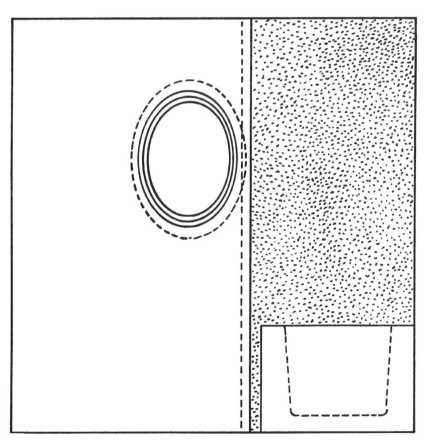

Ceramic Tile   265

☐ **640**
Check that the measurement and installation of prefabricated bathroom lavatory tops between two side walls will result in narrow caulking joints at each side. A 3/8-inch or 1/2-inch gap at each side, for example, results in a large hole that will have to be filled at the front corners of the countertop. When the caulking shrinks upon drying, the result is a concave unsightly area of caulking at each end of the countertop.

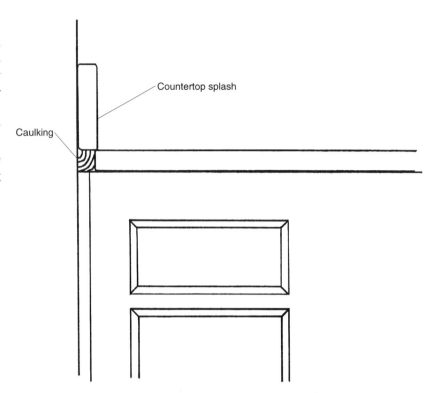

☐ **641**
When a countertop steps up to another height, check that the design of the higher cabinet is deeper than that of the standard-height lower cabinet to provide a vertical surface for the lower countertop bull-nose edge tile to butt into.

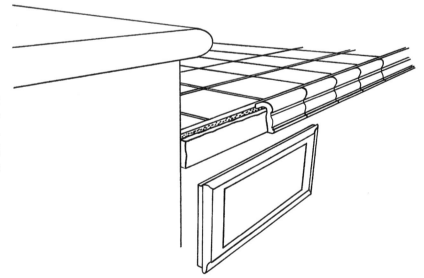

☐ **642**
Add slope to tiled shelves around bathtubs to allow the shelves to drain toward and into the bathtubs.

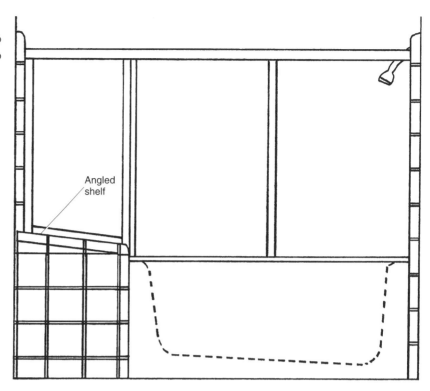

☐ **643**
When bull-nose drywall corner bead is used, check that the measurements for a wet-bar cabinet and a prefabricated countertop, for example, take into account the curvature of the bull-nose corner bead at the wall corners. You don't want the countertop splash to extend into the curvature of the bull-nose corner bead.

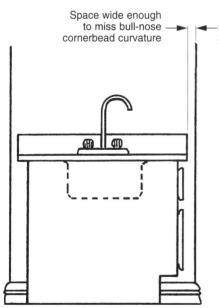

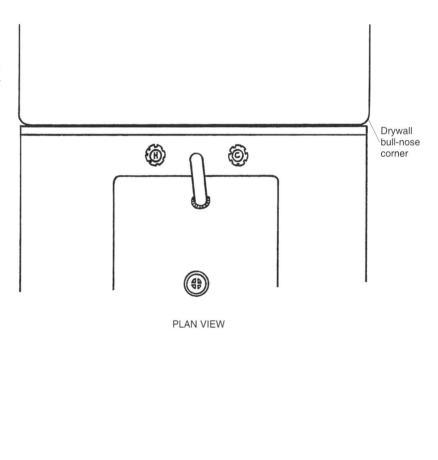

Ceramic Tile 267

☐ **644**
When imitation marble sheet panels are used at bathtubs and showers as a finish wall material, check that the panels extend all the way down to the floor during the initial installation. In this example, baseboard is not installed later to cover the gap at the floor, because wood baseboard trim normally is not used over marble.

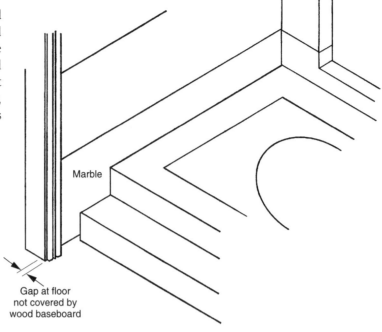

☐ **645**
For marble splashes at kitchen windows with wood casing trim, deepen the top edge of the marble splash at the window to create a stool effect to accept the bottoms of the wood casing trim.

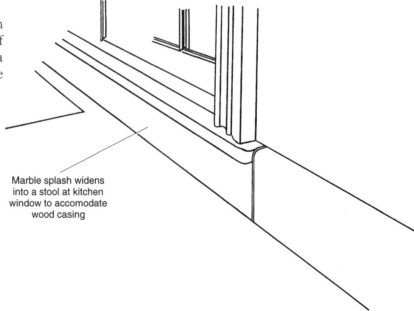

☐ **646**
For a detailed tile wainscot on a bathroom wall, for example, check that there is enough clearance from the wall corner to the edge of the door casing for the tile wainscot to fit. Cutting out the casing trim to fit around the tile wainscot creates the look of a design and construction mistake.

☐ **647**
For bathroom marble countertops, pieces that are spliced together, especially at the banjo area above the toilet, should get pieces that are similar in color and grain pattern. Marble can vary greatly in color shade and grain pattern, and two dissimilar pieces of marble next to each other can stand out visually.

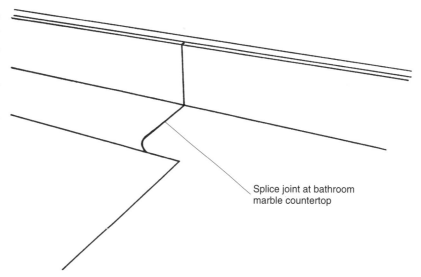

Splice joint at bathroom marble countertop

Ceramic Tile

☐ **648**
Check that a granite countertop, for example, at a kitchen pass-through bar top that joins a wall corner at a 45-degree angle does not extend in width over into a bull-nose drywall corner bead.

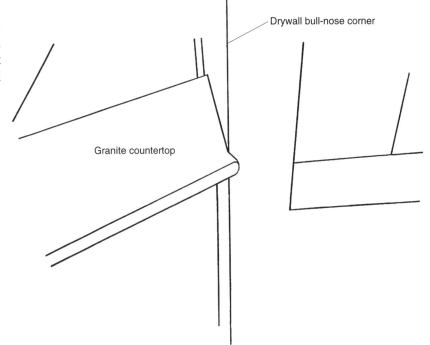

☐ **649**
In a bathroom with a sloping raked ceiling, for ceramic tile in a shower that is supposed to be installed at the full height up to the ceiling, stop the top row of tile a few inches short of the ceiling corner so that both sides can have an identical row of bull-nose cap tile. If the tile extends up to the ceiling, bull-nose tile cannot be used at the sloped ceiling section because the joint will look awkward.

CHAPTER 23

Insulation, Lath, and Stucco

☐ **650**
For preliminary insulation installed at locations that are subject to wind, use paper-backed insulation that can be stapled in place.

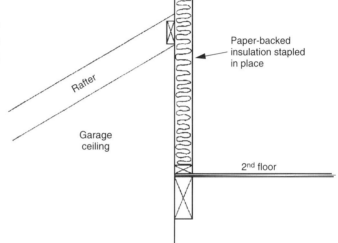

☐ **651**
Obtain a few bundles of insulation on the job site before the start of lathing; they can be used if necessary for preliminary insulation locations that were missed. Lathers can simply install this insulation before covering over the area.

☐ **652**
If wrought-iron handrailing is scheduled to be installed after lathing but before stucco plastering, install the wrought iron before the lath metal corner aid trim is nailed in place. The wrought-iron panels in this case will have been measured from wood-framed column to column, and the corner aid will prevent the panels from sliding into place.

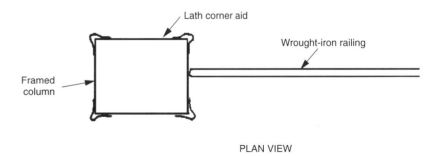

Continued on next page

Continued from previous page

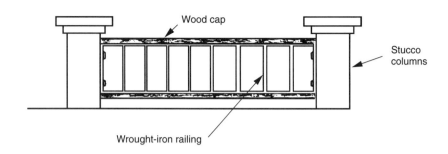

☐ **653**
Identify and mark broken windows with spray paint before the start of the scaffolding so that the costs of replacing windows that are broken as a result of scaffolding erection or plastering work can be identified and apportioned to the plastering subcontractor.

☐ **654**
Ask the plastering subcontractor ahead of time where on the job site deliveries of sand and bags of plaster should be placed. The placement of these materials determines the equipment setup location, and the plasterer may not be on the job site when these deliveries arrive.

☐ **655**
For masonry block walls finished with stucco, use both a scratch coat and a color coat. The grout joints may otherwise show through a single stucco color coat.

☐ **656**
Do not allow the stucco plastering to fill in window frame's drainage weep holes. This can lead to water seeping through the window frame and onto the interior drywall windowsills and walls.

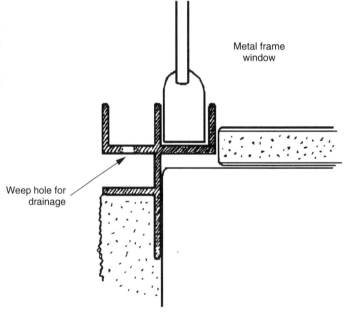

☐ **657**
Pull out the electrical wires from exterior outlet boxes before applying the stucco scratch coat. This prevents outlet boxes from being mistakenly buried by the lathing or stucco plastering because the wires are now sticking out from the wall. It also saves the cost of patching holes in the stucco after searching for buried electrical outlets.

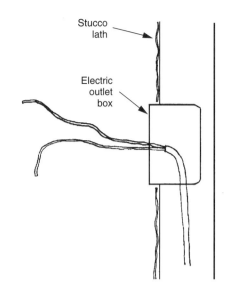

☐ **658**
Cover garage foundation screen vents before stucco plastering. This prevents having stucco sprayed through the vents and onto the garage stem foundation footing or the garage slab floor.

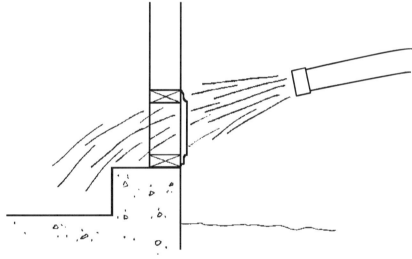

Insulation, Lath, and Stucco 273

☐ **659**
Check that stucco plaster is not sprayed into the exterior outlet for the dryer vent. This can create a lint trap and restrict the vent's airflow, causing the dryer to work less efficiently and thus take longer to dry clothes.

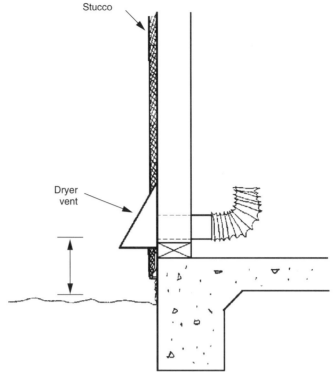

☐ **660**
Schedule the final stucco pickup repair work after the completion of the finish grading. This allows small tractors to bump into the wall corners before the final stucco repair work.

☐ **661**
Check that enough dimension is shown on the plans for two pieces of adjacent architectural foam trim.

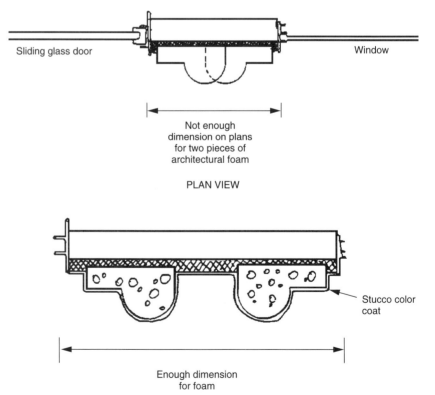

☐ **662**
Check that the roof overhang eaves have been washed and scraped clean of stucco overspray after the scratch and brown coats. If overhang eaves are spray-painted and have unremoved stucco overspray, the loose stucco will chip and break free, leaving bare wood exposed.

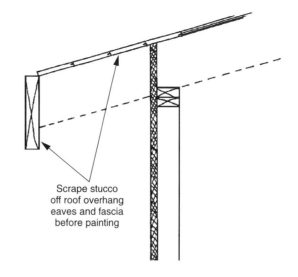

☐ **663**
Decide ahead of time whether ceramic tile address plaques should be recessed into and flush with the finish stucco surface or should stick out slightly with a buildup around the edges.

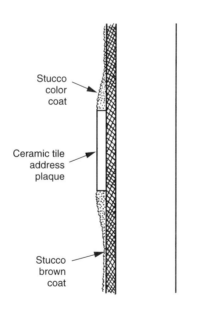
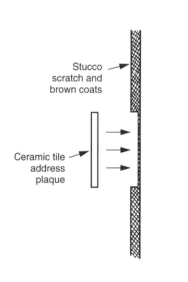

☐ **664**
Check that the plastering subcontractor has delivered to the job site and is applying the correct stucco color name and/or number. Stucco colors sometimes are changed during the design phase, and the plastering subcontractor may be working with old information.

Insulation, Lath, and Stucco 275

☐ **665**
For condominium projects with buildings that have stucco colors different from those of the sales models building, make sure the new home buyers understand that the exterior building color is not the same as that of the model building.

☐ **666**
Have the plastering subcontractor remove all scaffolding planks that end up inside the interiors of the units before the start of drywall taping, texturing, and painting. Carrying planks down a stairway may result in drywall dents and dings that require drywall repairs.

☐ **667**
Analyze the relationship on the schedule between the installation of the sliding glass doors and the erection of the scaffolding for stucco plastering. Install sliding glass doors first, before lathing, if the scaffolding will be in the way later.

☐ **668**
Have the plastering subcontractor thoroughly clean plastering overspray off roofing paper for the safety of the roofers who will install roof tiles later. The sand in the stucco plaster acts like small ball bearings at the edge of the roof, making the footing slippery. The plastering subcontractor should use a push broom or a power air blower to clean off the roofs.

☐ **669**
Check with the plastering subcontractor that enough scaffolding is free and available for the project and that its availability is not dependent on the progress and completion of the stucco work on another project.

☐ **670**
When a rough "skip-trowel" finish is applied during the stucco color coat phase, provide a flat, smooth surface around the rough outlet for the entry doorbell. The round doorbell cover will not sit flat against the rough skip-trowel stucco finish. Alternatively, have the stucco patch crew fill in the gaps around the doorbell cover during the final stucco pickup phase but include this patching work in the plastering subcontract, not as an extra cost.

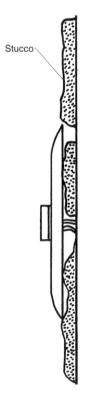

☐ **671**

For "Tudor"-style architecture with stucco applied in wall sections between $1^1/_2$-inch-thick wood plantons, check that the thickness of the stucco is correct at each wall section. You don't want a $^1/_2$-inch-thick or $^5/_8$-inch-thick stucco application in wall sections when the stucco is supposed to be $^7/_8$ of an inch thick.

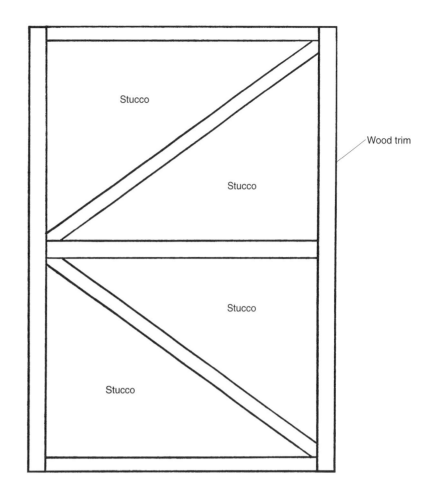

☐ **672**

For balcony decks with precast concrete handrail that is installed after the stucco brown coat but before the final color coat, check ahead of time that the legs, cross-braces, and wood planks of the scaffolding are not in the way of the installation of the precast concrete pickets and cap. This eliminates the need to leave out several pieces of the handrail, to be installed later, after the stucco is complete and the scaffolding has been removed.

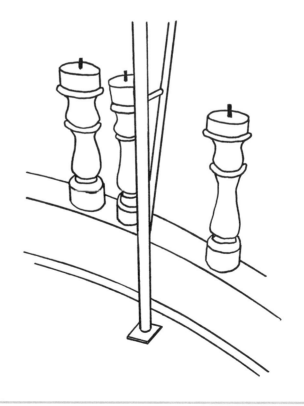

Insulation, Lath, and Stucco 277

☐ 673

For rooms that are cantilevered over a slope, check that there will be enough clearance underneath the underside of the cantilevered room to apply lath and stucco plastering, for example.

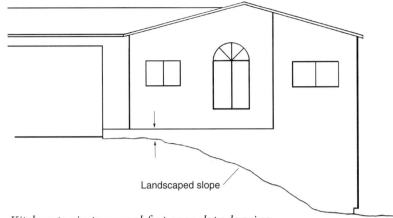

Kitchen projects several feet over slope leaving inadequate working space underneath kitchen for application of lath and stucco.

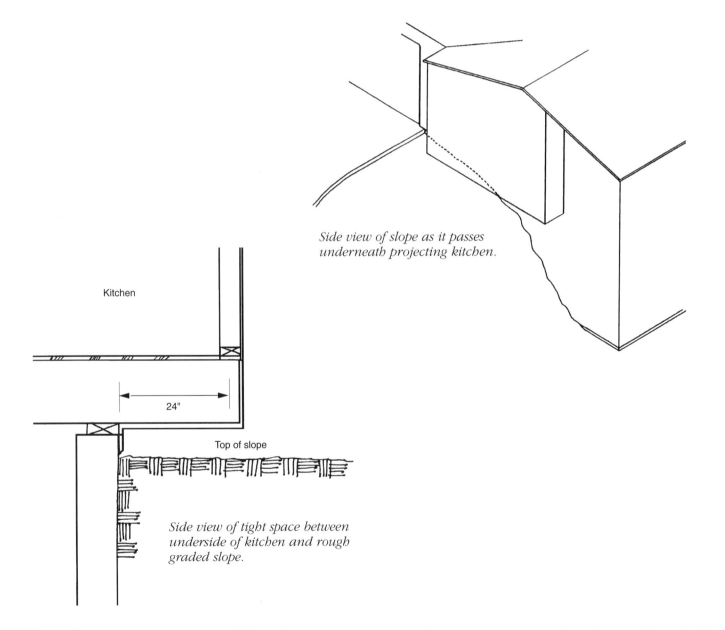

Side view of slope as it passes underneath projecting kitchen.

Side view of tight space between underside of kitchen and rough graded slope.

CHAPTER 24

Roofing

☐ **674**
The stocking of roofing materials on lower roof sections should leave ample working space for the stucco plasterers.

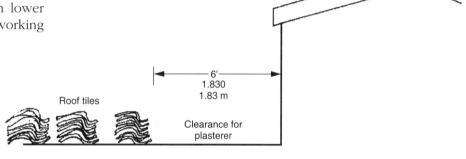

☐ **675**
For clay or concrete tile roofs, check that the roof tiles are installed with the correct amount of tile exposure in accordance with the manufacturer's recommendations. Roofing subcontractors can save money by lengthening the overall tile exposure slightly, saving a row of roof tiles or more per house.

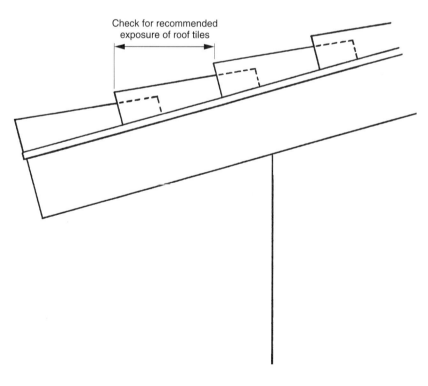

☐ **676**
For clay or concrete tile roofs, designate piles where broken tiles may be thrown off the roofs at the front and rear yards when there are narrow side yards between houses. This prevents broken tile pieces thrown off the sides of the houses from hitting the ground and then bouncing or rolling onto the lath paper, making small holes.

☐ **677**
For high-density condominium or apartment projects, discuss ahead of time the roof tile throw-off locations with the roofing subcontractor during the contract negotiations. It is time-consuming for the roofers when the throw-off locations are limited. At issue is the safety of other tradespersons working at ground level when the roofers are allowed to throw off broken pieces of tile from anyplace on the roof.

☐ **678**
Schedule the undercoat painting of roof fascia board to occur before the installation of roof paper. Paint overspray on the top of the roofing paper at the edge of the roof makes the roof slippery and unsafe.

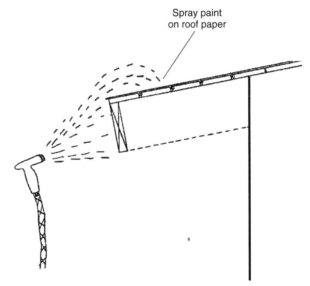

☐ **679**
Make an accurate count of the number, types, and sizes of sheet metal roof jacks for the vents through the roof and have all the jacks on the job site before the start of roofing. Consider organizing, labeling, and spreading the roof jacks per house just before the start of the roofing to achieve better materials management.

☐ **680**
For clay or concrete tile roofs, have the roofer sweep with a broom or blow off with a power air blower the dust that results from cutting tiles on the roof as part of each day's installation. If left on the roof overnight and moistened with dew, this cutting dust turns into a light paste that adheres to the roof tiles, is unsightly, and is difficult to remove later.

☐ **681**
Obtain the paint for the sheet metal roof vents from the painting subcontractor before the start of the roofing installation so that it can be used by the roofers to paint the vents.

☐ **682**

Make sure the roofers do not throw mud buckets from the roof onto the ground, splashing mud on the stuccoed walls. Colored cement mortar (mud) is used to fill in the ends of tile roof ridges, hips, and bird's mouths.

☐ **683**

For tile roofs, schedule the mudding of ridges, hips, and bird's mouths before the pouring of concrete walkways and driveways. Colored cement mortar that is accidentally spilled on the concrete below will stain the concrete.

☐ **684**

For large projects such as multiunit condominiums or apartments with two- or three-story roofs, consider hiring an independent roofing inspection company. This adds a level of technical consulting expertise to the job site to prevent roof leaks.

CHAPTER 25

Painting

☐ **685**
Get the interiors and exteriors 100 percent complete, with all the wood trim components and cabinets in place, before the start of painting.

☐ **686**
For production tract housing with phases and for condominium and apartment projects, keep a record of the "batch" numbers of the paint used for each phase or building so that the touch-up paint will match later. Different batch numbers of paint may not match perfectly for touch-up work.

☐ **687**
Coordinate the scheduling of the bathtub cleaning and the masking off of the bathtubs by the painter. Tub cleaning should be scheduled before or after the painting is done so that there is not a conflict between cleaning the tubs and putting the plastic sheet masking in place for the painting.

☐ **688**
Get the floors clean and free of dust and debris before the start of painting. Dust at the base of the walls can be picked up by a paint roller or lifted into the air by a paint spray gun and end up on the floor baseboard and the walls.

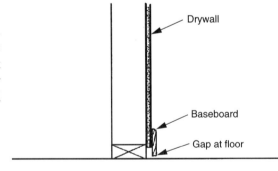

☐ **689**
Check that adequate care is taken when the painter sands nail holes through particleboard.

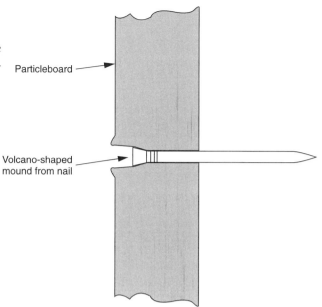

☐ **690**
Check that the finished factory edges of particleboard shelving are not rough and pitted; if they are, have the installer of the particleboard sand the rough edges smooth as part of the installation. Don't leave it for the painter to fix as if rough material were the responsibility of the painter.

☐ **691**
Have the painter caulk around the edges of HVAC registers during the caulk and putty painting prep phase.

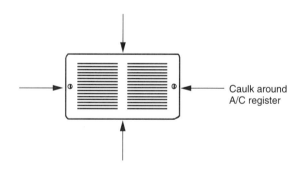

☐ **692**
Spackle, caulk, and paint the wood banjo cleat that supports the bath cabinet countertop at the toilet. This wood cleat is easily missed by the painter, yet if it is unpainted, it will be clearly visible by a homeowner sitting inside an adjacent bathtub.

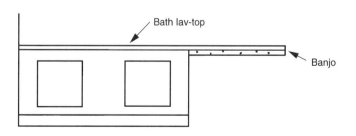

Painting 283

☐ **693**

Spray-paint the inside of sheet metal HVAC ducts with black paint as one of the first activities in the painting phase. This prevents seeing shiny bare sheet metal covered with drywall texture when one is looking through the register baffles.

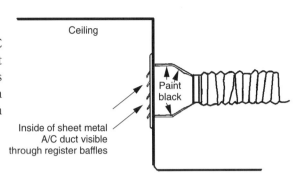

☐ **694**

If the painter removes cabinet drawers to be stood up vertically and spray-painted as a group in the center of a room after the drawers are painted or stained, have the painter come back and reinstall the drawers. This prevents the drawers from being accidentally knocked over and damaged by other tradespersons while the drawers are in the middle of a room and in the way of other work.

☐ **695**

Have kitchen breadboards in place during the staining and lacquering of the cabinets and then label and remove the breadboards to a storage location until they can be installed at the end of construction. This will get the breadboards to match the stain exactly in each individual kitchen and protect the breadboards from being stolen.

☐ **696**

Check that the tops of drywalled pot shelves above head height have been painted.

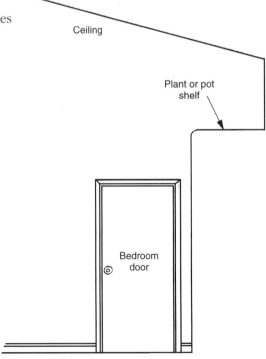

284   **Chapter 25**

☐ **697**
Paint the walls that are exposed within the interiors of cabinets.

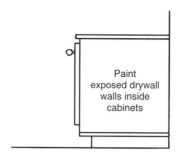

☐ **698**
Caulk and touch up with paint the edges around medicine cabinets during the final paint touch-up phase. Medicine cabinets are installed late in construction, after the painting is complete. This improves the appearance of the joint between the medicine cabinets and the drywall.

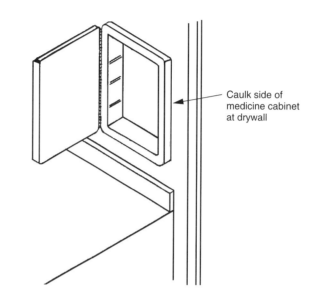

☐ **699**
Prepaint quarter-round wood trim for transoms above bedroom doors, for example. It is easier for the painter and results in better quality, because the painter does not have to cut in the wood trim at the surface of the glass. The builder must fit the prepainting of this quarter-round trim into the construction scheduling, however.

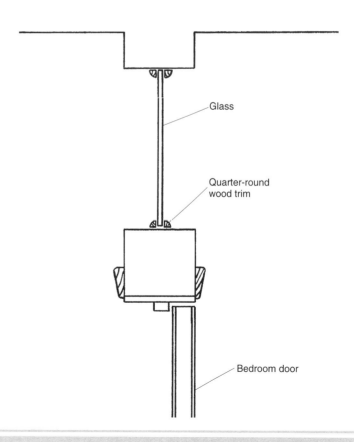

Painting

☐ **700**
Paint range hood sheet metal vents located in the interior of kitchen upper cabinets. Don't leave them as raw, unfinished sheet metal.

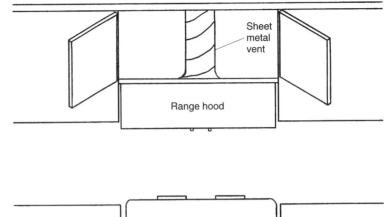

☐ **701**
Paint metal bumper posts located in front of water heaters and HVAC furnaces in garages. Don't leave them as raw, unfinished metal posts.

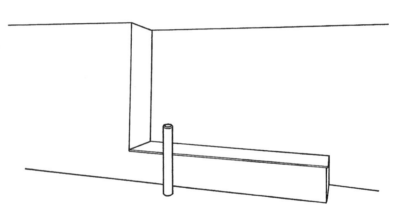

☐ **702**
Check that stairway treads are dusted off before the painter spray-paints the stair skirtboard. The spray-paint gun will kick up any dust on the stairways into the air, and the dust will mix with the paint and end up on the skirtboard or walls, creating rough areas.

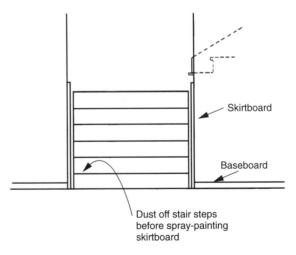

286 Chapter 25

☐ **703**
Check that the tops of door casings have been painted.

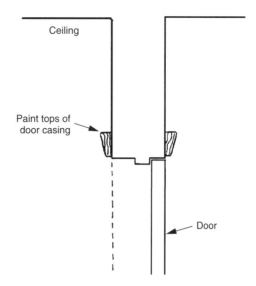

☐ **704**
When spray-painting or staining base shoe, have the painter elevate the base shoe off the floor to prevent dust on the floor from getting onto the base shoe. Dust mixed in with the paint results in a rough surface on the base shoe.

☐ **705**
Come up with a method for keeping paint overspray off stair wood handrail metal support brackets. Either have the painter mask off the brackets or have the finish carpenters remove the brackets before painting and then reinstall them after the painting is complete.

☐ **706**
Mask off, clean, and/or touch up with paint the black-painted metal surfaces of the fireplace firebox opening. Any paint overspray on these black surfaces should be left for the painter to remove or touch up.

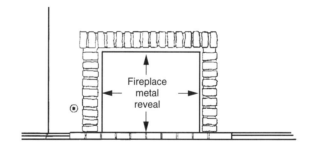

☐ **707**
For kitchens and bathrooms that get vinyl flooring, have the painter mask off the floor surfaces around the cabinets to keep the floor free of paint or stain and lacquer overspray. This allows the vinyl flooring glue to adhere properly to the subfloor surface and allows the flooring installers to avoid scraping off this overspray as part of floor preparation.

☐ **708**
Make the removal of paint overspray and drips a separate painting activity, not just part of the paint touch-up phase. Otherwise it will get mixed in with the other paint touch-up work and may not be done completely and thoroughly.

Painting 287

☐ **709**
Don't paint roof overhang eaves white. White magnifies the appearance of flaws in the wood.

☐ **710**
Consider painting the roof overhang eaves the same color as or a color a few shades away from the stucco color. This softens the transition at the joint between the stucco plastering and the wood overhang boards.

☐ **711**
Paint exterior doors soon after they are installed to seal and protect the wood. Don't allow the painter to postpone this activity and work it into the normal painting sequence for the sake of convenience. Warranties for wood doors require that they be painted within a short time after installation.

☐ **712**
Do not use rough-sawn wood for balcony deck handrail to match the rest of the exposed exterior wood work on the project. When paint is added, the rough and fuzzy areas on rough-sawn wood become hardened, resulting in sharp, sandpaperlike surfaces.

☐ **713**
Analyze the method for painting exposed sheet metal parapet wall caps and flashings. The paint must bond and stick to the metal or it will flake and chip off later, perhaps requiring an entire second repainting within a short time after the original painting. Consider cleaning solutions that prepare the metal surfaces for painting and look into "fuse-primer" paints that adhere to metal surfaces.

☐ **714**
When exterior wood is stained using a transparent stain, have the painter provide and use a matching color or standard paint for the sheet metal vents and flashing that cannot be stained.

☐ **715**
Have the painter supply roofjack paint in 1-gallon cans rather than 5-gallon buckets. Smaller cans are easier for the roofers to handle and use on the roof and eliminates the need for the roofers to transfer paint into smaller containers.

☐ **716**
When exterior wood is sandblasted before painting, have the balcony decks and other work areas clean and free of debris. The sandblaster wears a protective hood that limits peripheral vision, and you do not want the sandblaster to trip over loose pieces of wood or be distracted by having to think about stepping over debris.

☐ **717**
When roof fascia board and other wood trim are sandblasted before painting, have the sandblasting subcontractor clean off the sand from the roofs by using a push broom or power air blower.

☐ **718**
Prime all sides of wood siding boards before installation so that the boards do not cup or bow later as a result of uneven drying and shrinking.

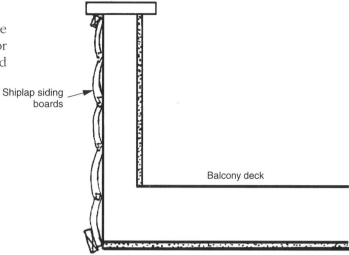

☐ **719**
When a mirror is inlaid in a masonry or stone facade above a fireplace, for example, pre-paint the white drywall at the inside edges of the stone facade a color that matches the color of the grout before the mirror is installed. This removes the contrasting white color of the bare drywall in areas where the mirror does not perfectly match the edge of the stone facade and grout.

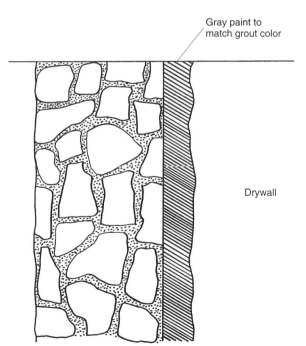

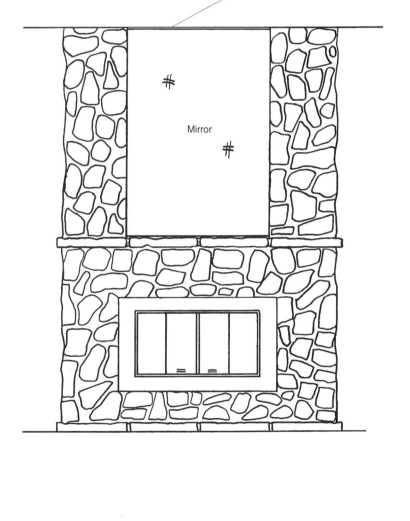

☐ **720**
When the apron wood trim for the window stool and apron has a hollow area on its backside, the painter should caulk each end of the apron trim during the painting prep phase.

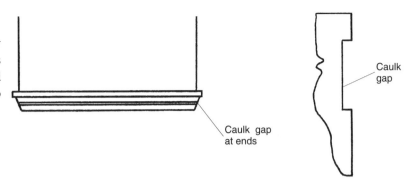

☐ **721**
When cabinets with particleboard doors and drawers in a laundry room, for example, are spray-painted, the edges of the doors and drawers will absorb the paint, requiring additional coats of paint brushed on by hand to achieve coverage. Each application of spray paint is thin to prevent paint runs on flat surfaces of the doors and drawers, and one therefore cannot apply enough paint material buildup at the door and drawer edges.

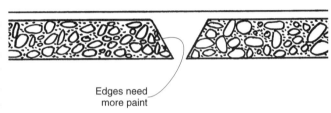

☐ **722**
Sand smooth the cut edges of roof rafter tails at the eaves before painting in single-story houses where the rafter tails are only a few feet above eye level. This removes the unsightly frayed edges at the cut ends and bottoms of the exposed rafter tails.

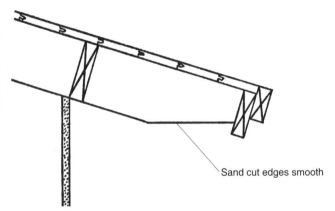

☐ **723**
For wood pot shelves underneath a kitchen window, for example, spray-paint the supporting wood corbels after the stucco brown coat along with the other exterior wood trim and prepaint the wood shelves at ground level before installation and after the stucco color coat. This gets all the surfaces of the shelf boards painted and removes the possibility of dripping paint on the stucco and concrete patios or walkways if the shelves were painted in place.

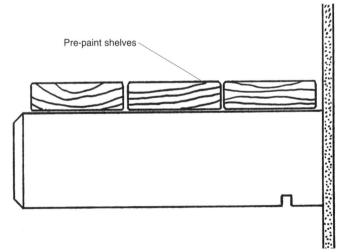

CHAPTER 26

Hardware

☐ **724**
For production tract housing, consider ease of assembly in choosing interior and exterior doorknob hardware. This can affect the construction schedule if the installation is difficult and time-consuming. It also can result in the finish carpentry subcontractor asking for an extra for the increased difficulty of installing the hardware if the brand of doorknob hardware was not selected and specified by the builder at the time of the contract.

☐ **725**
If hardware is purchased directly by the builder, break down the bulk packages of towel bars, towel bar ends, toilet paper holders, and towel rings and add them to the individualized hardware that is packed and labeled per lot or house number. This not only provides a check on whether there is the proper quantity of hardware on hand for the work but also organizes all the hardware pieces into individual boxes, labeled per lot number, that are ready for the finish carpenters each workday.

☐ **726**
Check that the entry door rails and stiles are wide enough for the type of entry door hardware being used.

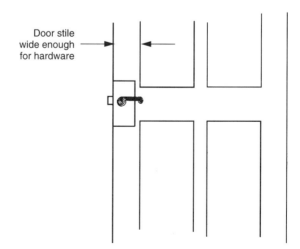

☐ **727**
Don't allow the installation of entry door hardware backset screws to split the edges of entry doors. Instead, predrill pilot holes for the screws.

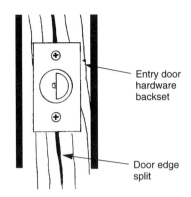

Entry door hardware backset

Door edge split

☐ **728**
Schedule the installation of thresholds and weather stripping to occur after the installation of doorknob hardware. This allows the threshold and weather stripping to be pushed up tightly against the door while the doorknob latch and deadbolt throw bolt are engaged, resulting in a snug fit without slack or play.

☐ **729**
For an apartment or condominium lobby with laundry and service doors that get repeated use, install round doorknobs rather than lever-type doorknobs. Round doorknob handles require a lot of gripping strength to overtwist to the point of breaking the doorknob, while lever-type doorknob handles require only downward force to reach the point of breaking the doorknob hardware.

☐ **730**
For production tract housing, check that the interior door hardware backsets match the precut mortices in the door edges. Round-cornered mortices require round-cornered backsets; otherwise the hardware installer must chisel out the corners to make them square.

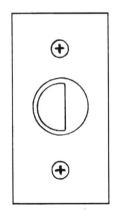

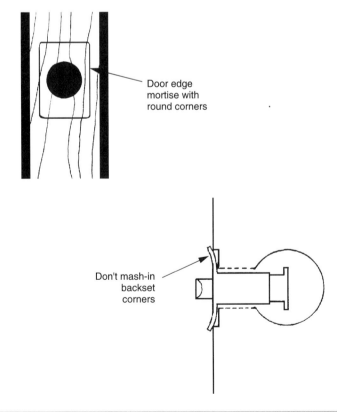

Door edge mortise with round corners

Don't mash-in backset corners

292 **Chapter 26**

☐ **731**
Swinging café-type doors leading into a kitchen, for example, should be attached to wood jambs, not attached directly over drywall. If the pivot hinges bind, they will twist, breaking the drywall paper and spilling out the crushed gypsum.

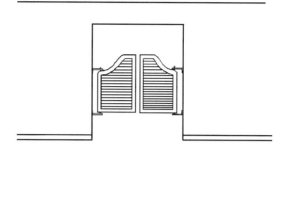

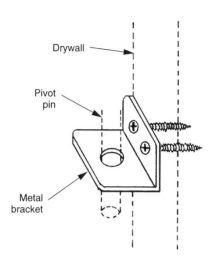

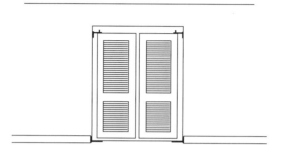

☐ **732**
Install pocket door latches perfectly flush with the edges of pocket doors so that the doors close tightly against the doorjambs.

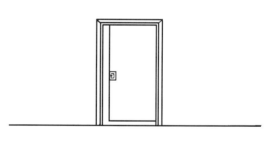

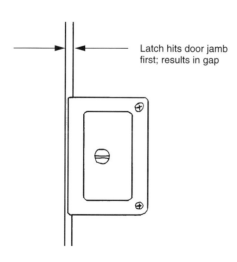

Hardware 293

☐ **733**
Check that the screws used to attach toilet paper holders to the sides of cabinets do not penetrate into the interiors of the cabinets. Have the hardware installer use shorter screws or cut off the ends of the screws inside the cabinets.

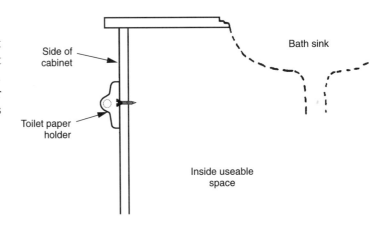

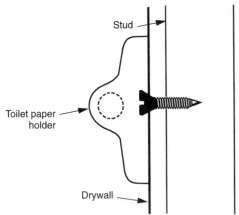

☐ **734**
Check that the toilet paper holders are installed with the correct distance between the two end posts for the standard width of a roll of toilet paper.

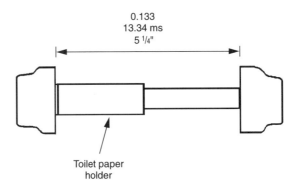

☐ **735**
For dummy-knob door handles attached to hollow-core Masonite veneer closet doors, use screws and inside-threaded metal studs to secure the doorknobs rather than putting screws directly into the Masonite veneer. Screws driven into the thin veneer material will not hold and will work loose over time.

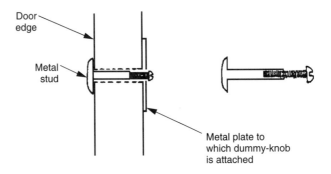

☐ **736**
Place wood backing in bathroom walls for the attachment of towel bars, toilet paper holders, and towel rings. This forces the hardware installer to use the wood screws that accompany the hardware rather than electing to put one molly-bolt fastener per end into the hollow wall areas for the sake of economy, resulting in the hardware coming loose later.

294 **Chapter 26**

☐ **737**
Check that the length of the screws that come with the bifold doorknobs matches the thickness of the wood bifold doors.

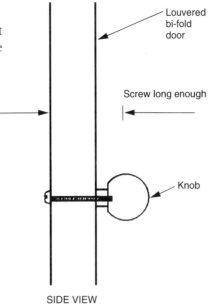

☐ **738**
Place wood backing in bathroom walls for expensive towel bars, toilet paper holders, and towel bar rings that come with their own matching colored screws that cannot be replaced with molly-bolt fasteners. This is also the case with clear plastic towel bars in which the screws are entirely exposed. One must use the screws that come with the hardware and therefore must have wood backing already inside the walls.

☐ **739**
Drill clean holes through entry doors for peephole hardware. Peepholes generally have flanges that are not wide enough to cover wood that has been chipped away as a result of drilling through a door from one direction only.

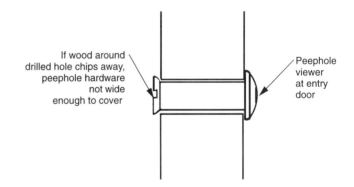

☐ **740**
For entry doors with glass panels at the top, delete the peephole hardware. The person standing outside can see the top of the head of the person looking through the peephole who might as well look out through the glass panels instead.

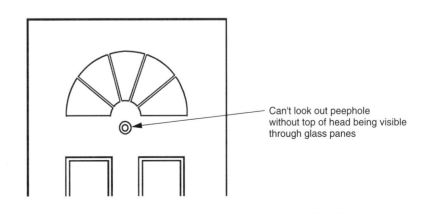

Hardware 295

☐ **741**
Check that there is enough dimension from the bottom of a light soffit to the bottom of a combination mirror and cosmetic box to create enough clearance to operate the faucet handle in a bathroom.

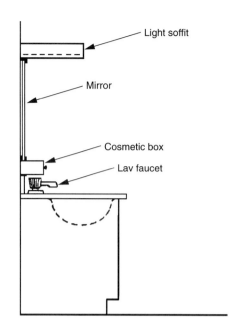

☐ **742**
For condominium and apartment projects with door closers on the entry doors, have the hardware installer use both the rough and the fine adjustments in setting the door closer tension. Entry doors should not slam shut as a result of the hardware installer using only the fine adjustment screw.

☐ **743**
Install medicine cabinets so that the backside joint with the drywall is tight. One can see this joint in the bathroom mirror.

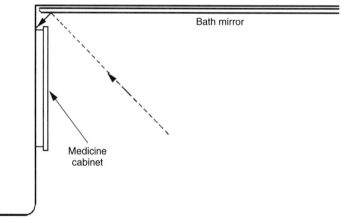

☐ **744**
Have the hardware installer adjust the doorknob striker plates to remove excess slack or play. Doors should close against the jamb stops with a solid feel.

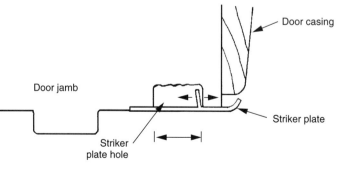

296 **Chapter 26**

☐ **745**
Schedule the installation of spring hinges on garage fire doors after the thresholds and weather stripping are installed. This allows the weather stripping installer to perform the work without having to struggle with self-closing doors.

☐ **746**
Have the builder's key cut to open not only the doors to the houses but also the locks to the common area gates, the recreation building, and the swimming pool area. This simplifies the keys for the project.

☐ **747**
Check that towel bars are not installed too close to shower doors that will be installed later, blocking the opening of the shower doors.

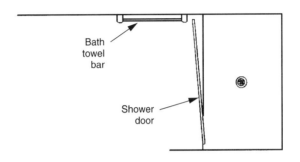

☐ **748**
Check that the size of address numbers meets the minimum size requirements of the police department and/or buildings department.

☐ **749**
Make the key to the swimming pool equipment room different from the homeowner's keys that open the gate to the swimming pool area. This prevents one homeowner from changing the pool or the spa water temperatures to suit himself or herself.

☐ **750**
When a builder's key is temporarily given to a tradesperson to gain access into houses that are locked, require that person to leave a monetary deposit so that he or she will return the key.

☐ **751**
Test the entry keys and the garage door opener transmitters before giving them to the new homeowners.

☐ **752**
Check that address plaques with four or more numbers will fit within the wall space or column pop-out shown on the plans.

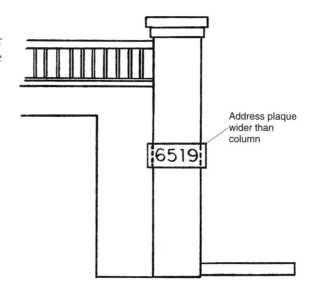

Address plaque wider than column

☐ **753**
If the builder purchases the hardware directly, consider ahead of time who will supply the hinge butts for the exterior and interior doors—the builder or the finish carpentry subcontractor. If the builder supplies the hinge butts as part of the hardware purchase, break up the hardware delivery and have the hinge butts delivered separately so that they are on the job site for the start of door hanging.

☐ **754**
Give the finish carpenters installing hardware $8\frac{1}{2} \times 11$-inch reduced floor plan cheat sheets that show the locations and dimensions for towel bars, towel rings, and toilet paper holders in bathrooms.

☐ **755**
Consider ahead of time the space conflict between a towel bar normally placed 60 inches above the floor and a light switch placed 48 inches above the floor in the same wall area. A towel draped over the towel bar will cover the light switch; either install the towel bar below the light switch or move the towel bar to another wall if possible.

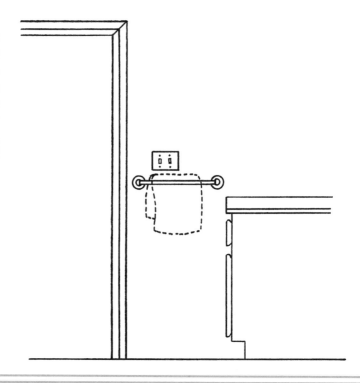

298 **Chapter 26**

☐ **756**
Predrill holes completely through wardrobe closet doors only if finger cups will be installed on the outside and inside surfaces of the closet doors. One can save time and money by drilling holes only partially through the front side of the closet doors and then installing the finger cups only on the front sides of the closet doors.

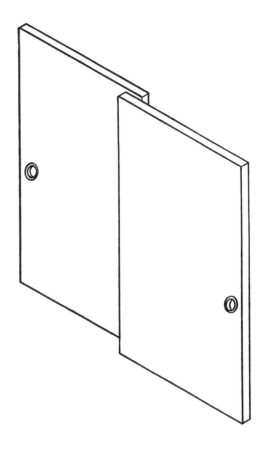

☐ **757**
Avoid bathroom floor plans in which the bath door fully opens at an awkward angle against adjacent towel bars, door stops, and side walls.

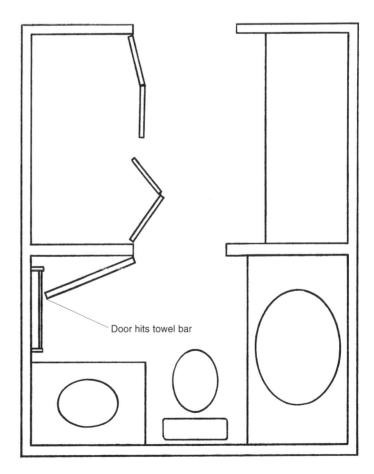

Door hits towel bar

Hardware

☐ **758**
When a bathroom door opens against an adjacent glass shower door, install the door stop simultaneously with the doorknob hardware so that someone does not accidentally open the bathroom door into the shower door, hitting the glass with the doorknob and breaking it.

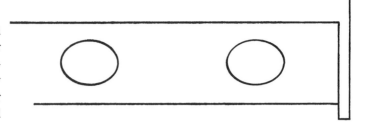

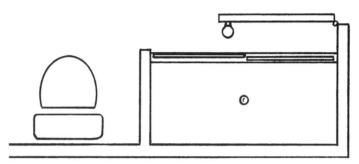

Without doorstop, doorknob hits shower glass doors

☐ **759**
For pocket doors between a kitchen and a dining room, for example, do not install the latch striker plate on the doorway jamb. The pocket door at this location does not have to be and should not be lockable for privacy like a bathroom door.

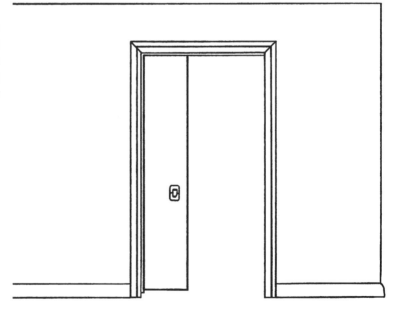

☐ **760**
For small bathrooms with angled walls that reduce the usable wall space, check that the location for a toilet paper holder is not directly within the path of the opening of a cabinet drawer.

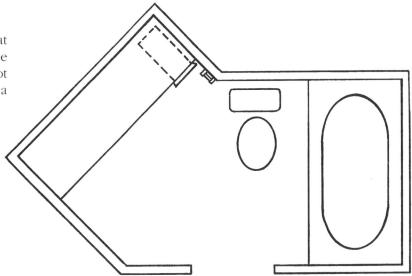

☐ **761**
When towel bars are installed before shower doors, make sure the towel bars will not be within the swing of the glass shower doors.

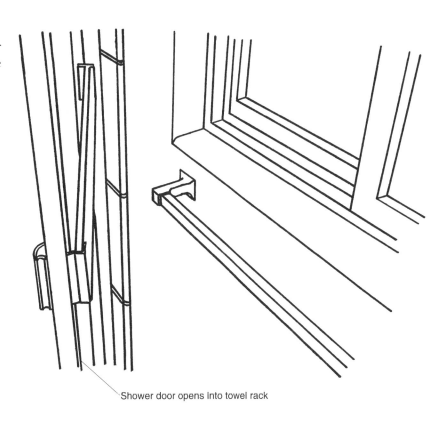

Shower door opens into towel rack

☐ **762**
Come up with a method for installing individual address numbers so that numbers are straight, aligned, and spaced uniformly. Address numbers are too important visually to install using the freehand "eyeball" method without the help of some type of straightedge or jig device.

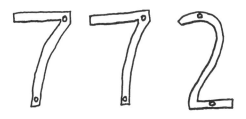

Hardware 301

☐ **763**

For medicine cabinets that have a definite top and bottom orientation, check that the correct numbers of right-hand and left-hand medicine cabinets are ordered and delivered to the job site. This is different from standard medicine cabinets for production tract housing, which can simply be turned over one way or the other because there is no top or bottom.

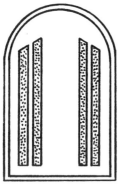

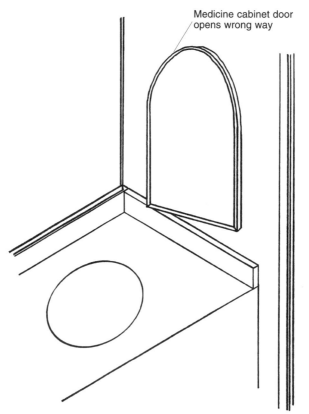

☐ **764**

For exterior doors that open out onto a second or third floor balcony deck or a flat roof area finished as a deck area, consider placing the doorknob 54 or 60 inches above the floor to place it out of the reach of small children. This prevents small children from letting themselves out onto the deck unsupervised.

☐ **765**
Don't combine a lever-type doorknob with doors that have ogee detailed wood trim placed too close to the edges of the door. The back of a person's hand will hit the ogee trim when the person is gripping the doorknob lever handle. The ogee trim should be farther from the door edges, or a round or oval doorknob should be used instead of a lever type.

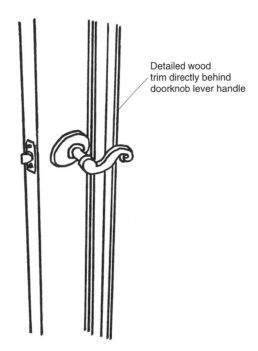

Detailed wood trim directly behind doorknob lever handle

CHAPTER 27

Finish Plumbing, Electrical, and HVAC

☐ **766**
Check that the toilet is not above the height of the bathroom lavatory's top banjo.

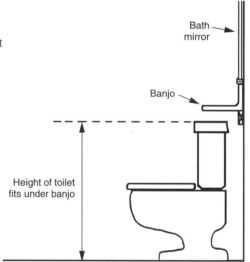

☐ **767**
Check the clearances for water heater vents, especially when a vent must travel horizontally to form a combination vent with the forced-air furnace unit before turning vertically to vent out through the roof. Use double or triple wall vents when clearances are too tight.

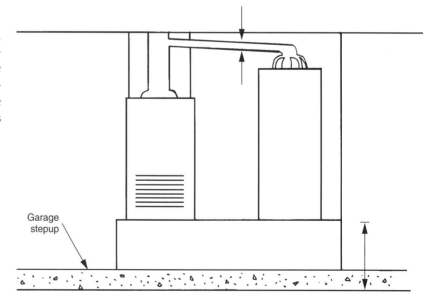

☐ **768**
For tract housing, use the same models of appliances such as garbage disposals, microwave ovens, dishwashers, and ranges for simplicity during construction. This prevents communication problems that result in the wrong appliances being spread and installed in the wrong houses.

☐ **769**
Order appliances with pigtail extension cords already attached; if they are not available, schedule the electrician to install the pigtails before the appliances are installed by the plumber.

☐ **770**
Fire sprinkler systems should have flowmeters with a delay mechanism, a backflow valve, and/or a 4- to 8-gallon-per-minute allowance. This prevents the fire alarm from going off as a result of a variation in the main-line water pressure caused by construction activities elsewhere in the project.

☐ **771**
For bathtub spouts that are installed by being twisted around a threaded valve, check that there is enough clearance around the valve. Faucet handles and other obstructions cannot be within the turning radius of the bathtub spout.

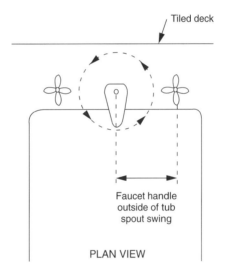

☐ **772**
Check that the bathroom lavatory faucets have the same number and spacing of the holes as the prefabricated lavatory tops.

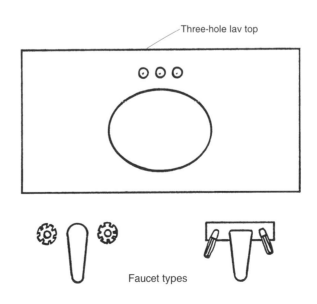

Finish Plumbing, Electrical, and HVAC

☐ **773**
Analyze the time period for the hot water to reach a bathroom far away from the water heater. Long waiting periods are a common complaint from homeowners in condominium projects when the master bathrooms are on the third floor and the water heaters are in the first floor garages.

☐ **774**
For large spa tubs in master bathrooms, check that the water heater's capacity is large enough. You don't want the spa tub to be only half full when the hot water begins to run out.

☐ **775**
Do not install pedestal sinks in bathrooms that will get hard surface flooring until after the flooring has been installed. They are too much work to install, remove, and then reinstall, especially in a tract house setting.

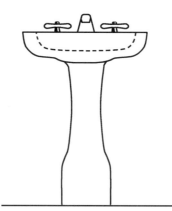

☐ **776**
Schedule finish plumbing before hardware installation. Towel bars and toilet paper holders may be in the way and may be accidentally hit and damaged while toilets are being installed in small bathrooms.

☐ **777**
Snake out the sewer line between the house and the street before laying first floor carpeting or hard surface flooring. Discover and remove possible blockages before the first floor is flooded by a sewer backup.

☐ **778**
During the final cleanup, when the electrical power is not yet available, spin the garbage disposal, using the hex head wrench that is provided, to discover and remove debris, nails, or tile grout inside the garbage disposal.

☐ **779**
Run water through all the faucets before the final cleanup to discharge any dirt, sand, and sediment in the plumbing water lines. This flushes out the water lines and gets this material into the sinks before the sinks are cleaned.

☐ *780*
Check that the bathtub and shower valves are in the off position and that bathtub and shower drains are free to avoid accidental flooding of the interior if a tradesperson turns the water to the building on for a construction use.

☐ *781*
If the builder purchases appliances directly, do not have the job site superintendents spread the appliances to the houses or units. It takes too much time and energy away from supervising the work.

☐ *782*
At the time of appliance delivery check that the model numbers are correct.

☐ *783*
If garbage disposals are purchased directly by the builder, schedule their delivery to the job site before the delivery of the other appliances. Garbage disposals can be installed by the plumber during the finish plumbing phase at the same time as the kitchen sinks, well in advance of the installation of the major appliances.

☐ *784*
Do not spread microwave ovens into the units along with the other appliances, because the ovens can be stolen. Microwave ovens are small and light enough to be easily lifted and hidden inside a worker's vehicle, unlike larger and heavier appliances. Store microwave ovens in a separate protected area.

☐ *785*
Collect appliance booklets and pamphlets shortly after the appliances are installed.

☐ *786*
Install the microwave ovens before installing the kitchen ranges. This provides easier access and working room without the ranges in the way, and sawdust will not fall on top of the ranges.

☐ *787*
Use the correct paddle size and shape in the flowmeter for a monitored fire sprinkler alarm system. It should match the water pipe's diameter and the water pressure.

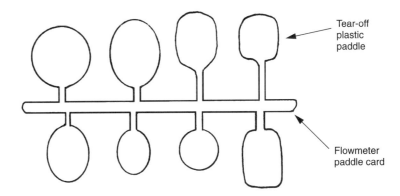

Tear-off plastic paddle

Flowmeter paddle card

Finish Plumbing, Electrical, and HVAC

☐ **788**
Remove and store elsewhere the handle to the fire sprinkler system hose bib so that someone does not mistakenly turn it on and activate the fire alarm.

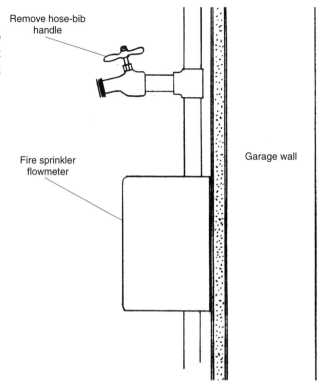

☐ **789**
Install registers with baffles facing upward at a stairway so that a person walking down the stairway will not see directly inside the A/C sheet metal duct.

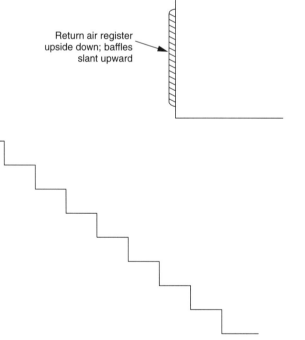

☐ **790**
For registers that are not close to a wall ceiling line or corner, use a torpedo level to get the registers level and plumb.

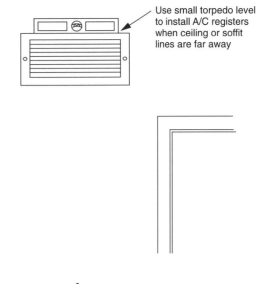

☐ **791**
Purchase exterior light fixtures that come with foam rubber gaskets for weather-resistant seals. This eliminates the need to caulk the outside edges of the light fixtures.

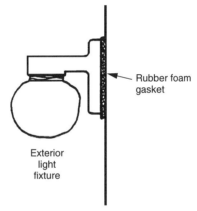

☐ **792**
Check the distance from the garbage disposal to the switched electrical outlet and provide a pigtail extension cord that will reach the outlet.

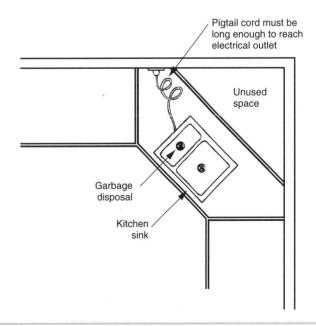

Finish Plumbing, Electrical, and HVAC

☐ *793*
When the builder purchases the light fixtures directly, have the light fixture company "house pack" the light fixtures at the time of delivery. Label the boxes for lot numbers and room locations and organize the light fixture boxes in groups per house lot number or address.

☐ *794*
Specify in the purchase agreement for the light fixtures that all light fixtures be fully preassembled and ready to install out of the box.

☐ *795*
When light fixture boxes are labeled and organized by the light fixture company upon delivery, label the boxes, using floor plan locations such as dining room, breakfast nook, and stair ceiling, instead of using obscure model numbers that require an ordering list to figure out where they go. This makes it easier for the builder and the electrician to spread and install the light fixtures.

☐ *796*
Divide the light fixture installation into two phases: the basic light fixtures during the normal finish electrical phase and the expensive hanging light fixtures later in construction, just before the final cleaning. This prevents expensive light fixtures from being stolen during the course of construction.

☐ *797*
Specify in the purchase agreement for the light fixtures that the light fixture company have one or two extras of each type of fixture in stock at its local warehouse for immediate replacement if the electrician opens up a box and discovers that the fixture cannot be installed because of a missing part or another flaw. This prevents the common occurrence of the builder having to wait two to four weeks for another light fixture while the electrician has completed the work and moved on to another project.

☐ *798*
Consider accurate placement of light fixtures in sales models and follow through with the same locations in the production units. Homeowners cannot complain later when light fixtures do not center over their individually differing dining room tables, for example. The builder's reply is, "Check the models."

☐ *799*
For large entries in expensive houses, consider making a full-scale cardboard or plywood mock-up of the hanging light fixture to check whether its proportional size compares well with the entry and the adjacent stairway, for example.

☐ *800*
Temporarily tie up hanging light fixtures above head height during construction to prevent them from being damaged by people walking into them.

☐ *801*
Install half-hot receptacles upside down, with the grounding hole at the top, so that new homeowners can identify a receptacle in a bedroom that is operated by the wall switch. This eliminates service complaints by unknowing homeowners that an electrical outlet does not work when in fact the outlet is half hot.

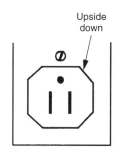

☐ *802*
Have the electrician install cover plates with screw slots uniformly turned vertical.

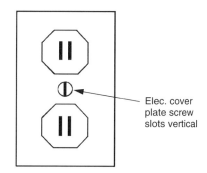

Elec. cover plate screw slots vertical

☐ *803*
Decide whether the electrician will supply and install all the light bulbs for the light fixtures or whether the new homeowners will purchase and install his or her own light bulbs.

☐ *804*
For condominium, apartment, and town house projects, include within the electrical subcontract a provision for the electrician to install light bulbs in all the common area exterior light fixtures as part of the installation. Some exterior light fixtures are placed high above the ground and should not be the initial responsibility of the new homeowners to install.

☐ *805*
For sales models and houses getting wallpaper, schedule the ring-out testing of the electrical wiring before the wallpapering so that any cut wires or shorts can be located, holes can be cut in the drywall, repairs can be made, and the drywall can be patched before the installation of expensive wallpaper.

☐ *806*
Have the electrical subcontractor label the circuit breaker panel using full-length words or the correct abbreviations in legible writing. The best method is to use an adhesive-backed label with the circuit breaker numbers and names typed on it.

☐ *807*
If the electrical cover plates will be wallpapered to match the surrounding walls, schedule the installation of the electrical cover plates ahead of the normal finish electrical, plumbing, and HVAC work so that the cover plates are in place for the wallpaper hanger.

Finish Plumbing, Electrical, and HVAC

☐ *808*
When electrical meters are set and power is provided, check in the unsold units that lights, bath fans, and appliances are not running because the switches were installed in the on position.

☐ *809*
For the electrical outlet inside the lower cabinet underneath the kitchen sink for the dishwasher and the garbage disposal, wire the outlet so that the continuously hot half of the outlet for the dishwasher is on the bottom and the half-hot switch outlet for the garbage disposal is on the top. The end of the appliance pigtail cord for the dishwasher is usually large and has a stiff connection where the cord meets the plug. It will block off access to the lower receptacle and make it difficult to plug in the extension pigtail cord for the garbage disposal.

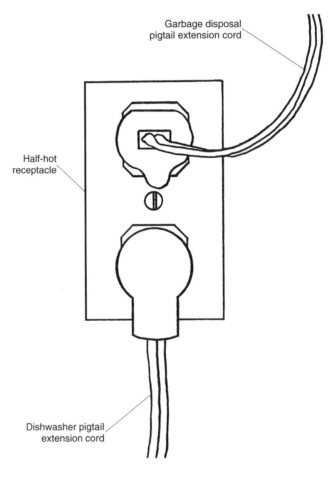

Garbage disposal pigtail extension cord

Half-hot receptacle

Dishwasher pigtail extension cord

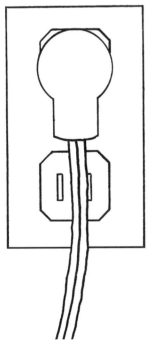

Stiff dishwasher pigtail blocks access to lower receptacle

312 **Chapter 27**

CHAPTER 28

Miscellaneous Finish

☐ **810**
Don't install medicine cabinets on two sides of the opening for a bathroom cabinet before the mirror has been installed. If the mirror is measured and cut to extend from wall corner to wall corner, the two medicine cabinets will be in the way and prevent the mirror from being installed.

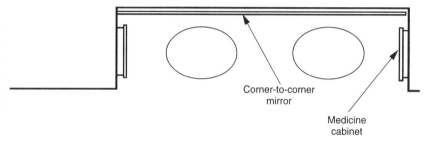

☐ **811**
When vanity strip light fixtures are installed vertically at the sides of bathroom mirrors, check that the bottom metal tracks for the mirrors are even with the edges of the mirror. Otherwise the bottom track may prevent the strip light fixture from being installed tight to the sides of the mirror, leaving a narrow wall reveal between the fixture and the mirror.

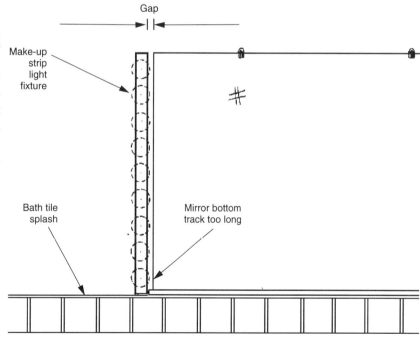

☐ **812**
Check that long lengths of bathroom mirrors will make it up stairways.

☐ *813*
Analyze bathroom mirrors when they are installed on walls with bull-nose drywall corner bead at windows, for example. Coordinate the dimensions of the mirrors with the round curvature of the bull-nose corners.

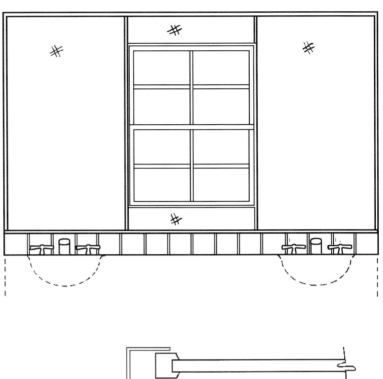

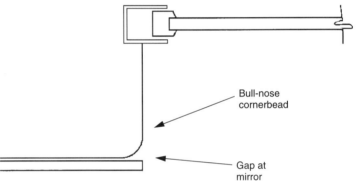

☐ *814*
Instruct the mirror subcontractor not to bother installing mirrors with chipped corners or desilvered edges, as these flaws eventually will be identified and the mirrors will be rejected by the builder.

☐ *815*
Schedule the installation of bathroom mirrors ahead of the installation of medicine cabinets. If the mirror installers are required to remove and reinstall medicine cabinets that are in the way of the mirror installation, the medicine cabinets sometimes will be reinstalled out of plumb and level, with the blame going to the finish carpenter, not the mirror installers.

☐ *816*
For pedestal sinks that are not installed until the hard surface flooring is selected by the home buyers and installed in the bathrooms, install the bathroom oval mirrors at the normal period during construction, using "cheat sheet" dimensions, at the correct height and centered over the future locations of the pedestal sinks.

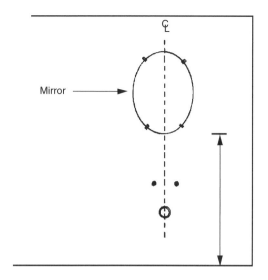

☐ *817*
Using a carpenter's pencil, lightly circle small dents and dings in the drywall requiring repair before the drywall final pickup phase. This establishes the level of quality required and removes these decisions from the discretion of the drywall repairpersons.

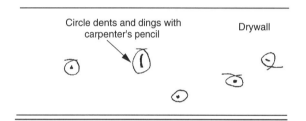

☐ *818*
Use color in the drywall texture at each repair so that the painter can identify and touch up the repairs afterward.

☐ *819*
Use drop lights and extension cords or flashlights to help identify small dents, nicks, and dings in drywall surfaces that need repair during the final drywall pickup phase.

☐ *820*
Get the drywall taping for walls that will have wallpaper perfectly smooth and without flaws before hanging the wallpaper. Some defects in the drywall underneath the wallpaper will be clearly visible and will detract from the wallpaper's appearance.

☐ *821*
Apply wall sizing or enamel paint to the wall surfaces before installing wallpaper. This creates a barrier between the wallpaper glue and the drywall surface paper. If a section of the wallpaper must be removed later, the drywall surface won't be damaged.

Miscellaneous Finish **315**

☐ *822*
Have the wallpaper hanger thoroughly wash or clean off the glue from the surrounding door casing and baseboard wood trim. If some types of glue are left to harden as a thin film over painted wood trim, they are difficult to remove later.

☐ *823*
For exterior balcony decks, a 6-inch on-center spacing of 2-inch by 2-inch nominally thick wood handrail spindles will not result in the required maximum open spacing gaps of 4 inches; 2-inch wood is only $1^1/_2$ or $1^5/_8$ inches thick. Use a $3^{15}/_{16}$-inch spacer block to install the handrail spindles instead of the 6-inch on-center method.

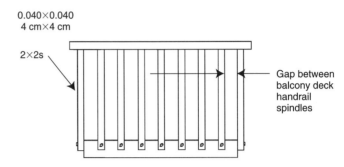

☐ *824*
Have the weather stripping installer cut off the bottoms of exterior doors and install the thresholds with a small amount of excess play at the bottom. This prevents a customer service request later if the door swells or settles slightly and rubs against the threshold, sticking in place in the closed position.

☐ *825*
Do not allow subcontractors to use the garages of houses under construction to store materials and equipment. Locked garages create logistics problems for the builder when other tradespersons need to perform work inside the locked garages. Subcontractors should maintain storage bin containers on the job site.

☐ *826*
Have the garage door installer adjust the steel tension rods at the top of the garage doors so that the doors close tightly against the garage door header.

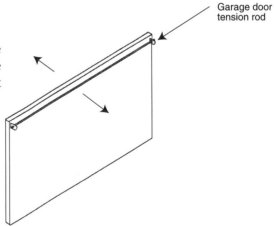

316 **Chapter 28**

☐ **827**

Avoid a sharp 90-degree corner for stairway wrought-iron handrail that makes a transition from a full-height wall to a half-height pony wall partway down the staircase. Make the transition corner a 45-degree angle or a softly contoured round corner for safety and liability reasons.

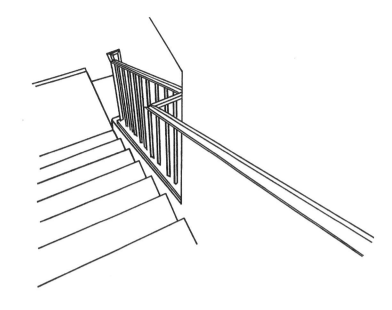

☐ **828**

For wrought-iron handrailing with knuckles, balance the number and spacing of the spindles so that the knuckles are uniformly distributed, especially for short lengths of handrail. You don't want an even number of spindles for a 4-foot-long handrail section, for example, with the spindle on one end having a single knuckle and the spindle on the other end having two knuckles.

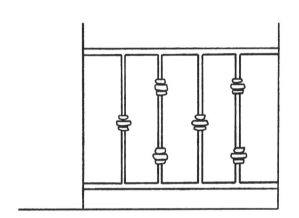

☐ **829**

Check that the raised hearth for a fireplace is not deeper than the dimensions of an adjacent wall to prevent a small section of the end of the raised hearth from being unintentionally exposed. It would then have the appearance of a design and construction mistake.

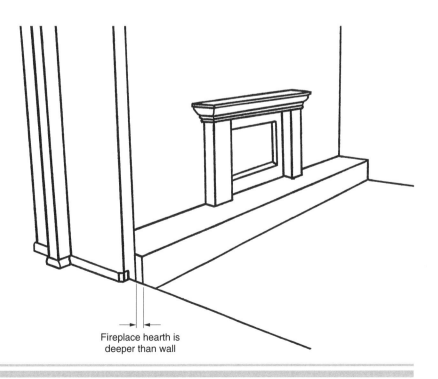

Fireplace hearth is deeper than wall

Miscellaneous Finish

☐ **830**
Check that the fireplace mantel is not longer than the width of the wall space, corner to corner. If it is, decide ahead of time how to finish the ends that extend beyond the wall corners.

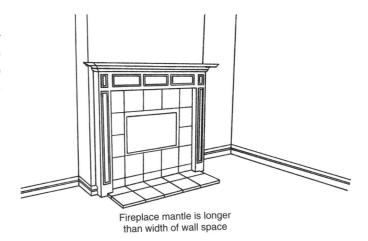
Fireplace mantle is longer than width of wall space

☐ **831**
Make sure a prefabricated fireplace mantel unit with vertical legs is not wider than the wall space at a fireplace.

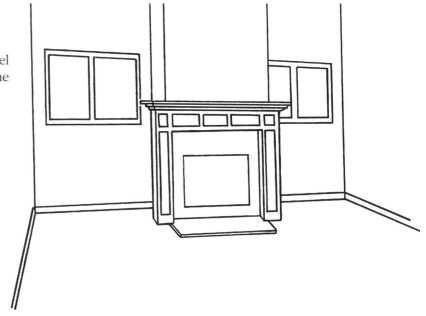

318 **Chapter 28**

☐ **832**
For fireplaces placed at a 45-degree angle within a wall corner, check that the surrounding windows, sliding glass doors, and stairways will allow the fireplace to be symmetrical and balanced on each side.

☐ **833**
For a fireplace placed at a 45-degree angle within a wall corner with 45-degree angled hearths at each end, check that the front corners of the legs of a prefabricated fireplace mantel unit do not extend beyond the edges of the angled hearth.

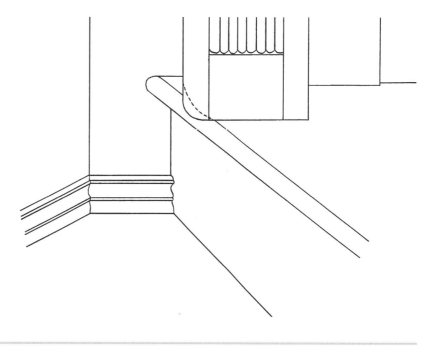

☐ *834*

For a bathtub with a small gap on the floor at the front side of the tub resulting from a subfloor that is not flat or level, consider covering the gap between the bottom edge of the bathtub and the vinyl flooring with plastic trim pieces that match an ivory white bathtub. You can't nail wood base shoe trim to the bottom of a bathtub, but it is fairly easy to glue plastic trim to the bottom of a bathtub by using latex caulking.

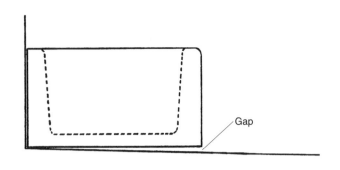

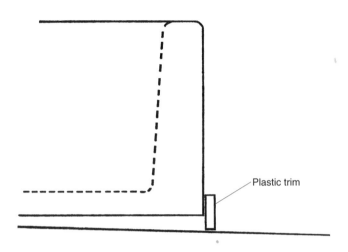

☐ *835*

For a microwave oven that sits on top of an upper cabinet shelf within an opening that is enclosed on all four sides, remember to order the accessory trim kit along with the oven. The trim kit covers the gaps between the microwave and the sides and top of the cabinet opening. It does not come with the microwave but must be ordered separately.

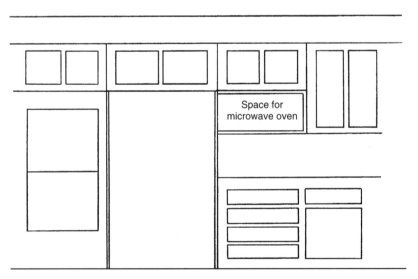

☐ **836**

Check on the plans that the kitchen cooktop appliance has the same length as a microwave oven above it and/or that the vertical separation between the cooktop and the microwave oven meets the minimum clearance requirements of the cooktop manufacturer. A 30-inch microwave oven over a 36-inch cooktop must be at least 18 inches above the cooktop, because there is 3 inches of upper cabinet (considered combustible) directly above the cooktop on each side. A 30-inch microwave oven over a 30-inch cooktop, however, need be only 15 inches above the cooktop, because the microwave oven is not considered combustible. Check the local codes

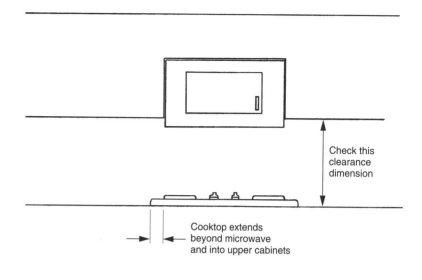

☐ **837**

For exterior doors overlooking a scenic view, install window blinds at the high point at the top of the door so that when they are opened, the blinds will clear the top panels of glass and not obscure the view.

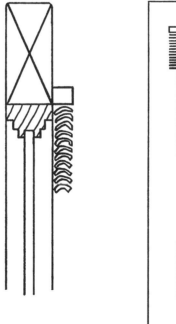
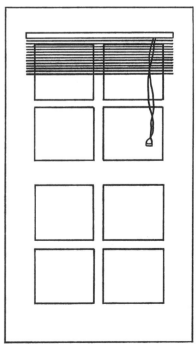

Miscellaneous Finish *321*

☐ **838**
Check that a prefabricated, one-piece marble fireplace face, for example, does not extend past the curvature of the adjacent wall corner's bull-nose drywall corner bead. The premeasurement before fabrication is not to the wall corner but to the start of the curvature of the bull-nose corner bead.

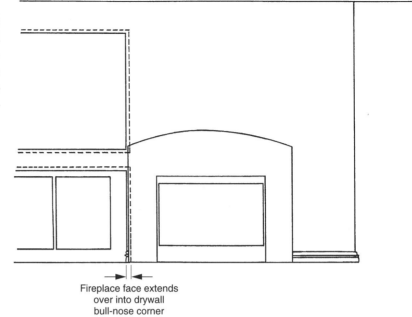

Fireplace face extends over into drywall bull-nose corner

☐ **839**
Design and build balcony deck pony walls to be equal to or wider than the precast concrete handrail cap. This allows the handrail cap to butt into the end of the pony wall without a wide grout joint on each side of the cap to soften or taper the corners of the cap.

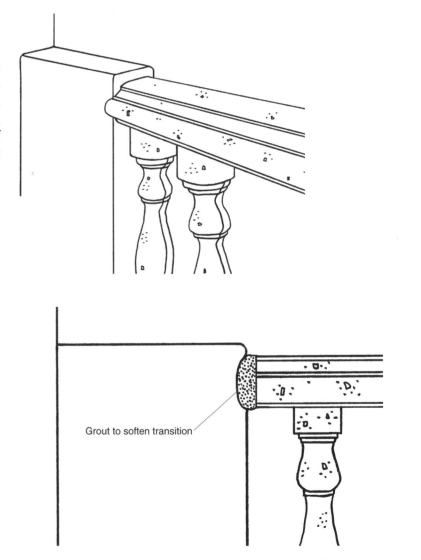

Grout to soften transition

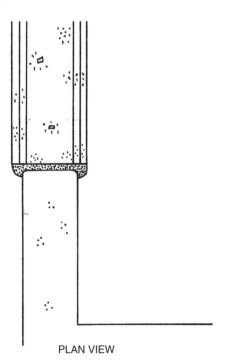

PLAN VIEW

322   **Chapter 28**

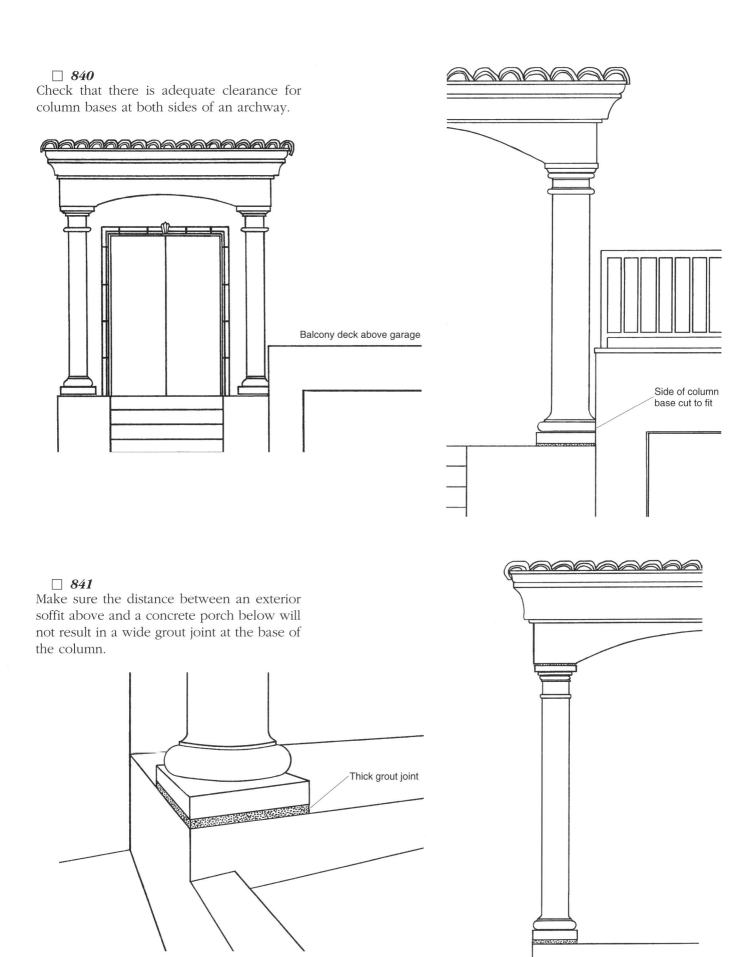

☐ **840**
Check that there is adequate clearance for column bases at both sides of an archway.

☐ **841**
Make sure the distance between an exterior soffit above and a concrete porch below will not result in a wide grout joint at the base of the column.

☐ *842*
Check that balcony deck pony walls are higher than the precast concrete handrails. The top of the handrail cap should die into the pony wall equal to or below the pony wall; otherwise part of the precast cap ends will be exposed, giving the appearance of a design and construction mistake.

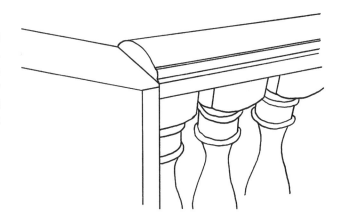

☐ *843*
Make sure precast concrete handrails and the surrounds around a sliding glass door for an exterior balcony deck do not meet at the same place. If they do, order specially shaped precast pieces to match the contours of the adjoining piece.

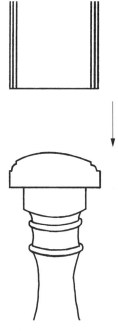

☐ *844*
When dissimilar pieces of precast concrete trim join together, order specially shaped pieces that match the contours of the adjoining piece or backcut on the job site at a 15-degree or 30-degree angle the ends of the thicker precast members. The partially exposed square ends of the thicker pieces sometimes look like a design and construction mistake.

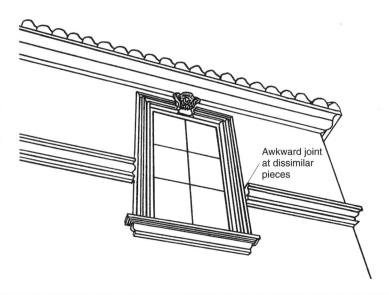

324   **Chapter 28**

☐ **845**
Check that a prefabricated column encasing a structural wood post will not extend beyond the edge of a split-level step at the interior entry, for example. If the structural post is located at the edge of the split-level step, the thicker column will project into the lower floor area.

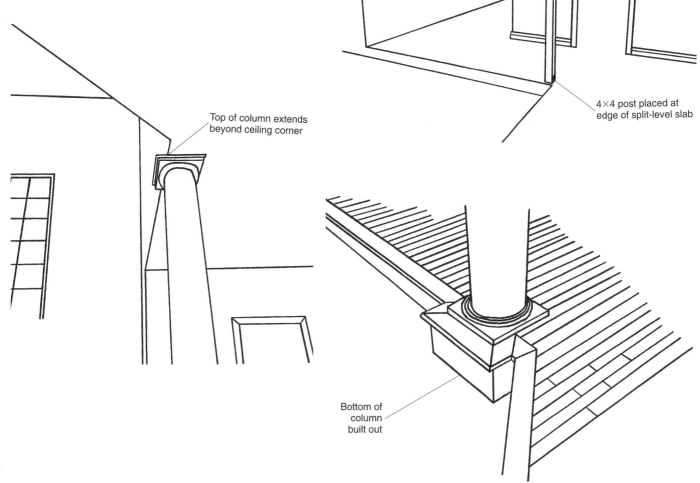

☐ **846**
Construct a wooden ramp that protects the new metal thresholds at entry doors during the delivery of appliances that are rolled into house interiors on appliance dollies. The ramp can be picked up and carried from house to house for each delivery of appliances.

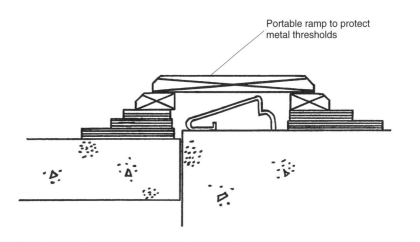

Miscellaneous Finish 325

☐ **847**
For roof overhangs that have rafter tails without fascia board, install rain gutters without a downward slope toward a downspout or add L-metal flashing over the tops of the rafter tails before installing the rain gutters to cover the gap that is created above the rain gutter as it slopes downward toward the low point at the downspout.

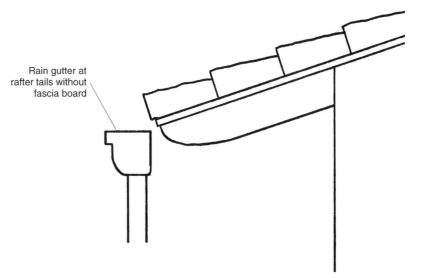

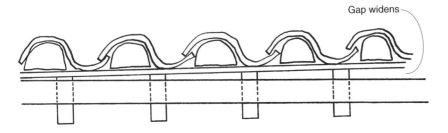

☐ **848**
For wrought-iron handrailing with a separate bottom rail, turn the attachment flange to face upward to clear the floor baseboard; otherwise the baseboard must be cut out around the attachment flange.

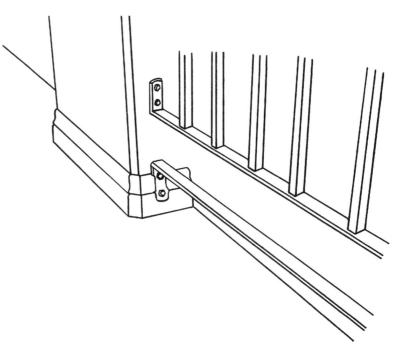

☐ **849**
For wrought-iron handrailing with a separate bottom rail, anticipate areas where the bottom rail will have to be precut to fit around hardwood skirtboard trim at a stairway landing, for example.

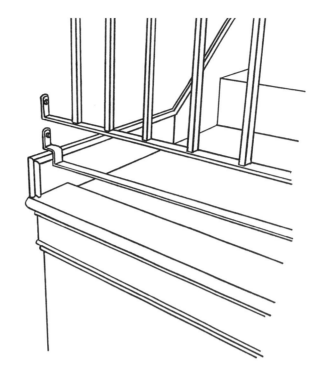

☐ **850**
A method to provide a flat surface for light fixtures on exterior walls with wood cedar shingles or bevel siding, for example, is to install a finish wood plaque around the rough electrical outlet box and then cut the shingles or siding to fit around the plaque.

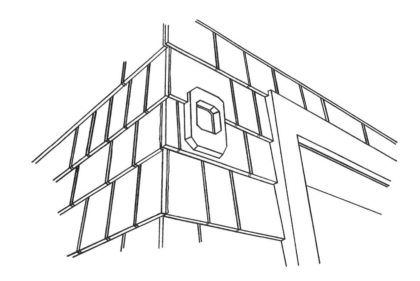

☐ **851**
Lay out the locations of windows so that balcony deck wrought-iron handrailing does not attach to the wall through foam plastic architectural trim.

☐ **852**
Lay out the locations of windows so that exterior light fixtures do not attach to the wall through foam architectural trim or, during the rough electrical phase, find other locations for the light fixtures that will have flat, open stuccoed wall surfaces.

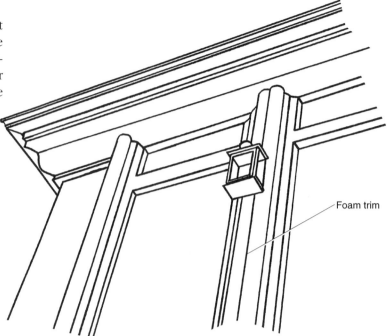

☐ **853**
For architectural foam trim that is finished with a stucco color coat, make sure there is enough clearance for everything to fit. If the architectural foam trim must be cut to fit around surrounding members such as windows, balcony decks, and other foam trim at roof eaves, this can look like a design and construction mistake.

Continued on next page

Continued from previous page

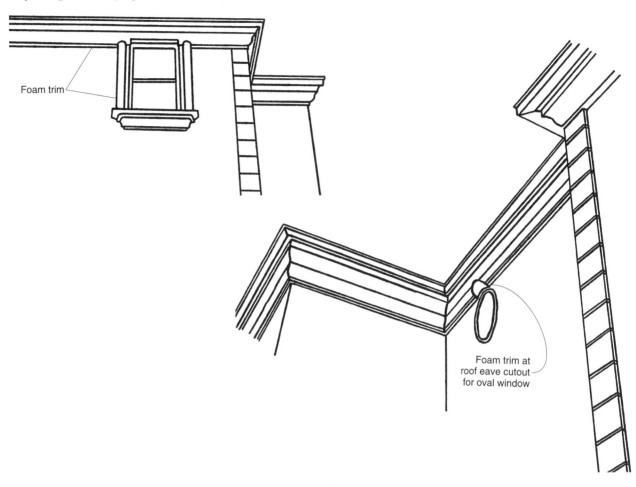

☐ **854**
Check that there is enough clearance from the top of an arched window to the roofline for precast concrete surround trim to fit, especially when rain gutters are installed.

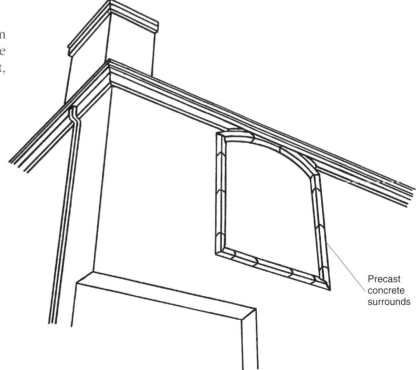

Miscellaneous Finish

☐ 855
Make sure the depth of architectural foam trim at the roof eaves does not bury the top reveals of the second floor window frames.

☐ 856
Analyze ahead of time the joint between two dissimilar precast concrete members, in this case between a balcony handrail cap and a wainscot. The exposed end of the wainscot piece will look unfinished and ill planned.

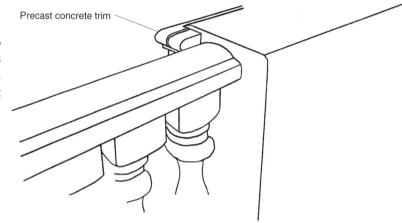

☐ 857
Avoid installing bathroom mirrors that fill the entire wall space above a countertop splash from wall corner to wall corner and from the splash to the ceiling. It is difficult to get the wall corners and ceiling perfectly straight and square to match the mirror.

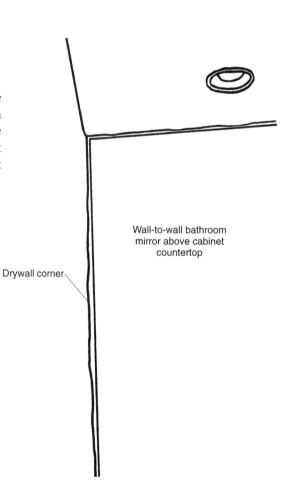

CHAPTER 29

Flooring

☐ *858*
Walk, mark, and screw off plywood subfloor squeaks late in the construction, after the box-out and sweep phase that follows the completion of the final trades. This is the best time to correct floor squeaks because the interiors of the units are free of boxes and debris, the subfloors have been swept, and the flooring installation will follow shortly afterward.

☐ *859*
Don't allow the quality-control supervisor from the flooring subcontractor to walk through the houses and spray-paint on the floors the corrections needed on the subfloors before the start of flooring. The builder may not agree with the level of quality requested by the flooring supervisor, and the new homebuyers will see these markings on the subfloors and consider these areas to be actual problems to be repaired, when in fact they will not be noticed after the flooring is in. The flooring supervisor should use a scratch pad instead and give the punch list to the builder without marking the floors.

☐ *860*
Include the grinding down of minor plywood subfloor high spots within the normal floor preparation work required from the hardwood flooring installers. This gets minor problems repaired on the spot by the flooring installers to suit their individual requirements.

☐ *861*
Have the plumber reinstall toilets that have been removed for the installation of hard surface flooring. This gets the toilets correctly reinstalled by plumbers rather than the flooring installers.

☐ **862**
Lay out floor tile in bathroom toilet rooms, for example, so that the cut row of tiles is against the baseboard at the back wall of the room. This provides a finished factory tile edge at the tile-to-carpet joint at the front of the toilet room rather than a cut tile edge.

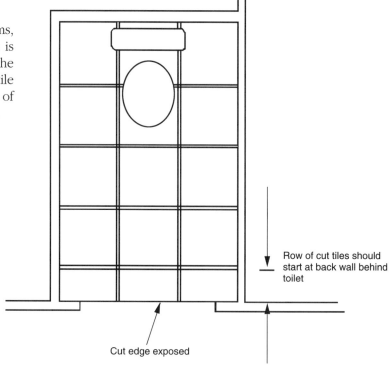

☐ **863**
Don't grout the flooring next to wood siding boards. The moisture from the tile grout will seep up the wood and create stains that match the color of the tile grout. Install the tile flooring ahead of the wood siding boards or thoroughly seal the bottom of the boards.

☐ **864**
Analyze the relative thickness of hard surface flooring and carpeting to prevent tile edges and mortar or underlayment from being exposed because of thin carpeting.

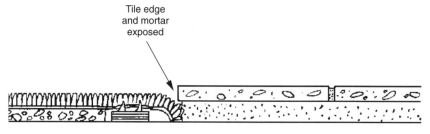

☐ **865**
Have the vinyl flooring installers trim off the edges of vinyl flooring at the wood baseboard. This prevents the finish carpenter installing the base shoe from having to trim off the excess vinyl flooring.

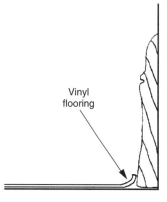

☐ **866**
Analyze the different thicknesses of two types of hard surface flooring materials joining together in a doorway, for example, and install some type of sloping ramp to provide a transition between the two dissimilar floor coverings.

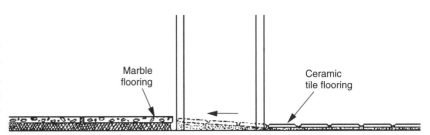

☐ **867**
When paper is used as walkway runners to protect new wood flooring from damage from foot traffic, cover the entire wood floor area. Sunlight coming through the windows will discolor the uncovered wood flooring areas slightly over time so that when the paper runners are removed, the wood flooring will have two noticeably different colors.

☐ **868**
Order vinyl flooring in 12-foot-wide rolls instead of 6-foot rolls to eliminate seams in kitchens and large bathrooms.

☐ **869**
Have the vinyl flooring installers caulk the joints between the vinyl flooring and the interior doorjambs.

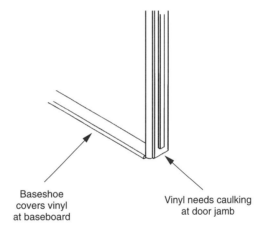

☐ **870**
If the hard surface flooring will be installed all the way underneath appliances, discuss and include this in the flooring subcontract so that the extra square footage does not become an extra cost requested by the flooring subcontractor.

☐ **871**
Have carpet layers spackle the hammer divots in the floor baseboard that result from misses while installing the tack strip.

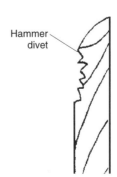

Flooring 333

☐ *872*
Have the carpet layers use the metal nailing bar tool to install the tack strip all the way back in hard-to-reach cabinet toe kick areas.

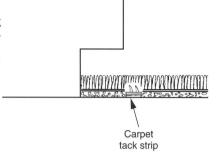

☐ *873*
If the carpet layers stack interior doors in bathtubs to get them out of the way during carpet installation, require that the doors be placed on top of carpet padding to protect the bathtubs.

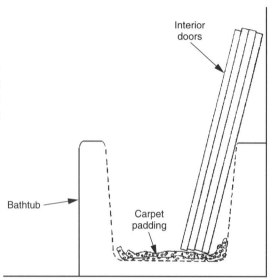

☐ *874*
Analyze ahead of time how to finish areas where stair carpeting meets hard surface flooring on upper floor levels. Either continue the hard surface flooring down the riser at the top step for the carpet to butt into or provide some type of overhang at the upper floor level for the carpet to butt into.

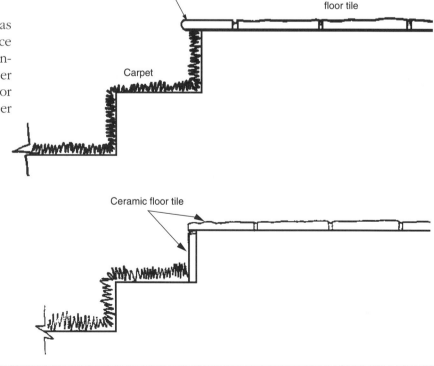

334 Chapter 29

☐ **875**
Provide and install a metal trim shoe to finish the edge of the subfloor and vinyl flooring at the joint with the stairway carpeting.

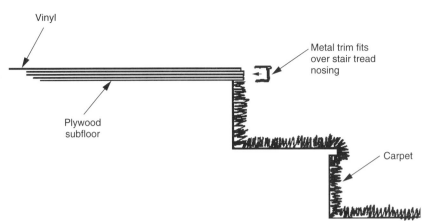

☐ **876**
Use tapered wood shingles to provide a sloping ramp at the transition from a thin Berber® or woven-type carpet to a thicker hard surface flooring.

☐ **877**
Choose between a "waterfall" and an "upholstered" method of carpeting at stair tread nosings and build the stair treads accordingly with either angled risers or treads with a standard 1-inch nosing projection. You don't want to have to add wood trim to the fronts of the treads as an afterthought during carpet installation to achieve upholstered stair steps.

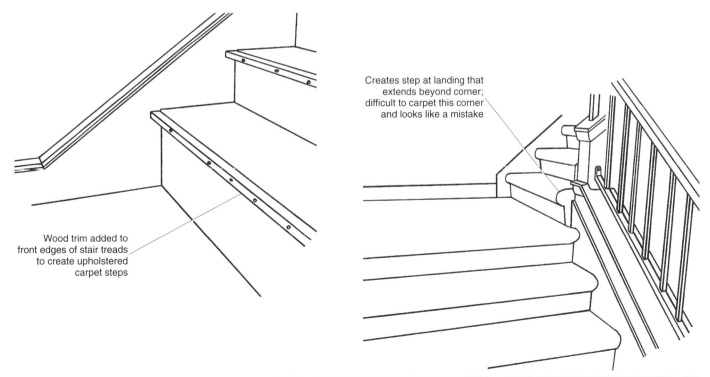

Flooring 335

☐ **878**
For Berber-type carpeting in stairways, consider the grain pattern of the carpeting in estimating the square yardage required. Stair carpet must be cut lengthwise from the roll of carpeting to keep the grain pattern running in the correct direction. Otherwise the Berber-type carpet will bird's-mouth as it bends around each tread nosing.

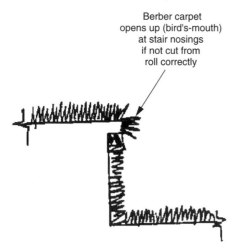

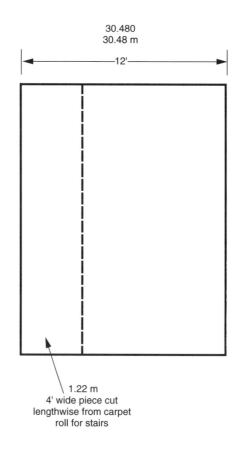

☐ **879**
For floor plans with 45-degree doorways at bedrooms, add more yardage to the carpet estimate when using patterned carpet that must be matched and aligned at the doorways.

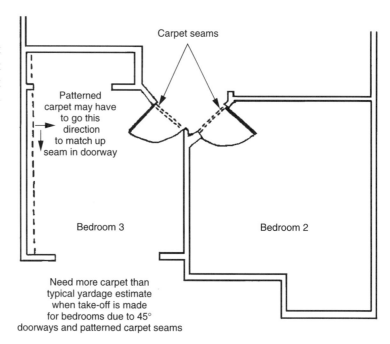

☐ **880**
Check toilets for cracked bases after the installation of the tack strip. Require the carpet layers to use the metal nailing bar tool to attach the tack strip around toilets.

☐ **881**
Include within the painting subcontract a provision to touch up the baseboard after carpet installation. A pencillike line usually results from the carpet layers' trimming tools running along the baseboard as a guide.

☐ **882**
Don't scribe cut vinyl flooring around toilets. Remove the toilets beforehand and reinstall them after the vinyl flooring has been installed.

☐ **883**
Composition board underlayment beneath a fireplace hearth must be even with or slightly inside the edge of the hearth tile so that it does not interfere with the carpet installation. If the underlayment projects beyond the hearth tile, it will prevent the carpet from reaching the hearth tile.

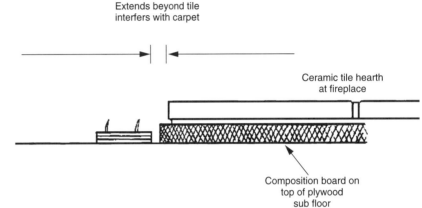

☐ **884**
Analyze the relative thicknesses of fireplace hearths and carpeting and determine in the subcontracts who will be responsible for installing ramps and floating at the transition from carpet to hard surface flooring—the hard surface tile installers or the carpet layers.

☐ **885**
Have stair handrail spindles go on top of a wood cap border so that carpeting can tuck into the wood cap rather than having to be cut individually around each handrail spindle. This looks better and saves a lot of time for the carpet layers.

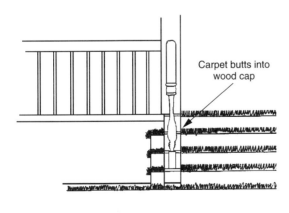
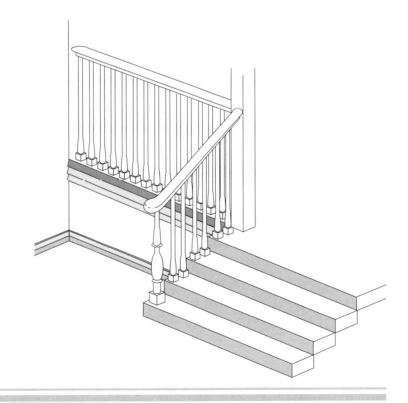

Flooring

☐ **886**
Check that the thickness of the hard surface flooring adjacent to a wood cap for the handrail spindles is not above the level of the wood cap; otherwise the top edge of the tile grout will project above the cap and be noticeable.

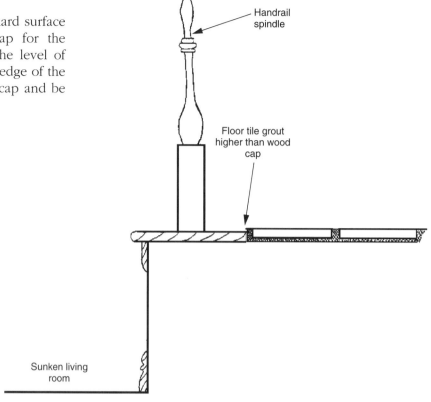

☐ **887**
Save leftover pieces of carpet, vinyl flooring and hard surface tile, and marble or wood for the new homeowners. Use scrap carpet pieces as temporary doormats at the entry door and the garage fire door until the homeowners move in.

☐ **888**
Have the floor tile installer mix the tiles from all the boxes to blend the shade variations from box to box. You don't want color variations from one box to uniformly end up in the same area on the floor.

☐ **889**
Design and install wood borders on stair treads and risers so that after the handrailing is installed, the carpeting is centered between the handrailing and the opposite wall. The width of the wood borders should be equal after the handrailing is installed on one side of the stairway.

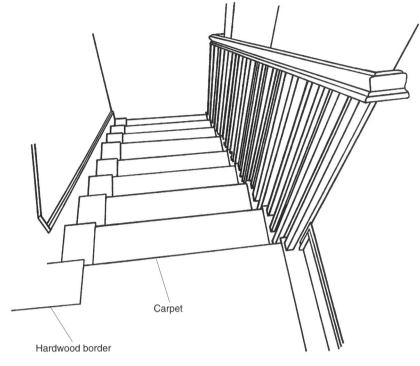

☐ **890**
For powder baths and toilet rooms with marble flooring and marble baseboard, for example, make sure there is enough wall space at the door casing for the marble baseboard to at least slip behind the edge of the casing. If the marble baseboard butts into the casing, there may be an unfilled gap at the wall corner.

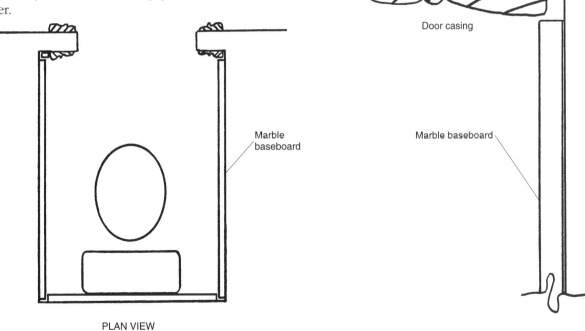

☐ *891*
When a second or third floor kitchen gets vinyl flooring, for example, decide ahead of time how to trim out the edge of the kitchen floor at the adjacent stairway when this open space gets a wrought-iron handrail.

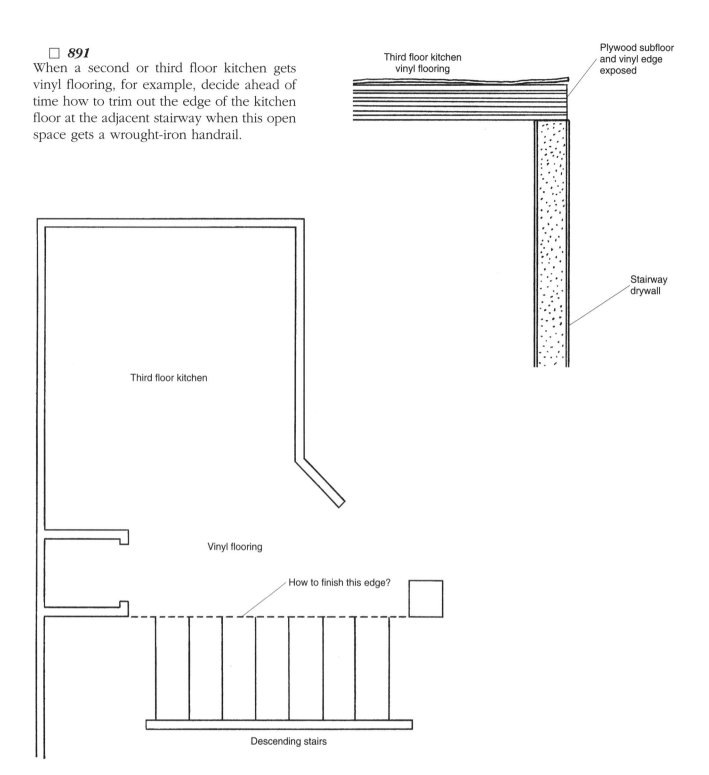

☐ **892**
When stairs gets hard surface tile flooring, check that the wrought-iron handrail is installed at the correct distance above the stair steps to allow for the thickness of the tile flooring that will be installed later.

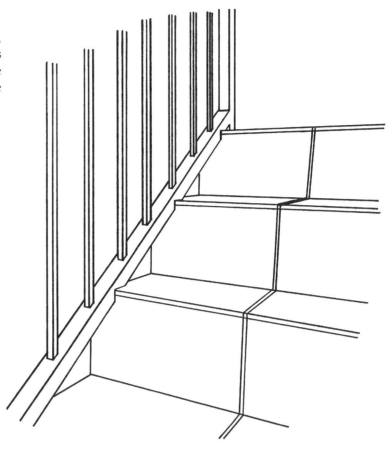

☐ **893**
Lay out floor tiles so that a row of cut tiles starts at a wall; that way, full-size tiles can be installed out to the middle of a room at the joint with the carpeting. Don't start with a row of full tiles at the wall and then end up with a row of 2-inch-wide tiles in the middle of the room at the transition from hard surface to carpet at a hallway, for example.

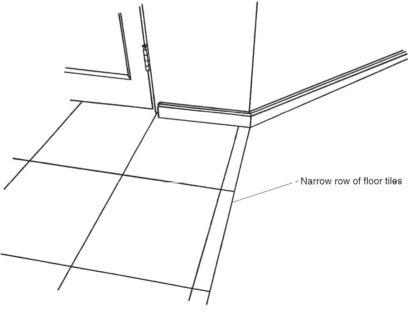

Narrow row of floor tiles

☐ **894**
Analyze ahead of time the effects of mortar floated hard surface ceramic, marble, or clay floor tiles on the surrounding members, such as cabinets, HVAC registers, and bathtub access doors.

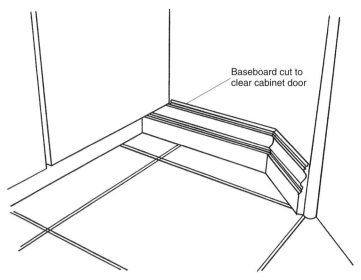
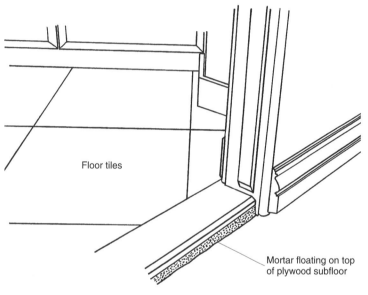
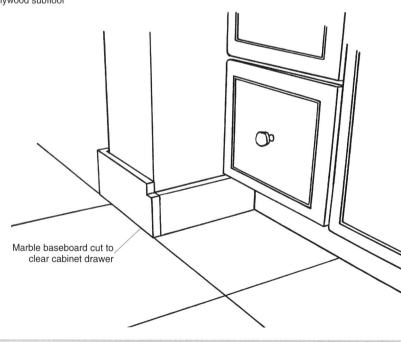

☐ **895**
For hard surface floor tiles that are installed over a floated mortar base, stop the floor tiles short of a sliding pocket door. The tiles may be higher than the carpeting, interfering with the opening of the pocket door.

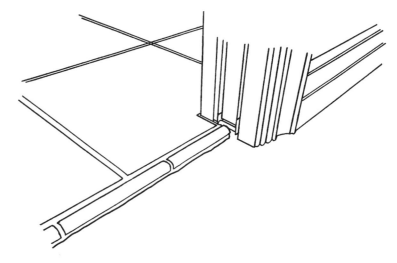

☐ **896**
Check that the stair framing anticipates the different thicknesses of the finish flooring so that no stair step risers vary in height more than ³/₈ of an inch after the flooring is installed.

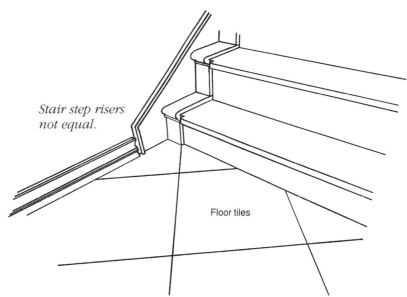

Stair step risers not equal.

Floor tiles

☐ **897**
Include the installation of 1×2-inch wood carpet returns in the flooring subcontract or the finish carpentry subcontract.

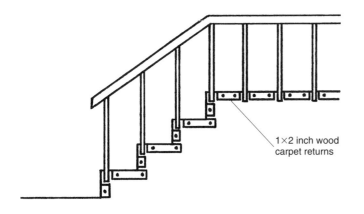

1×2 inch wood carpet returns

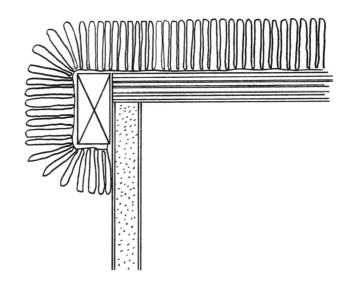

Flooring 343

☐ **898**

If the gap between a raised concrete slab and the adjacent wall framing is too wide for normal-width carpet tack strip, tell the carpet layers ahead of time so that they can bring 2-inch-wide, commercial-type tack strip to the job site for the carpet installation.

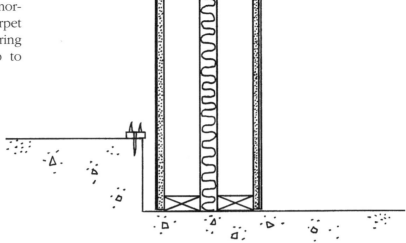

☐ **899**

For stairways with a wrought-iron handrail that prevents the carpet layers from reaching the outside edges of the stair steps, remember to tell the carpet layers ahead of time to bring ladders to the job site so that they can finish the stair step carpeting from a hallway below, for example. This prevents the carpet layers from having to stand unsafely on top of several stacked toolboxes to reach those areas.

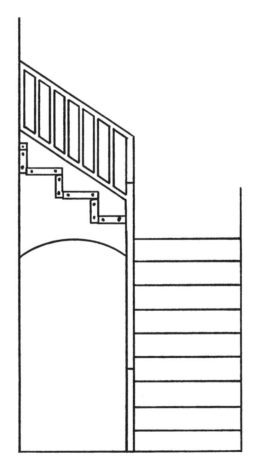

CHAPTER 30

Concrete Walkways and Driveways

☐ **900**
During the design phase of the project check the rise and run of the concrete stair steps from the sidewalk or street elevation up to the entry doors to determine whether the stairs and walkways will actually fit as designed.

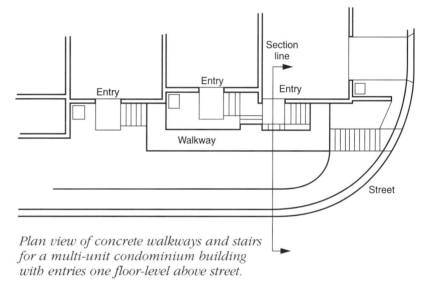

Plan view of concrete walkways and stairs for a multi-unit condominium building with entries one floor-level above street.

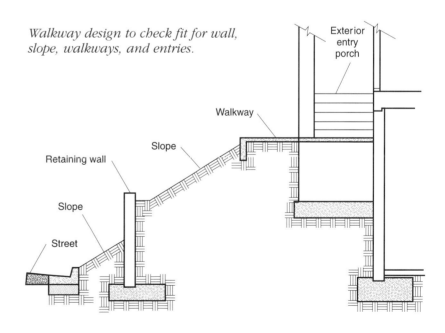

Walkway design to check fit for wall, slope, walkways, and entries.

Concrete Walkways and Driveways *345*

☐ *901*
For high-density condominiums and apartment projects, check that the dimension between an entry door and the side of a garage is large enough to accommodate the rise and run of the stair steps plus the minimum width of an entry porch landing and the inclusion of legal handrails.

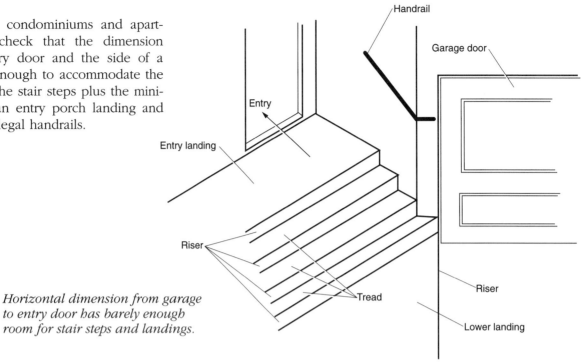

Horizontal dimension from garage to entry door has barely enough room for stair steps and landings.

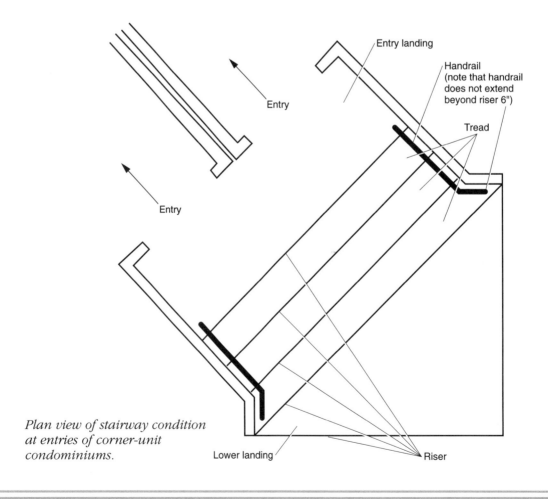

Plan view of stairway condition at entries of corner-unit condominiums.

☐ **902**
For high-density condominium and apartment projects, make sure there is enough space between the buildings for common area walkways, entry porches, stair steps, and handrailing to fit.

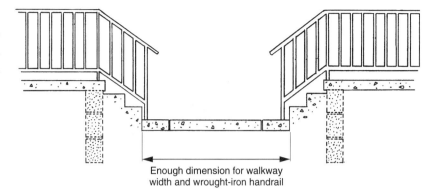

End view of common-area walkway between two entry porches.

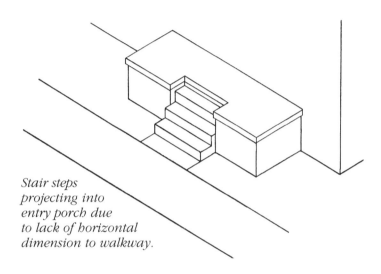

Stair steps projecting into entry porch due to lack of horizontal dimension to walkway.

☐ **903**
Design and pour entry concrete walkways so that tile that is added later dies into the wall surfaces rather than extending beyond the wall corners. Also, get the stucco weep screed installed during the lathing phase to match the slope of the stair steps that are poured later.

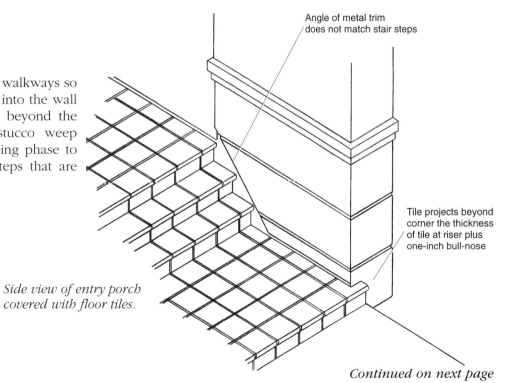

Side view of entry porch covered with floor tiles.

Continued on next page

Concrete Walkways and Driveways 347

Continued from previous page

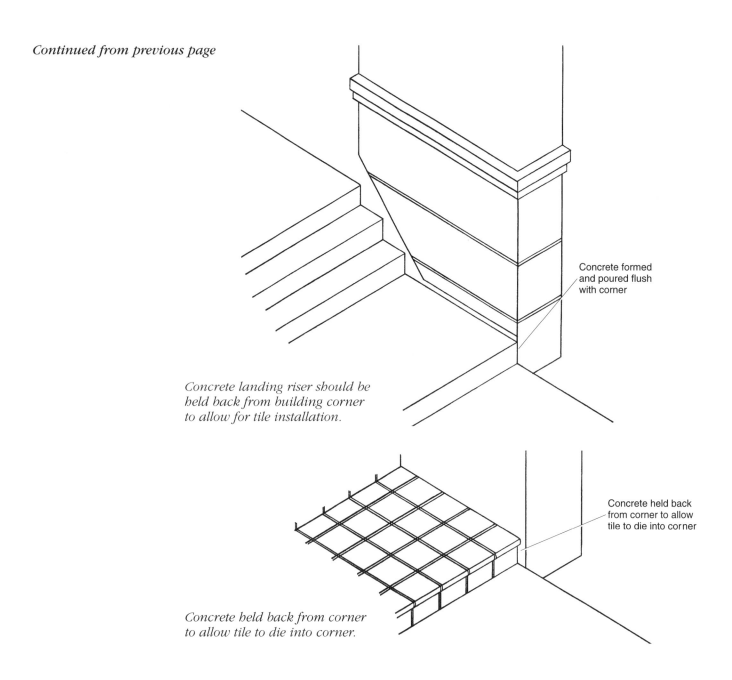

Concrete landing riser should be held back from building corner to allow for tile installation.

Concrete held back from corner to allow tile to die into corner.

☐ **904**
For concrete walkways at street corners, drop the elevation at the rear of the walkway to match the drop at the curb for a handicap ramp. This prevents the cross-slope of the walkway from becoming too steep at the handicap ramp.

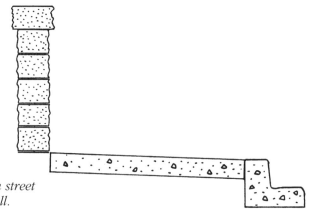

Concrete walkway between street curb and block garden wall.

Continued on next page

Continued from previous page

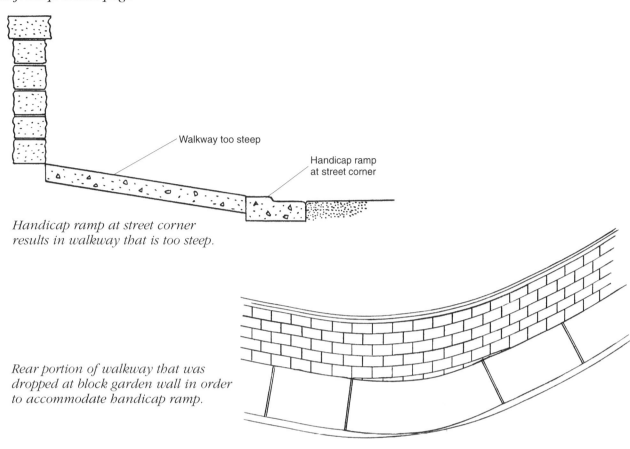

Handicap ramp at street corner results in walkway that is too steep.

Rear portion of walkway that was dropped at block garden wall in order to accommodate handicap ramp.

☐ **905**
For concrete walkways, configure the steps to prevent a severe cross-slope.

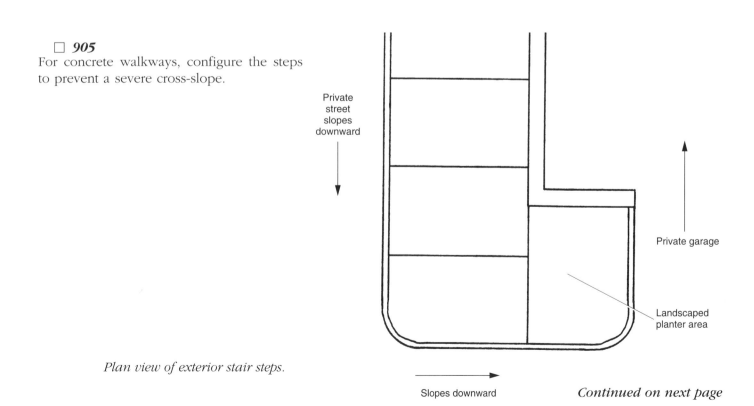

Plan view of exterior stair steps.

Continued on next page

Concrete Walkways and Driveways

Continued from previous page

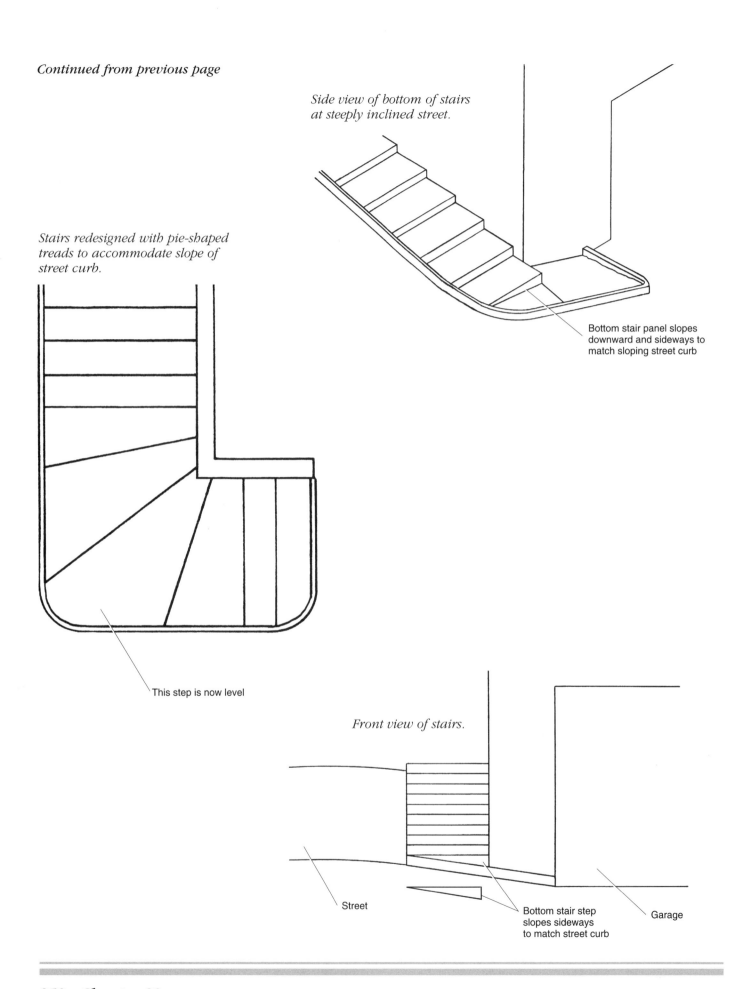

Side view of bottom of stairs at steeply inclined street.

Bottom stair panel slopes downward and sideways to match sloping street curb

Stairs redesigned with pie-shaped treads to accommodate slope of street curb.

This step is now level

Front view of stairs.

Street

Bottom stair step slopes sideways to match street curb

Garage

☐ **906**
Avoid narrow, unused spaces between concrete stair steps and a building's exterior walls. They become traps for dust and trash. Pour the steps all the way over to the building.

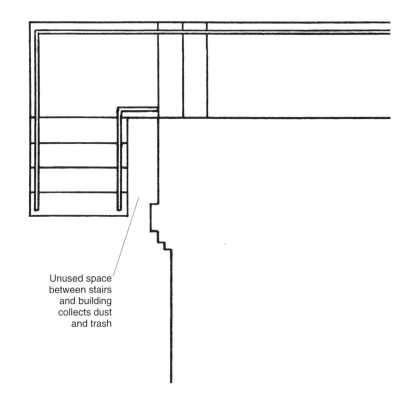

Plan view of concrete stairs.

☐ **907**
Check for architectural banding on exterior walls that project into and reduce the size of a minimum-width walkway.

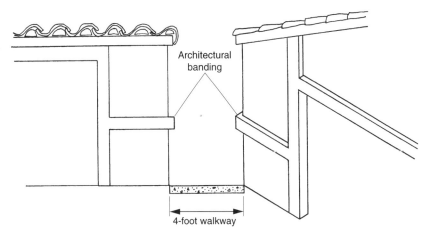

Architectural banding projects into concrete walkway.

Concrete Walkways and Driveways 351

☐ **908**
For condominium and apartment projects, avoid combining an electrical meter area elevated above a trash bin area with an adjacent concrete stairway going up to the entry door level, for example. This is an unsafe condition for entering and exiting from the electrical meter area from the stairway because there is no flat landing area.

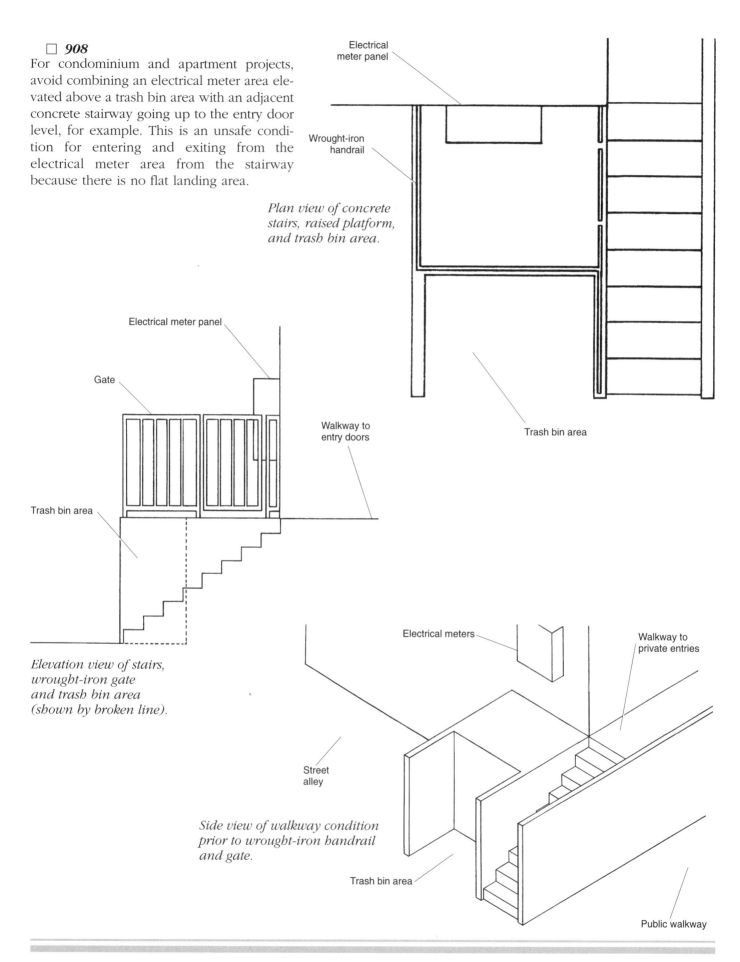

Plan view of concrete stairs, raised platform, and trash bin area.

Elevation view of stairs, wrought-iron gate and trash bin area (shown by broken line).

Side view of walkway condition prior to wrought-iron handrail and gate.

352 **Chapter 30**

☐ **909**
For condominium and apartment projects with subterranean garages, check that there is enough space between the side yard property line and the building corner for stair steps down to the subterranean garage, a side yard walkway, a fence, and a gate.

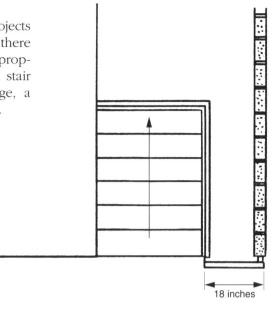

Leftover space between property-line block wall and stairs leading to parking structure.

☐ **910**
Check that the dimensions for concrete patios as shown on the plans will coordinate with the widths of precast columns covering structural wood posts that support second floors above.

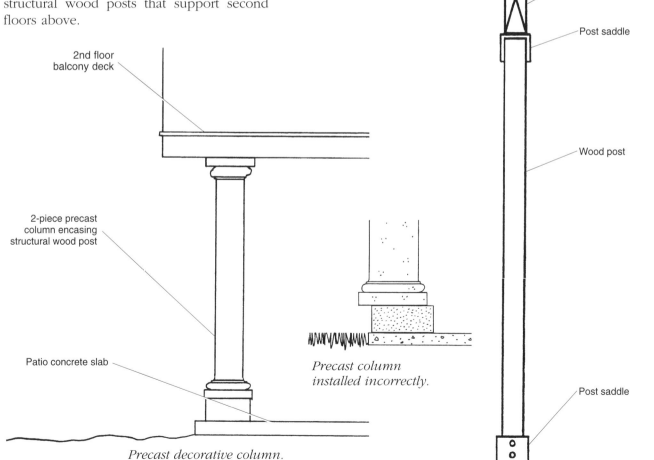

Precast column installed incorrectly.

Precast decorative column.

Concrete Walkways and Driveways

☐ **911**
Check that all the structural connections and the dimensional locations shown on the plans for concrete footings and the entry concrete porch result in the placement of a precast column at the building corner that works out aesthetically with the surrounding architecture. The finished appearance of the architecture should determine the positioning and alignment of the structural members and concrete rather than the other way around.

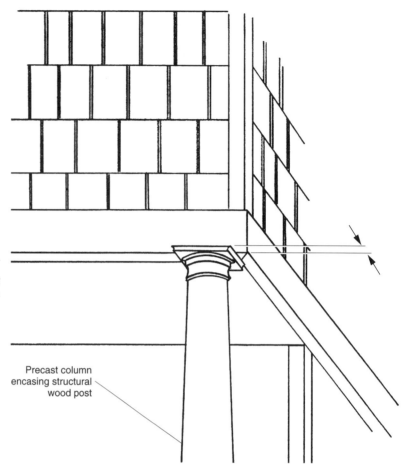

Precast column at front entry not aligned with building corner due to structural conditions below.

☐ **912**
For garage slabs that are lower than the adjacent side yards, check the plans and make sure during construction that there is enough garage width for a concrete landing with steps and a doorway that swings inward into the garage.

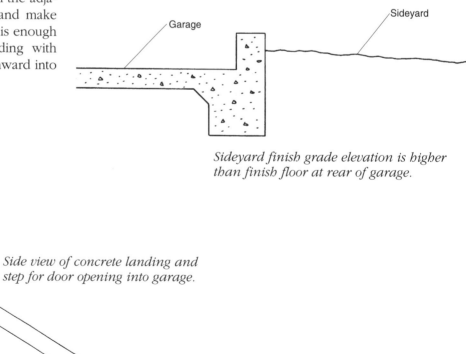

Sideyard finish grade elevation is higher than finish floor at rear of garage.

Side view of concrete landing and step for door opening into garage.

Continued on next page

Continued from previous page

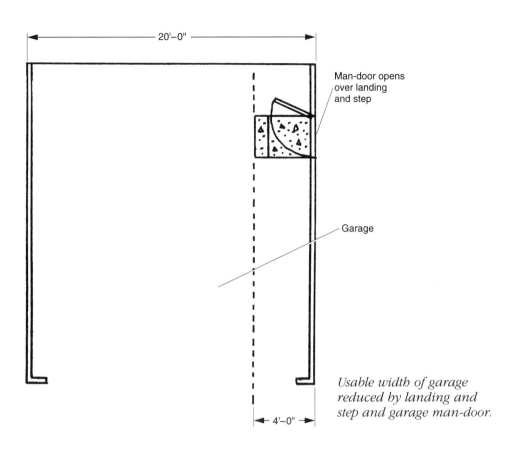

Usable width of garage reduced by landing and step and garage man-door.

☐ **913**
Pour concrete walkways to accommodate the thickness of floor tile that is added later.

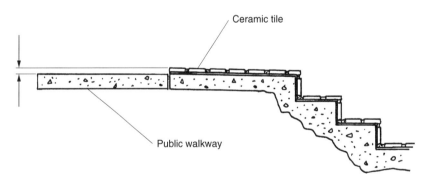

Tile added to top landing and stairs projects above public sidewalk.

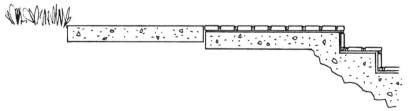

Concrete top landing lowered the thickness of mortar and tile, so finish surface is flush with public sidewalk.

Concrete Walkways and Driveways *355*

☐ **914**
Consider foot traffic patterns in designing common area walkways for condominium and apartment projects. Displace grass lawn areas with concrete at inside corners where pedestrians will take shortcuts.

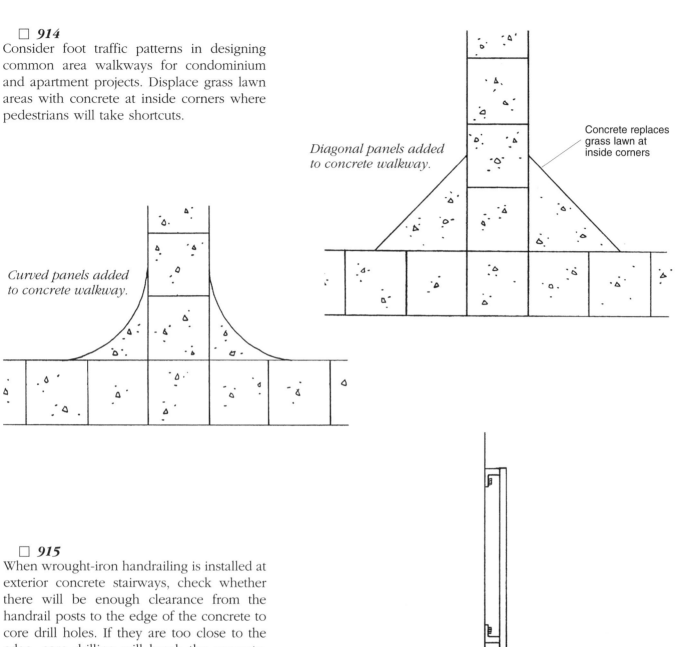

Diagonal panels added to concrete walkway.

Concrete replaces grass lawn at inside corners

Curved panels added to concrete walkway.

☐ **915**
When wrought-iron handrailing is installed at exterior concrete stairways, check whether there will be enough clearance from the handrail posts to the edge of the concrete to core drill holes. If they are too close to the edge, core drilling will break the concrete. Attachment flange base plates with exposed anchor bolts will have to be used instead, and such plates do not look good.

Concrete stairs widened at ends for wrought-iron handrail.

3–4 inches so edge of concrete step doesn't break off

Continued on next page

Continued from previous page

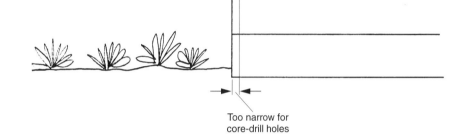

Metal flange and lag screw bolt are used to install handrail in stairs not designed and built for core-drilled holes.

☐ **916**
Check that the design and construction of a masonry block retaining wall match those of an adjacent concrete stairway to eliminate bottom steps that are unsafe or require additional handrailing.

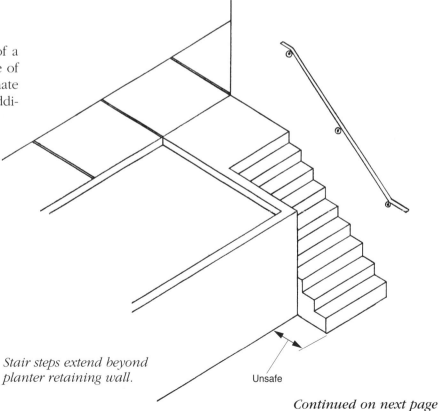

Stair steps extend beyond planter retaining wall.

Continued on next page

Concrete Walkways and Driveways

Continued from previous page

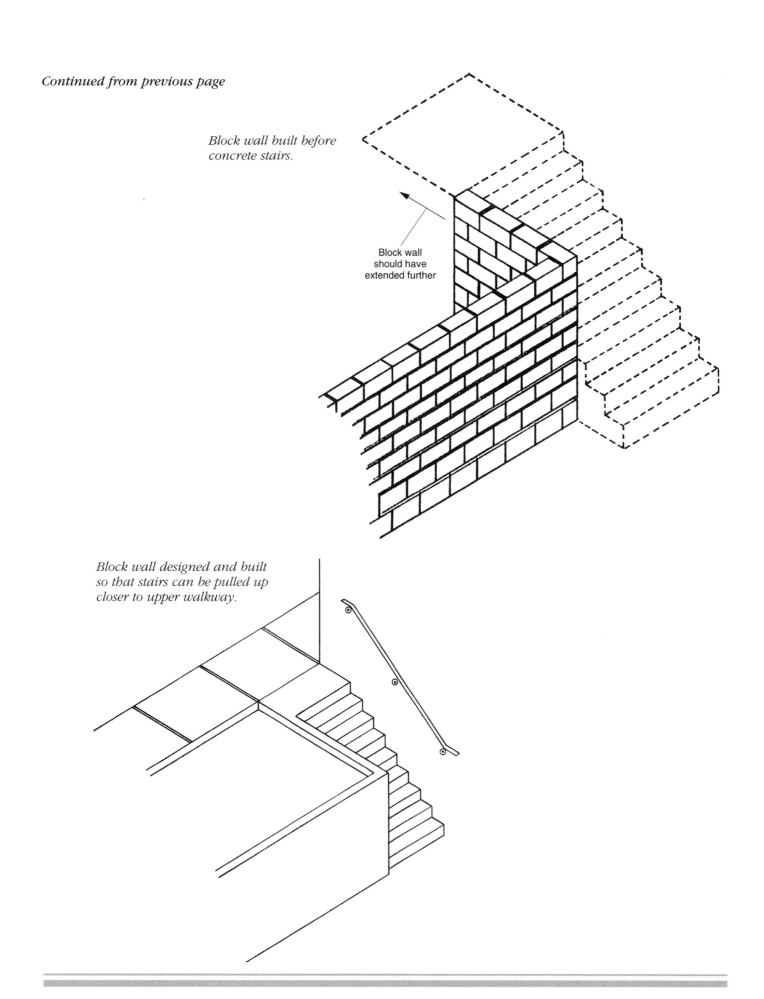

Block wall built before concrete stairs.

Block wall should have extended further

Block wall designed and built so that stairs can be pulled up closer to upper walkway.

☐ **917**

Analyze the elevation differences between finish grade and entry doors, the number of stair steps required, the width of the stairways, and the handrailing that may be required in terms of the architectural aesthetics of the entries and the cost of the handrailing in the budget for condominium and apartment projects.

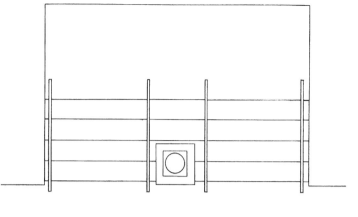

Plan view of wrought-iron handrail at entries to two condominium units.

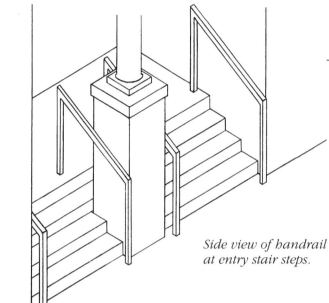

Side view of handrail at entry stair steps.

☐ **918**

Consider using a concrete grade beam footing at common area wrought-iron gates for expansive soil to provide a uniform resistance to the expansive forces of the soil, preventing differential uplift at each side of the gate and thus preventing the gate latch from requiring continual adjustment.

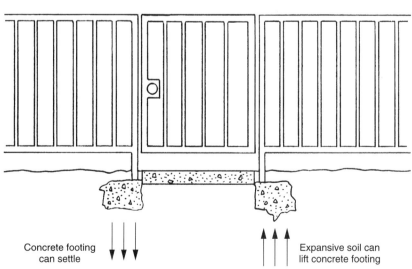

Wrought-iron fence gate supported by independent footings.

Continued on next page

Concrete Walkways and Driveways

Continued from previous page

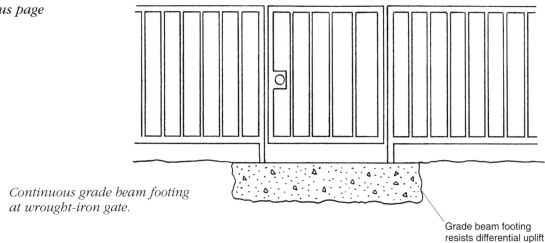

Continuous grade beam footing at wrought-iron gate.

☐ **919**
Consider upgrading the thickness and/or concrete mix for entry porch concrete slabs to prevent cracks. This should be in addition to the normal proper design, presaturation, and compaction of the subgrade underneath the concrete.

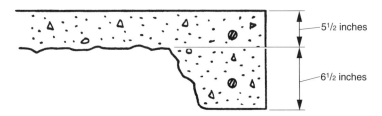

Thickened slab and deepened perimeter footing help prevent cracks in concrete.

☐ **920**
Make sure the design of exterior entries allows water to drain away.

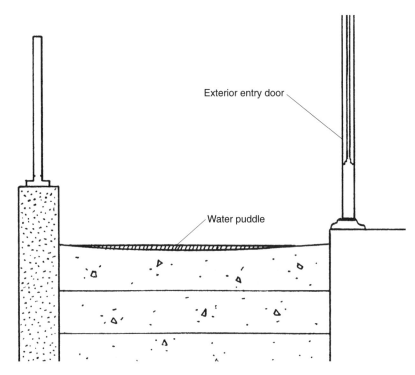

Low spot in concrete walkway results in water puddle outside entry door.

Continued on next page

Continued from previous page

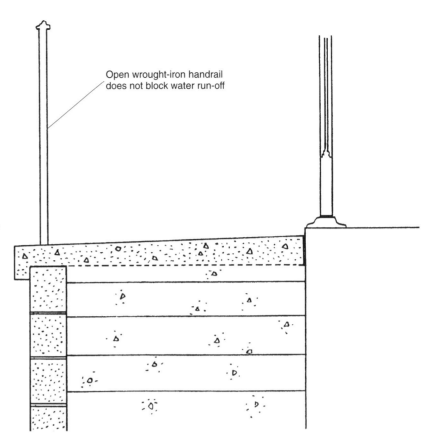

With this design, water drains over the edge of the sloped walkway.

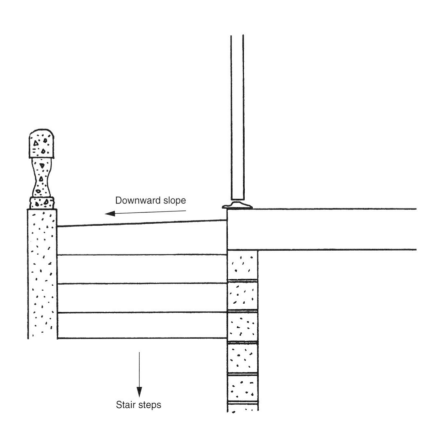

Provide adequate drainage slope away from building for concrete entry porch.

921
When concrete walkways are contracted for with a unit-price method such as linear feet, physically measure the actual linear footage of the walkways installed to verify the figures given on the concrete subcontractor's invoice.

☐ 922
Patch flaws in gray concrete street curbs to get the edges straight before pouring colored concrete next to these curbs. If the edges of the curbs are not straight, the colored concrete will flow over into the imperfections, resulting in a joint that is noticeably crooked as a result of the contrasting colors of concrete.

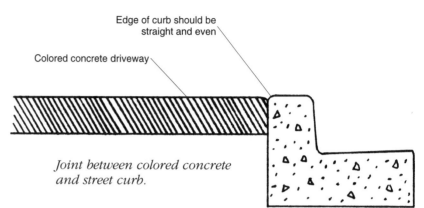

Joint between colored concrete and street curb.

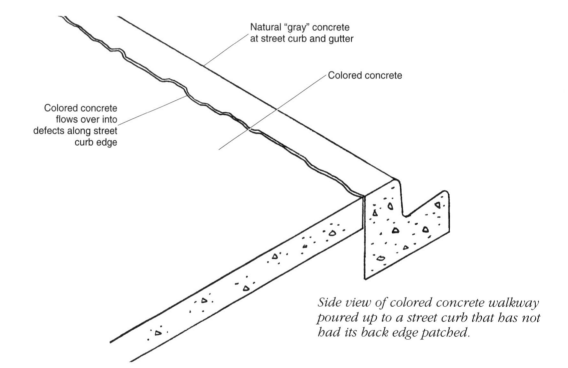

Side view of colored concrete walkway poured up to a street curb that has not had its back edge patched.

☐ 923
Make sure there is not an elevational conflict between a dryer vent daylighting out through the concrete foundation and the top of the elevation of a concrete walkway.

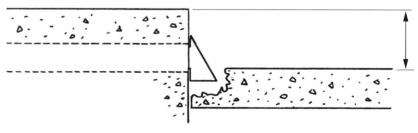

Concrete walkway must be chipped out for vent that is too low.

Continued on next page

Continued from previous page

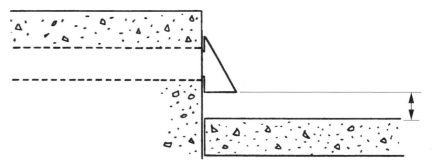

Difference between elevations of slab and walkway provides enough clearance for vent.

☐ **924**
Make sure there is not an elevational conflict between a vent to an island cabinet cooktop daylighting out through the concrete foundation and the top elevation of a concrete walkway.

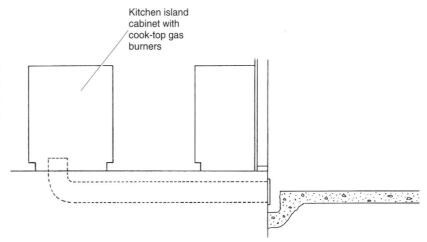

Vent for cook top exits through foundation that is too low in relation to exterior concrete walkway.

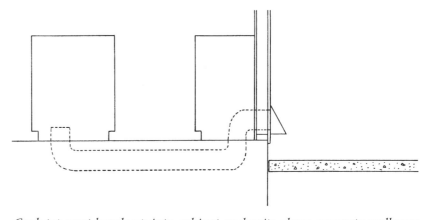

Cook top vent bends up into cabinet and exits above concrete walkway.

Concrete Walkways and Driveways

☐ **925**
For safety and liability reasons, check that there is enough level standing space at the garage laundry area after a washer and dryer are installed.

☐ **926**
Wrought-iron handrail should not project into a minimum-width common area walkway in a condominium or apartment project. Add another short panel of concrete walkway beyond the lowest stair step to maintain the walkway's width.

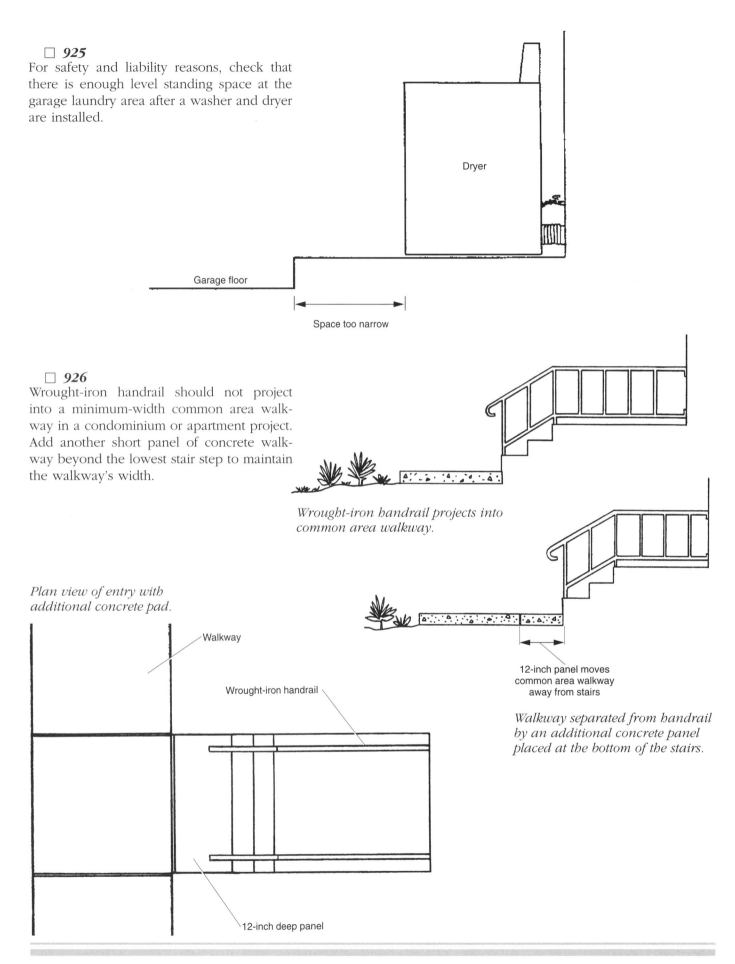

Plan view of entry with additional concrete pad.

Wrought-iron handrail projects into common area walkway.

Walkway separated from handrail by an additional concrete panel placed at the bottom of the stairs.

364　**Chapter 30**

☐ **927**
In using prefabricated plastic or concrete pads for A/C condenser units, make sure there is enough dimensional space for the pads to fit between the buildings and the concrete walkways.

Prefabricated concrete or plastic pads for A/C condenser should fit between building and concrete walkways.

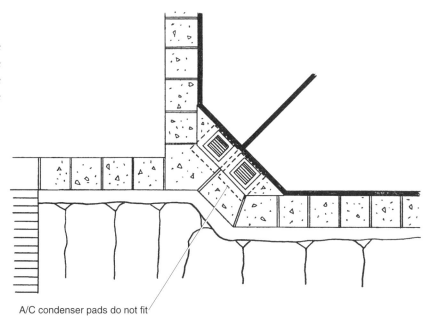

A/C condenser pads do not fit

☐ **928**
Analyze entries built on top of compacted backfill at retaining walls, for example. Consider extending the footings all the way down to undisturbed competent soil to avoid future settling of the entry porches.

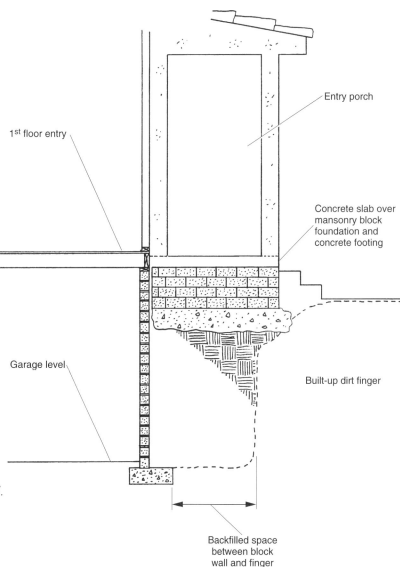

Entry porch over compacted backfill.

Entry porch

1st floor entry

Concrete slab over mansonry block foundation and concrete footing

Garage level

Built-up dirt finger

Backfilled space between block wall and finger

Concrete Walkways and Driveways 365

☐ **929**
Analyze the connection between concrete stoops and the buildings with the goal of preventing future settling or movement of the stoops and thus preventing cracks from appearing at the joints between the buildings and the entry stoops.

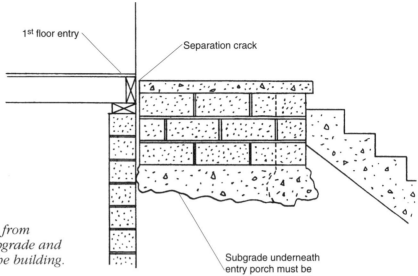

Entry stoop may come loose from improper compaction of subgrade and inadequate connection to the building.

☐ **930**
Check that the locations of garage screened vents do not conflict with concrete steps and flatwork that will be poured later.

Check locations of screed vents on garage walls in relation to concrete flatwork to be poured later in construction.

☐ **931**
Place plastic sleeve conduit underneath concrete walkways and driveways as needed for landscape irrigation and lighting that will be installed later.

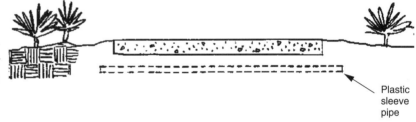

☐ **932**
Protect newly poured colored concrete walkways from stains caused by gray grout spilled from adjacent masonry block garden walls built after the concrete has been poured. Water blasting and other methods cannot always remove these types of stains.

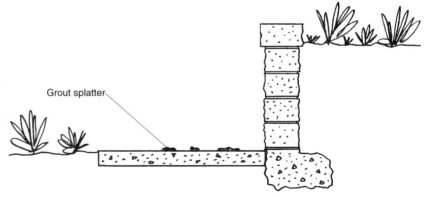

Grout splatter from masonry wall stains concrete walkway.

☐ **933**
Use a mock-up of an automobile to check for driveway high points that will hit the bottoms of automobiles. Change the contour of the driveway or add a second high point.

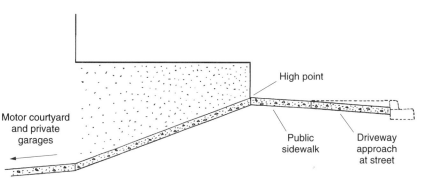

Side view of concrete driveway leading down into lower motor courtyard.

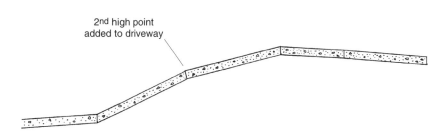

Driveway with a second hinge point or high point added to soften the first high point.

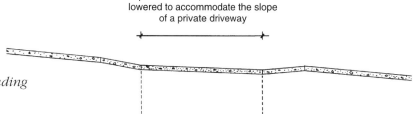

View of public sidewalk while standing in the street and looking at the driveway (broken line).

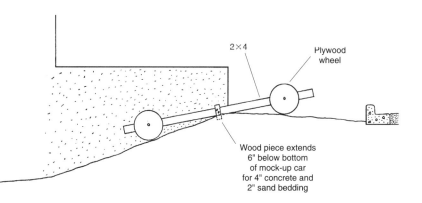

One method of testing driveway slopes and high points.

Continued on next page

Concrete Walkways and Driveways

Continued from previous page

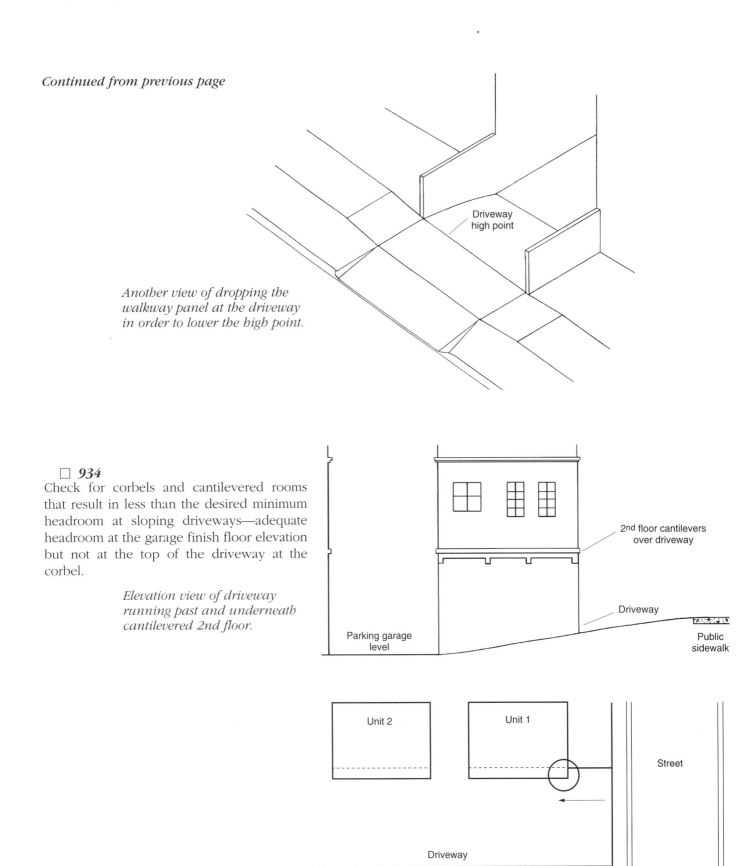

Another view of dropping the walkway panel at the driveway in order to lower the high point.

☐ **934**
Check for corbels and cantilevered rooms that result in less than the desired minimum headroom at sloping driveways—adequate headroom at the garage finish floor elevation but not at the top of the driveway at the corbel.

Elevation view of driveway running past and underneath cantilevered 2nd floor.

Plan view of first corbel location at cantilevered 2nd floor.

Continued on next page

Continued from previous page

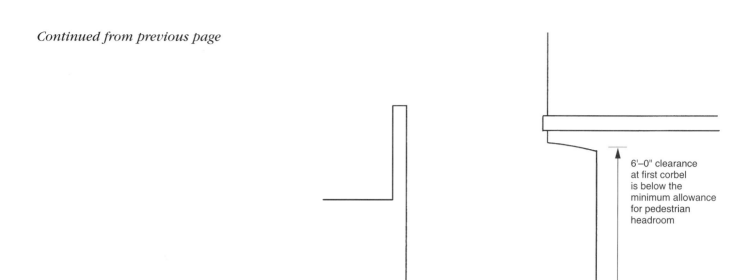

Side view of driveway while standing on public sidewalk.

☐ **935**
Check that driveways have the minimum turning radius needed to enter and exit from the garages, especially when the orientation of a garage has to be changed because of the location of an underground utility vault in the sidewalk at the street.

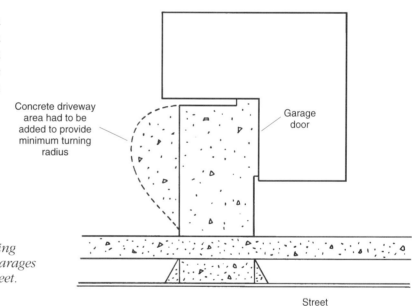

Check for adequate turning radius on driveway for garages that run parallel with street.

☐ **936**
Do not allow tradespersons to park their automobiles on freshly poured concrete driveways or inside garages. Their cars may drip oil onto the concrete surfaces, making brand-new concrete look old and used.

Concrete Walkways and Driveways

☐ **937**
For subterranean parking garages, consider using plastic sleeve conduit below the driveway concrete for the garage door opener sensor wire to avoid having unsightly saw-cut grooves filled with caulking.

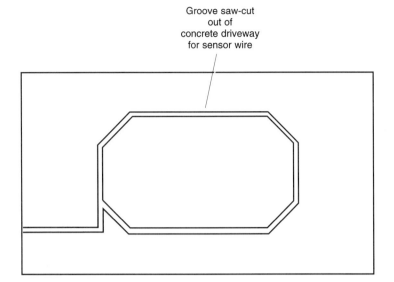

Saw-cut pattern and caulking exposed in driveway section with garage door opener sensor wire.

☐ **938**
Try to locate trash bin areas in subterranean garages that allow easy removal of the bins. If they are placed in a difficult location, the surrounding walls may become dented and dinged as a result of the bins hitting the walls.

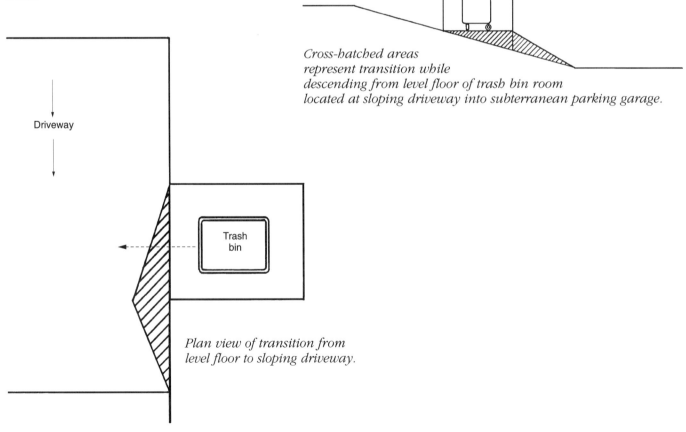

Cross-hatched areas represent transition while descending from level floor of trash bin room located at sloping driveway into subterranean parking garage.

Plan view of transition from level floor to sloping driveway.

☐ **939**
Get vaults for plumbing cleanouts in driveways placed before the driveways are poured, especially for colored concrete with a stamped pattern. Saw cutting for the cleanout vault after the concrete has been poured looks like an afterthought.

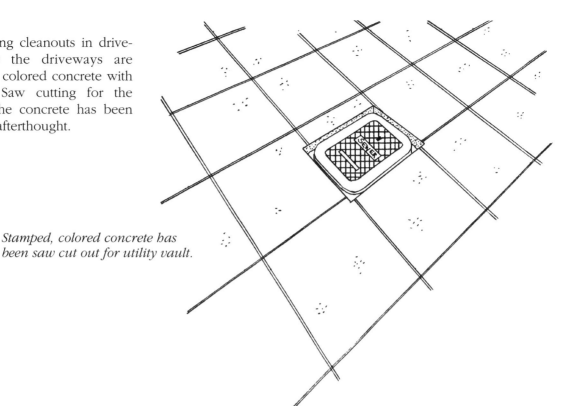

Stamped, colored concrete has been saw cut out for utility vault.

☐ **940**
For houses on steeply inclined streets, check that the elevation of the driveway at the garage door is higher than the low point at the driveway approach at the street. There must be some vertical fall between these two points for a diagonal swale across the driveway to work. If this has not been done, you must know this early in order to change the city sidewalk and approach slightly to accommodate the drainage swale or raise the garage finish floor elevation.

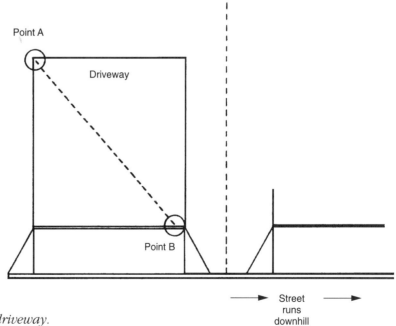

Drainage flow-line across driveway.

Continued on next page

Concrete Walkways and Driveways *371*

Continued from previous page

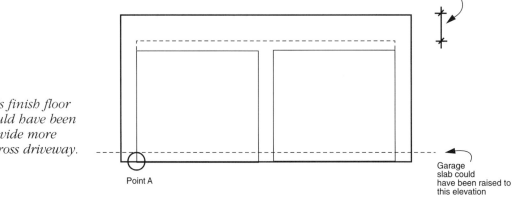

Elevation view of street and garage.

Garage slab's finish floor elevation could have been raised to provide more drainage across driveway.

☐ **941**
For steep driveways that lead down into a motor courtyard for condominiums or town houses, come up with a surface finish on the concrete that is rough enough to prevent automobiles from losing traction and spinning out of control when the surface is wet.

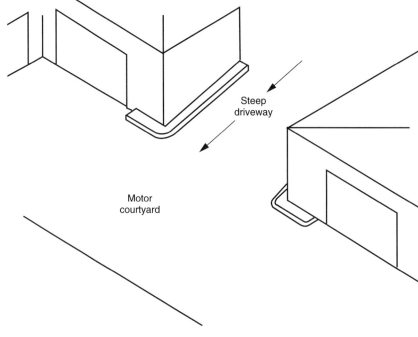

Steep driveway leading down to a motor courtyard for condominium units.

☐ *942*
For projects with a mix of two- and 3-car garages, check with the city or county for restrictions on driveway width. A 50-foot-wide lot with a 40 percent driveway width restriction would allow only a 20-foot-wide driveway, which would be too small for a 3-car garage. One can't move around floor plans for marketing reasons without considering driveway restrictions.

☐ *943*
Provide in the project budget for the engineering surveyors a provision to stake the property line locations and corners of irregularly shaped, commonly shared driveways.

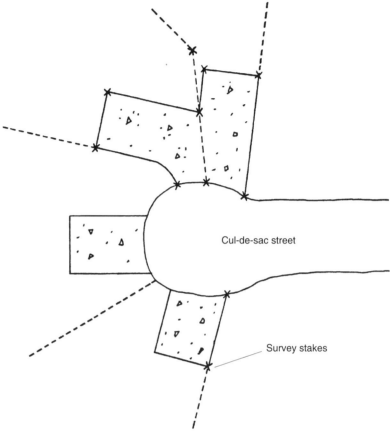

Have civil engineering survey crew stake critical points at property lines for irregularly-shaped driveways.

☐ *944*
Check that the design mix of the concrete in the driveway within the city and/or county right-of-way setback from the street meets the required public standards. Either upgrade the entire driveway to meet the public standards or pour the public portion of the driveway along with the sidewalk and the driveway approach. Pour the private portion of the driveway separately with a lower-quality concrete mix.

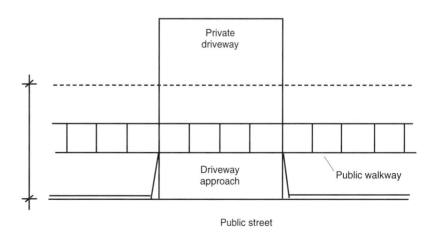

City or county standards for concrete design mix may be required within the setback easement area (broken line).

Concrete Walkways and Driveways *373*

☐ **945**
Coordinate the pouring of the garage slab concrete so that the edge of the driveway's colored concrete ends up midway underneath the garage door. This prevents 2 or 3 inches of the gray garage slab concrete from being exposed beyond the garage door and contrasting with the colored and/or stamped driveway concrete.

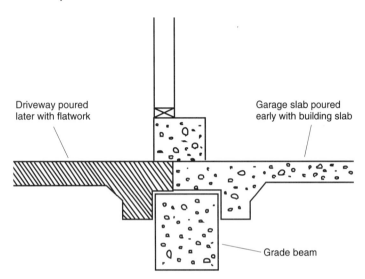

Concrete grade beam at front of garage lowered to allow for colored concrete driveway.

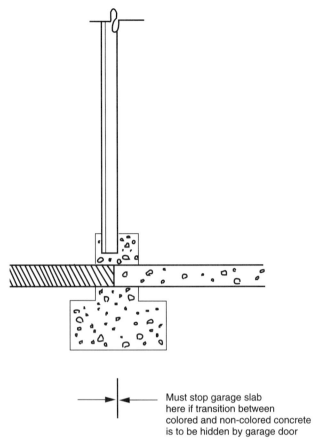

Joint between colored concrete in driveway and standard gray-colored concrete inside garage is hidden by garage door.

☐ **946**
For curved driveways, use the tooled control joints as an opportunity to create an aesthetically pleasing pattern.

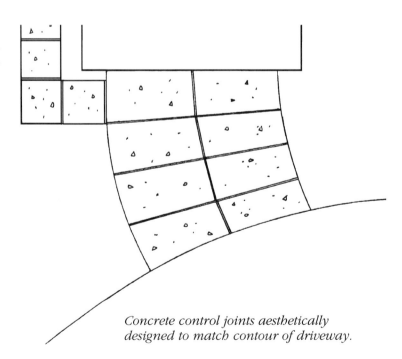

Concrete control joints aesthetically designed to match contour of driveway.

☐ **947**
For driveways poured over expansive soil, in addition to the normal design and construction precautions to prevent uplift, consider using a "key" connection between the garage slab and the driveway to prevent differential uplift at this location.

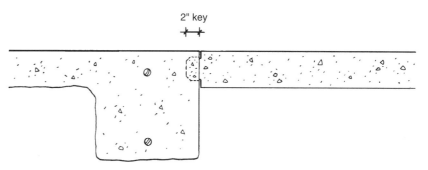

One method used to resist differential displacement between a garage slab and driveway.

☐ **948**
Check that the number of guest parking spaces on the conditions of approval for condominium projects matches what is shown in the plans.

☐ **949**
When the subgrades for driveways are not cut to the correct elevation during the rough grading phase, consider ahead of time what to do with the spoils late in the construction, when this subgrade material will be removed in preparation for forming and pouring the concrete driveways.

☐ **950**
For a drainage swale across a concrete driveway, have the precise grading plans show the dimensions and elevations for the swale.

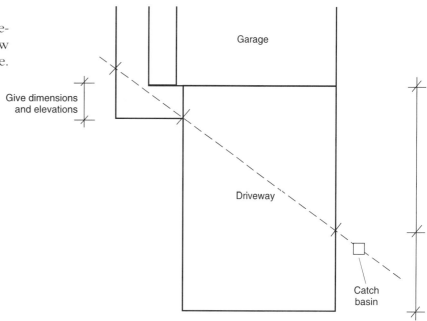

Drainage swale across a concrete walkway and driveway.

Concrete Walkways and Driveways

□ *951*
For projects that include houses with both two- and three-car garages, when the product mix is changed during the design phase and houses are switched for marketing considerations, check that a conflict is not created in regard to the locations of storm drain catch basins at the street curb when a house with a three-car garage is moved to a lot that previously had a house with a two-car garage.

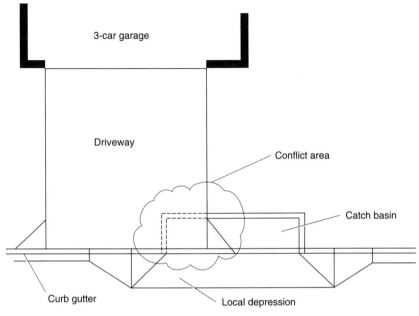

Storm drain catch basin built in street conflicts with concrete driveway.

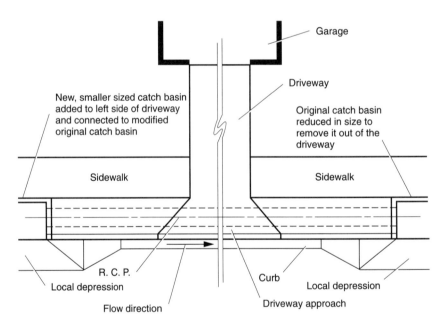

Expensive modification of storm drain catch basin to make room for driveway leading to a 3-car garage.

☐ **952**
Check that the storm drainage path for a steeply inclined street does not result in a driveway approach being located where the street has a curve. Water running down the curb and gutter will jump over the lower driveway approach section and flow across the driveway and into landscaped areas and hillside slopes, causing erosion, for example. One solution is to build a two- or three-course-high block garden wall at the downhill side of the driveway to stop and redirect the water back out into the street curb and gutter. The cost of this extra garden wall should be included in the project budget.

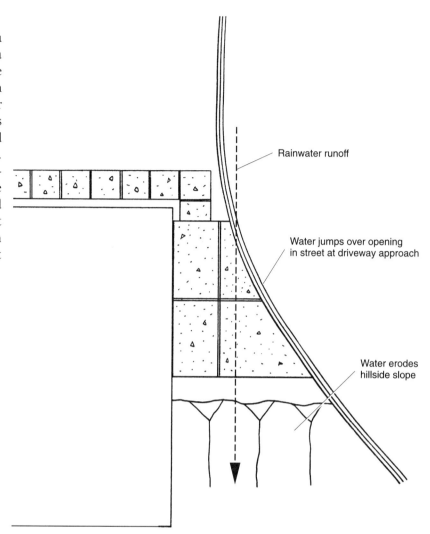

Plan view of rainwater path down steep street, over the driveway approach and driveway, and down landscaped hillside slope.

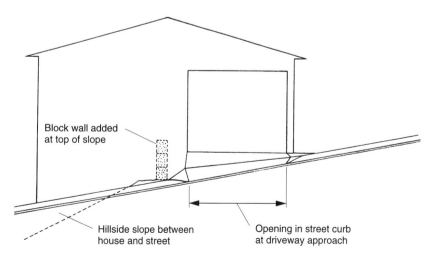

Side view of driveway approach and sideyard slope.

Concrete Walkways and Driveways

☐ *953*
For a condominium or apartment building with a row of adjacent units with the same finish floor elevation and with entry doors in close proximity to the ribbon gutter in a private street within the project, check that the slope of the ribbon gutter does not result in stair steps and short walkway sections that are too steep at the low point of the ribbon gutter.

Finish floor elevations of 4-unit building are the same, but street curb and gutter flows downward for storm drainage.

Differences in elevation between finish floor and street at each entry.

CHAPTER 31

Walls, Fences, and Gates

☐ **954**
As the finish grade drops for landscape drainage, check that the tops of the footing elevations for block walls also drop to remain below the finish grade.

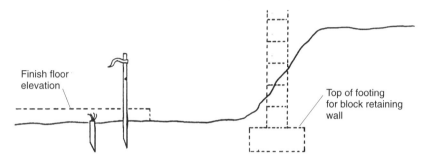

Relationship of finish grade at sideyard to top footing for a block retaining wall and the finish floor elevation of the house.

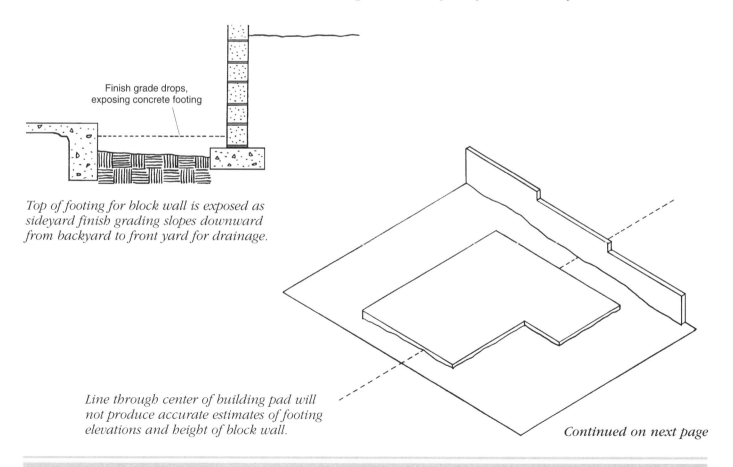

Top of footing for block wall is exposed as sideyard finish grading slopes downward from backyard to front yard for drainage.

Line through center of building pad will not produce accurate estimates of footing elevations and height of block wall.

Continued on next page

Walls, Fences, and Gates 379

Continued from previous page

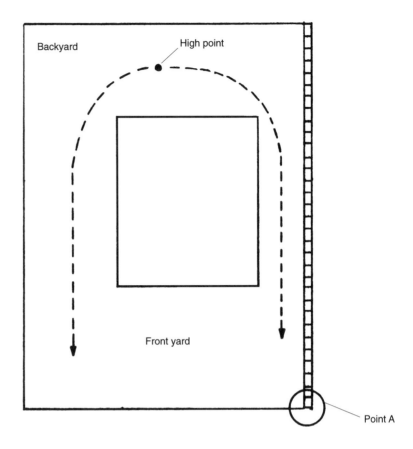

Block retaining wall must go deeper as finish grade lowers from backyard to front yard.

☐ **955**
Check that the top of the footing elevation for block walls is coordinated with the vertical fall required to achieve the correct slope for back-of-wall subdrain pipes. If necessary, lower or step the footing and add courses of block to provide the vertical fall.

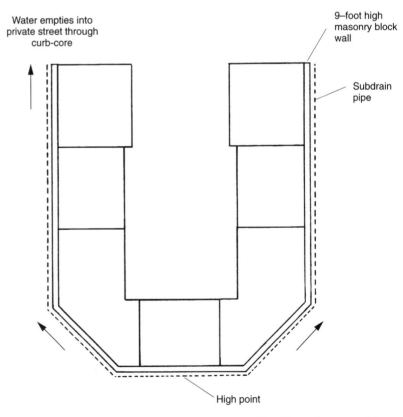

Subdrain at base of perimeter block wall for three-story condominium building.

Continued on next page

380 **Chapter 31**

Continued from previous page

Block wall and footing elevation must provide adequate fall for back-of-wall subdrain pipe in relation to the finish floor elevation of the interior units.

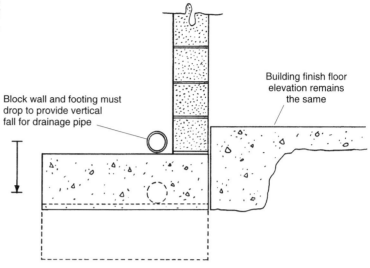

Side view of block wall footings that step down to provide vertical fall for subdrain pipe.

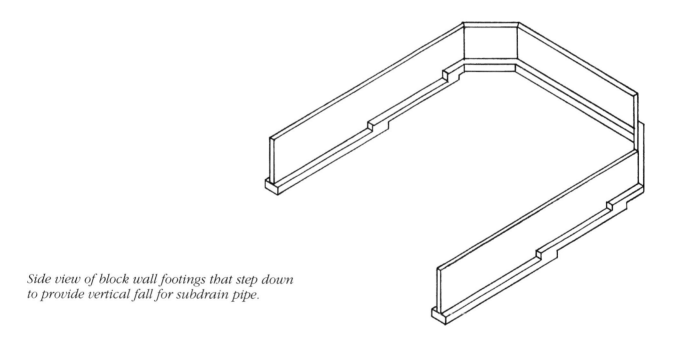

☐ **956**
Make sure the block wall footings are low enough or are retained at each side by formboards to provide an open space for the underground gas pipes to a row of gas meters.

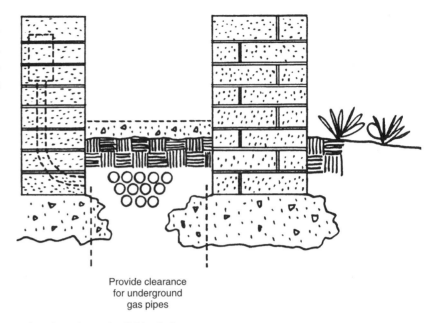

Either formboard-off block footing at entrance into gas meter areas or lower footing by adding two or three courses.

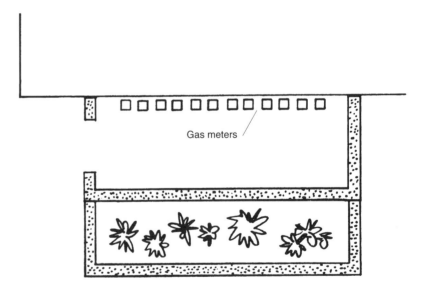

Plan view of a row of gas meters behind a planter area.

☐ **957**

For expansive soil, presaturate the trench soil underneath a concrete footing supporting the block column at a project entrance gate in accordance with the soil engineer's recommendations. This helps prevent the footing and column from lifting and throwing the entrance gate out of adjustment.

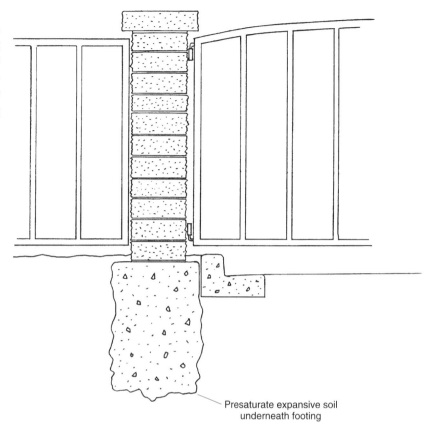

Block column supporting project entrance gate.

☐ **958**

Check that the overlap between a block garden wall and an adjacent stairway retains the landscaped slopes. If it does not, the slope will erode onto the stair steps as a result of landscape irrigation watering, creating safety and maintenance problems.

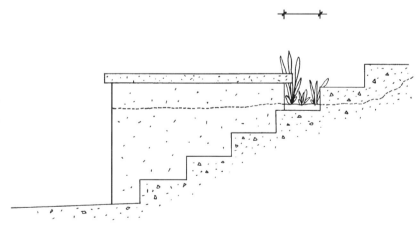

Gap between stair step and planter wall.

Continued on next page

Continued from previoous page

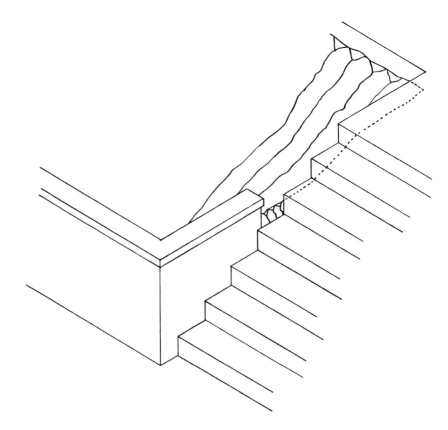

Side view of slope that is not retained by a wall at the center of the stairway.

☐ **959**
Make sure the top elevations of block walls at stairways are not overbuilt in terms of retaining the stairways or architectural aesthetics. You can save money by stepping down and eliminating portions of the block walls.

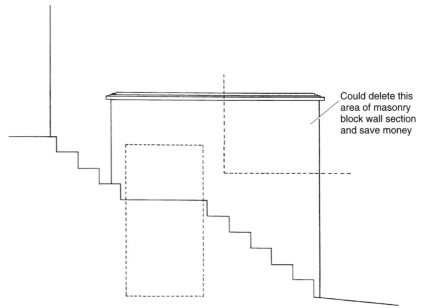

Could delete this area of masonry block wall section and save money

Side view of block wall behind exterior concrete stairs.

☐ **960**
Check that the locations of block retaining walls do not encroach on the turning radius and access into garages.

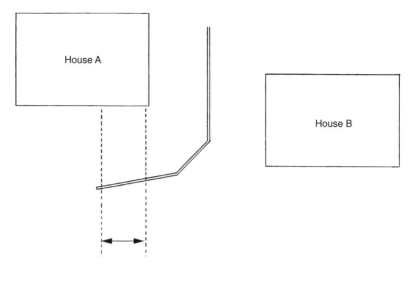

Retaining wall extends in front of garage.

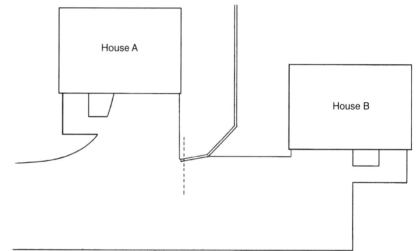

Bottom of steepened driveway to house B is now at broken line, and wall length has been shortened to provide better access into house A garage.

☐ **961**
Make sure small retaining walls are shown on the plans and included in the scope of work to create level areas for A/C condensers on rear yard or side yard slopes.

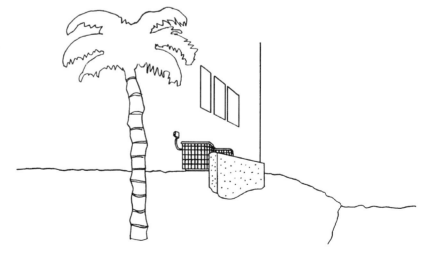

Retaining wall required to provide flat area for A/C condensers at top of rear yard slopes.

Walls, Fences, and Gates 385

☐ **962**
Check that block retaining walls are high enough to retain the slopes before the time when landscaping is installed. Erosion before and after landscaping can wash over the top of the block wall and stain the stuccoed finish surface.

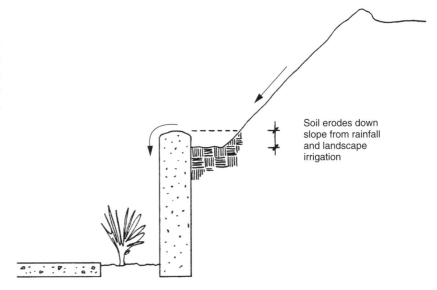

Erosion on slope fills up space behind, flows over the top of, and stains stuccoed retaining wall.

☐ **963**
For driveways leading down into subterranean parking garages, stop the block retaining wall at the point where the finish grade elevations are the same on both sides of the wall. This will prevent motorists from driving into or over the short wall when entering or exiting from the parking garage.

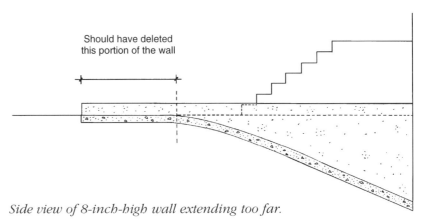

Side view of 8-inch-high wall extending too far.

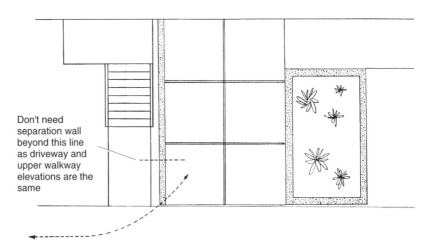

Automobiles drive over 8-inch-high wall entering and exiting subterranean parking garage.

Continued on next page

Continued from previous page

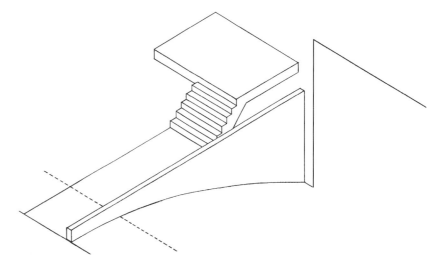

Another view of block wall separating driveway from walkway.

☐ **964**
Check that one- or two-course-high garden walls are not designed and built too close to a parking stall so that the bottom of a car door hits the wall when the door is opened.

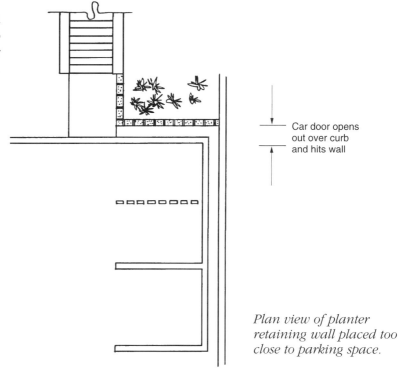

Plan view of planter retaining wall placed too close to parking space.

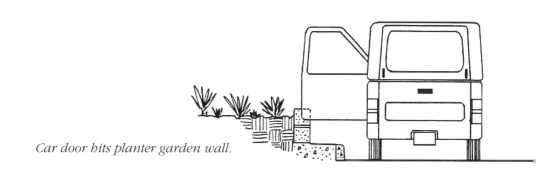

Car door hits planter garden wall.

Walls, Fences, and Gates 387

☐ **965**

Make sure the elevation at the top of the block wall does not create a tripping hazard at a concrete walkway. The block wall should be even with the finish grade at the sidewalk or run several courses above the finish grade, not 2 or 3 inches only.

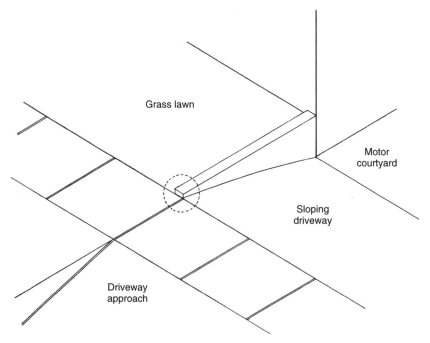

End of stuccoed block wall extends several inches above grade creating tripping hazard.

☐ **966**

For block garden walls that do not retain planter areas, consider building the walls with a 1-inch or 2-inch air space between the garden wall and an adjacent building wall. This will eliminate the possibility of separation cracks by removing the connection altogether.

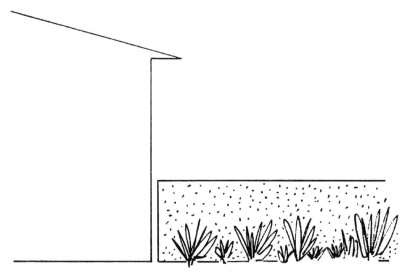

Two-inch separation gap between garden wall and house.

☐ **967**
Design a good connection at the joint between a planter retaining wall and the building to prevent future hairline cracks or wider gaps. You need more than just a masonry block wall built tight to a stuccoed exterior wall and then finished with stucco.

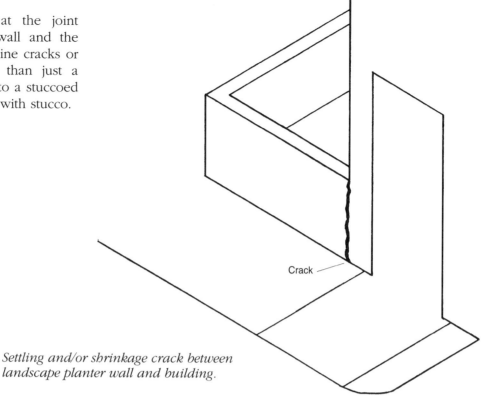

Settling and/or shrinkage crack between landscape planter wall and building.

☐ **968**
Coordinate the elevation of the top of the footing for a block wall retaining a concrete drainage swale, for example, with the elevation of the catch basin drainpipe daylighting out through a street curb cored hole.

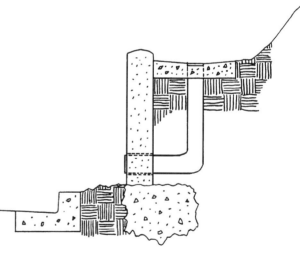

Storm drainage pipe from swale empties directly onto narrow landscaped area.

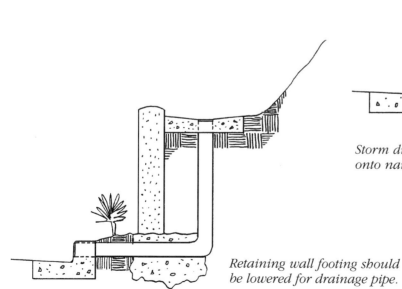

Retaining wall footing should be lowered for drainage pipe.

Continued on next page

Walls, Fences, and Gates

Continued from previous page

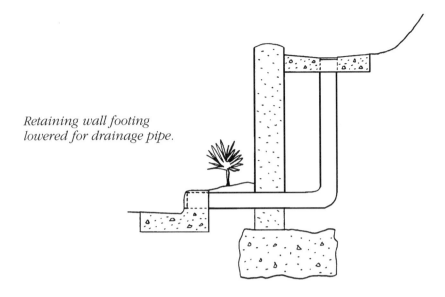

Retaining wall footing lowered for drainage pipe.

☐ **969**
For decorative slump-stone garden walls at the entrance to a project, for example, check that the design and installation of the waterproofing and subdrain prevent water from leaking through the wall and causing efflorescence.

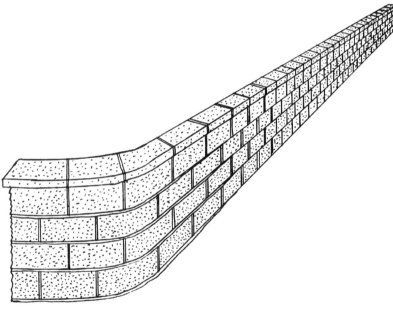

Slump-stone masonry garden wall at project entrance is least desirable location for water leaks and efflorescence stains.

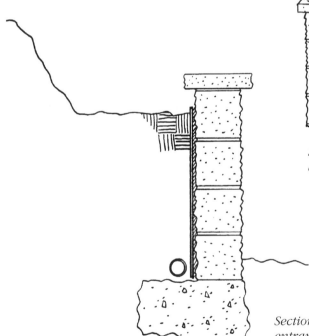

Section view of waterproofing at entrance street slump stone wall.

☐ **970**
For block walls with two different size blocks, create an angled splay by using mortar at the flat shelf transition to prevent water from collecting on the shelf and possibly leaking through the wall.

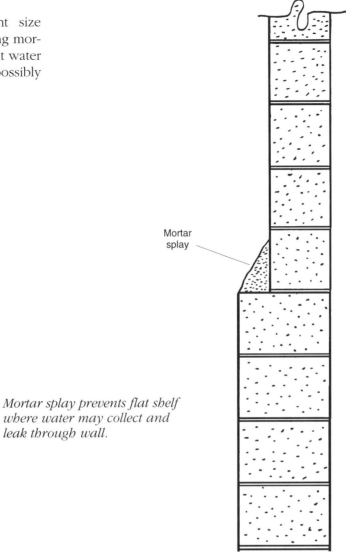

Mortar splay prevents flat shelf where water may collect and leak through wall.

☐ **971**
When a trench for a block wall is mistakenly overexcavated, check that any soil put back in the trench is processed and compacted properly; otherwise part of the block wall may settle later, resulting in cracks along the grout lines.

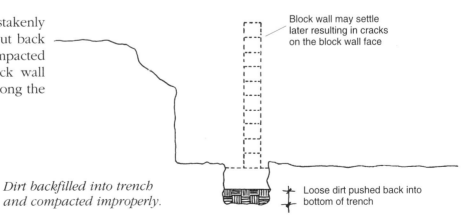

Dirt backfilled into trench and compacted improperly.

☐ **972**
Check that the soil excavated for a swimming pool is not casually spread out over the yard area to save the cost of exporting the soil, changing the elevation of the finish grading and creating a layer of unprocessed and uncompacted soil as a subbase for the concrete flatwork, possibly surcharging a block wall that is not designed to retain soil.

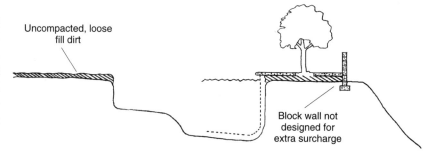

Dirt generated from swimming pool excavation spread out loosely in rear yard, resulting in uncompacted subgrade under concrete flatwork.

☐ **973**
For high-density condominium and apartment projects with common area landscaping, check that the side yard slopes throughout the project are retained on all sides.

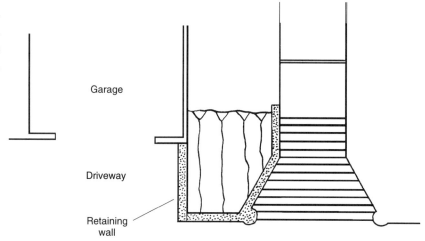

Sideyard slope adjacent to concrete stairs.

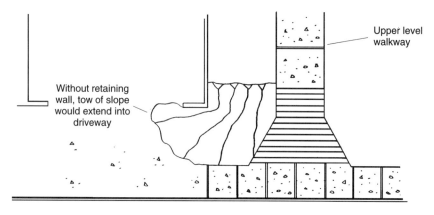

Retaining wall added at left side of landscaped slope.

392   **Chapter 31**

☐ **974**
When designing wood fencing for projects with common area landscaping, come up with a design that eliminates a positive side and a negative side of the fence. The positive side shows the fence boards only, and the negative side shows the posts and horizontal supports. Consider using a "picture frame" design that allows both sides of the fence to be positive.

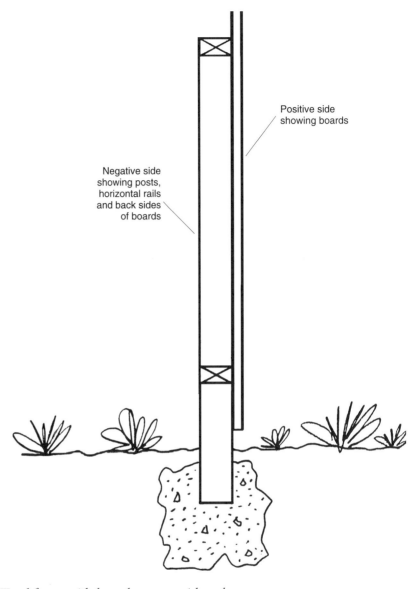

Wood fence with boards on one side only.

Continued on next page

Continued from previous page

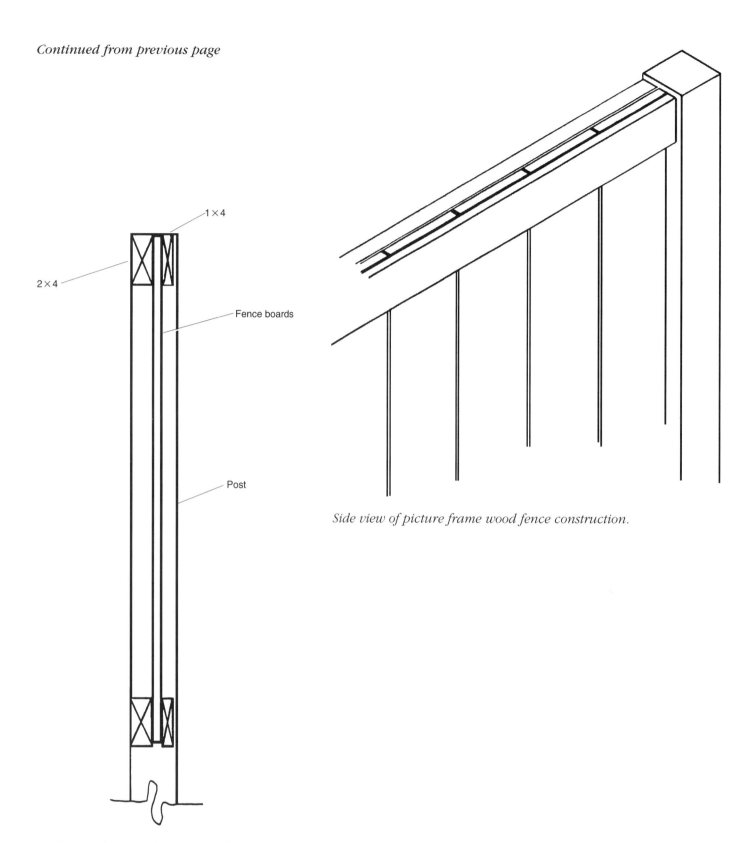

End view of picture frame wood fence construction, where both sides are positive.

Side view of picture frame wood fence construction.

□ **975**

In designing wood fencing mixed with stuccoed columns, consider the locations of the ends of the wrought-iron handrails at stairs and design the locations of the columns so that the handrail ends attach to the columns, not to the wood fencing boards.

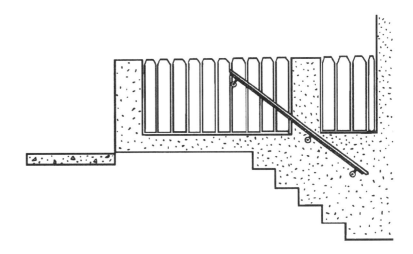

Wrought-iron handrail ends up in middle of wood fencing boards.

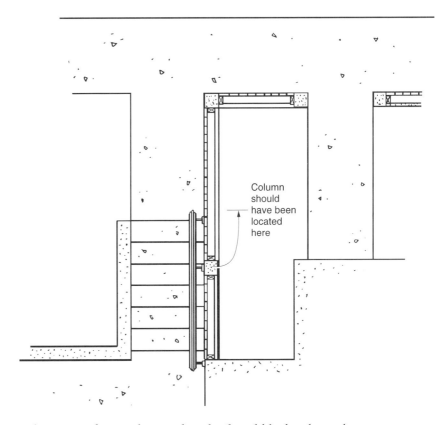

Plan view of wrought-iron handrail and block column layout.

Walls, Fences, and Gates 395

☐ **976**
When wood fence panels are attached to masonry block columns, come up with a design that provides a strong connection between the fence and the columns, such as threaded steel bolts set in grouted cells projecting out from the columns for the attachment of the fence panel ends. Fence panels that are attached to masonry block columns with powder-driven shot pins may not be strong enough to resist wind loads.

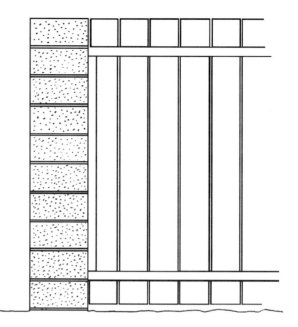

Wood fence panels attached to masonry block columns.

☐ **977**
Attach the 2×4 gateposts on the ends of a masonry block wall so that the gate can open fully without hitting the block wall corner. Otherwise this may pull the gatepost loose from the wall because of the fulcrum condition.

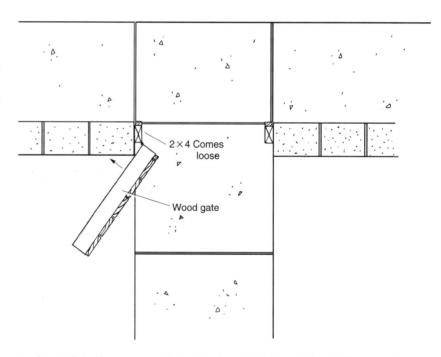

Wood gate hits the corner of the block wall before being fully opened.

Continued on next page

Continued from previous page

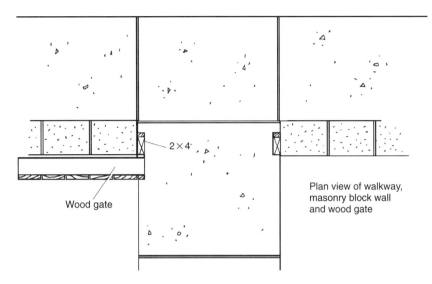

Vertical 2 × 4 board attached to block wall at edge.

☐ **978**

Use the correct galvanized nails for wood fencing. Rough-surfaced, hot-dip coated galvanized nails resist rusting and will not leave brown rust stains running down the fence boards. Another option is to use stainless steel nails.

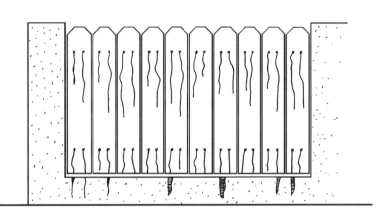

Wood fencing nail heads rust leaving dark stains on boards and stucco.

☐ **979**

Check whether entries require handrails because of the number of steps. If they do, have a design that includes the handrail as an architectural feature of the entry, not just as an afterthought at one side of the entry to meet building code requirements.

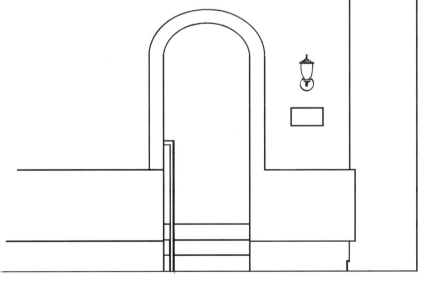

Stairsteps at entry require handrail.

Walls, Fences, and Gates

☐ 980

On large, high-density condominium and apartment projects with wrought-iron fencing between the yards, consider designing, measuring, and installing the fencing so that it does not connect to any of the exterior walls. This allows the measurement and installation of the fencing to be less than exact because of the 2- to 4-inch gap at the exterior walls.

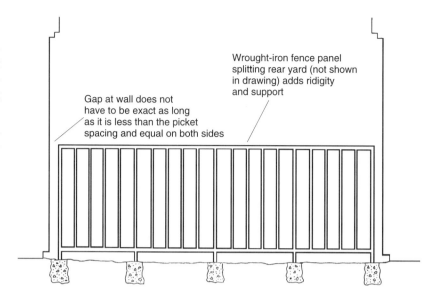

Wrought-iron fence panel placed between two walls is not attached to either wall.

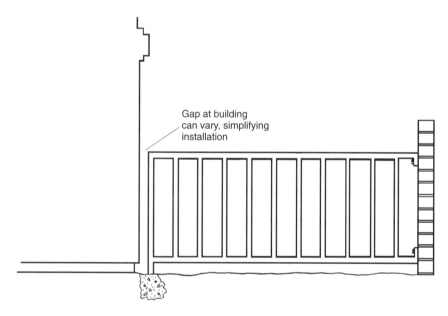

Wrought-iron fence panel attached to block wall for support but not attached to building.

☐ **981**

For small backyards with a short distance between the exterior wall of the house and a masonry block wall, for example, upgrade the dimensions of the top and bottom rails of a wrought-iron fence panel to allow it to span the entire distance without a center post. This eliminates the corrosion and rusting problems that result when a wrought-iron post is embedded into a concrete footing at ground level.

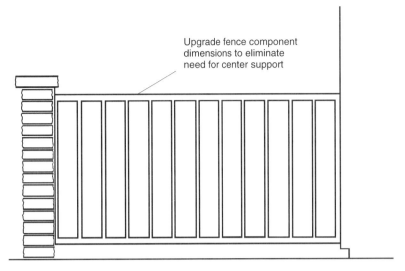

If wrought-iron fence panel is to be attached to wall surfaces at both ends, fence panel components can be thickened so as to eliminate the need for center support-post.

☐ **982**

When wrought-iron handrail posts are embedded into core-drilled holes in concrete walkways and stairs, apply two applications of concrete mortar to set the posts. The first application of concrete mortar will shrink, leaving a depression around the wrought-iron posts that collects landscape irrigation water, resulting in rusting and deterioration of the wrought-iron.

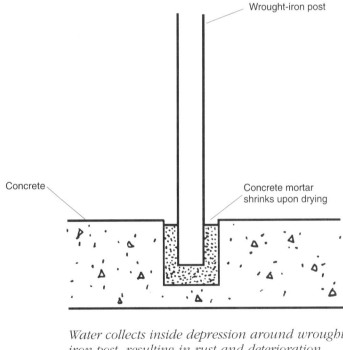

Water collects inside depression around wrought-iron post, resulting in rust and deterioration.

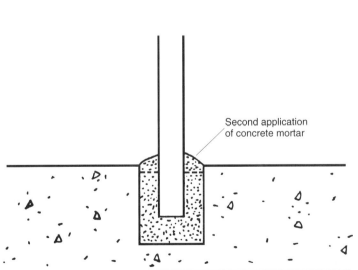

Second application of concrete mortar adds sloped surface to shed water away from base of wrought-iron post.

Walls, Fences, and Gates

☐ **983**
Weld all four sides of the bottoms of wrought-iron handrail pickets. This keeps moisture and landscape irrigation water from running underneath the pickets and causing rust stains and streaks on the bottom rail.

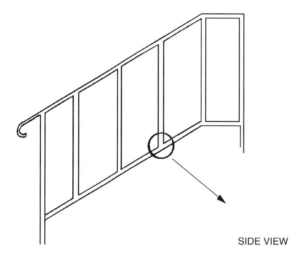
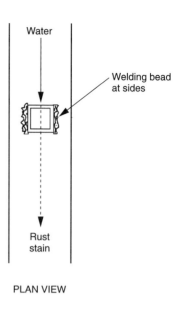

☐ **984**
Consider the minimum required height of the gate and the wrought-iron fence panels at a swimming pool when you are designing and building masonry block columns. Otherwise you may have to change the tops of the wrought-iron fence panels and gate to a more expensive design to wrap around the column caps. If the gate must be 6 feet high to meet building codes or health department requirements, the column caps must be higher than 6 feet.

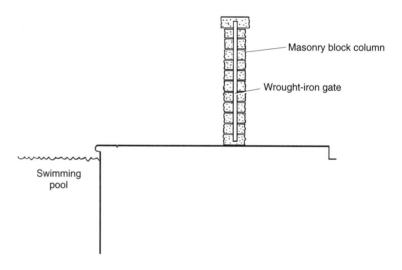

Side view of wrought-iron gate for swimming-pool area.

Continued on next page

Continued from previous page

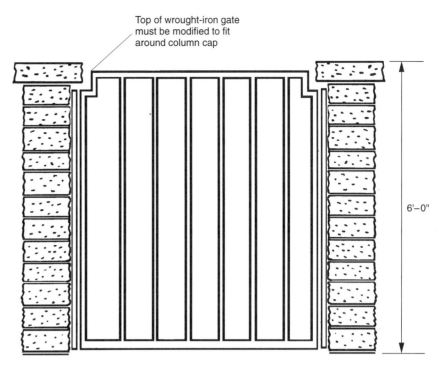

Masonry block columns are not high enough for minimum required gate height.

☐ **985**
For a security gate into a project, check that the adjacent wrought-iron fence panel is not so short that it can easily be bent or shifted over enough to release the gate latch.

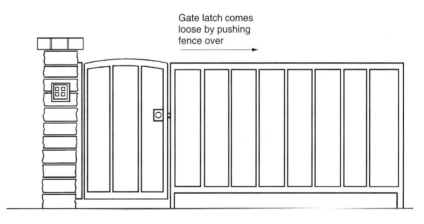

Left side of gate is securely attached to block column, but the fence panel on the right side is flexible.

Walls, Fences, and Gates 401

☐ *986*

Design concrete steps to be wide enough so that holes can be core drilled into the stair treads for a wrought-iron handrail. This eliminates the need to attach the wrought-iron handrail to awkward wall surfaces such as columns with architectural banding, which ends up looking like an afterthought.

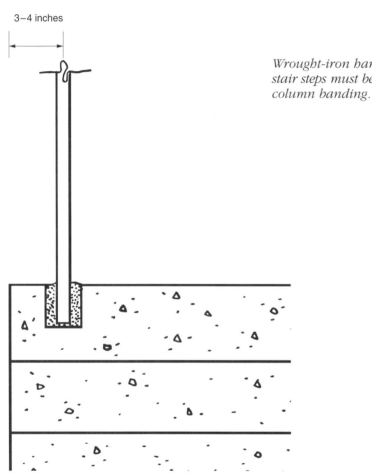

Widened stair steps with enough clearance for core-drilled holes.

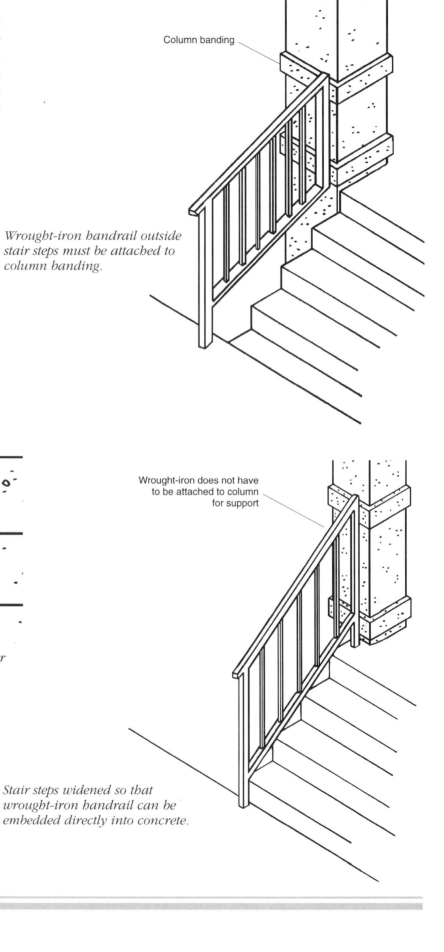

Wrought-iron handrail outside stair steps must be attached to column banding.

Stair steps widened so that wrought-iron handrail can be embedded directly into concrete.

402 **Chapter 31**

☐ **987**
Check that wrought-iron handrailing does not interfere with the outward swing of a nearby casement window.

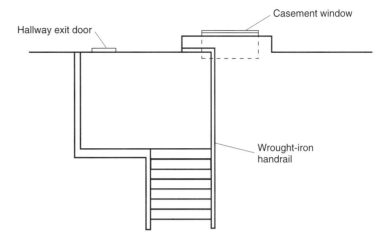

Plan view of wrought-iron handrail at exterior stair landing that interferes with the opening of the casement window.

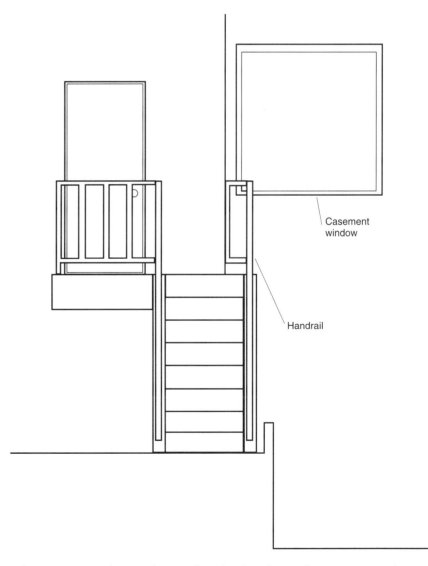

Elevation view of wrought-iron handrail in front of casement windows.

☐ **988**

Use high-quality ball bearing hinges for project entrance gates.

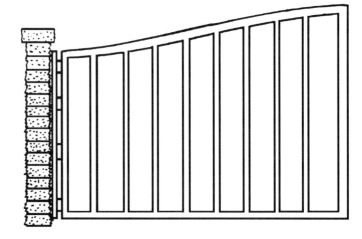

Use high-quality hinges at project entrance gates.

☐ **989**

For doors to trash bin enclosures, consider using rugged and more durable corrugated steel with hinges welded to metal shoes embedded in the masonry block enclosures.

Plan view of corrugated steel gate with hinges welded into metal shoe.

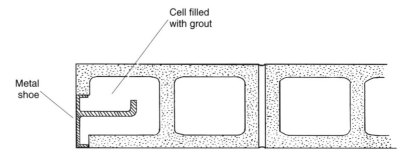

Plan view of custom-made metal shoe embedded in masonry block cell.

☐ *990*
Consider the design layout of common area wrought-iron fencing and access to utility meters and fire sprinkler standpipes in condominium projects.

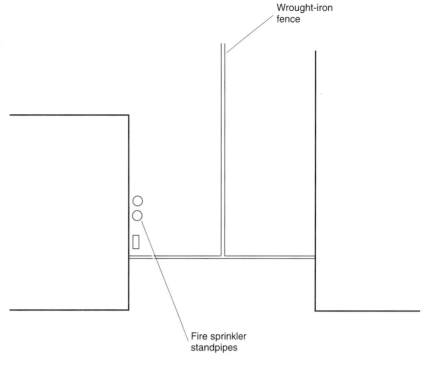

Utility meters and fire sprinkler standpipe access blocked by wrought-iron fence.

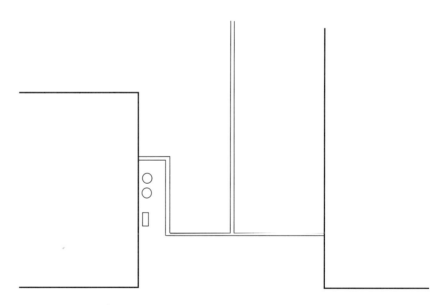

Wrought-iron fence design changed to provide direct access to utilities.

Walls, Fences, and Gates

☐ 991

Order precast concrete handrail cap with angled ends that match the slope of the stairs or come up with a design that allows the last pieces of cap at the bottom step to be horizontal. Cap ends that are cut square look odd.

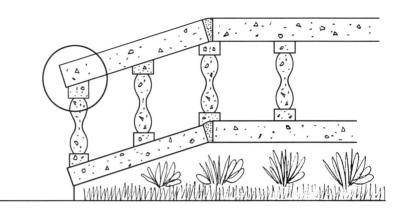

Precast concrete handrail cap without angled ends.

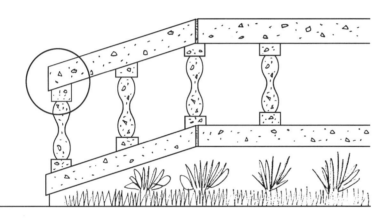

Precast concrete handrail cap with angled ends to match slope of stairs.

This method eliminated aesthetic problem of angled ends for precast handrail.

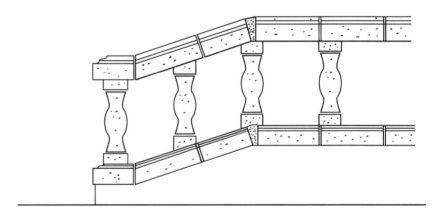

☐ **992**

For precast concrete handrails, check the layout of the pickets to determine whether the spacing will be uniform for each section of the handrail. Pickets that have different spacing at short sections of handrail compared with the rest of the handrail can look awkward and out of place. Check the building codes for the allowable maximum spacing in relation to the height of the handrail above the ground.

Inches   Spaces (Inches)

12   [1][2][3]   0
      [1] [2]   4

16   [1][2][3][4]   0
      [1][2][3]   2
      [1] [2]   8

18   [1][2][3][4]   0.66
      [1][2][3]   3
      [1] [2]   10

24   [1][2][3][4][5][6]   0
      [1][2][3][4][5]   1
      [1][2][3][4]   2.67
      [1][2][3]   6

36   [1][2][3][4][5][6]   2.4

48   [1][2][3][4][5][6][7][8]   2.29

50   [1][2][3][4][5][6][7][8]   2.57

120   [1] . . .   [ ]   2.10

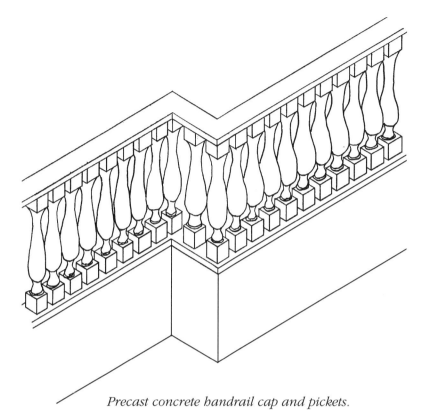

Precast concrete handrail cap and pickets.

Examples of possible handrail picket spacing using 4 × 4-inch pickets. The numbers at the left are the lengths of the wall section, and the numbers on the right are the spaces between pickets.

Walls, Fences, and Gates 407

☐ **993**

For wood handrail cap installed over a wrought-iron handrail, at splice joints drill holes through the wrought-iron on each side of the wood cap close to the splice joint so that lag bolts will keep both ends of the wood cap firmly in place. Don't rely on the predrilled, uniformly spaced holes to match the layout of the wood cap splice joints.

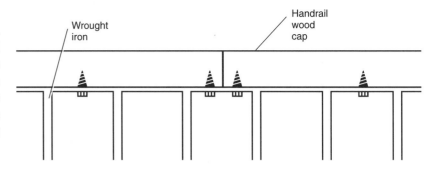

Lag-screw bolts placed close to each side of splice joint.

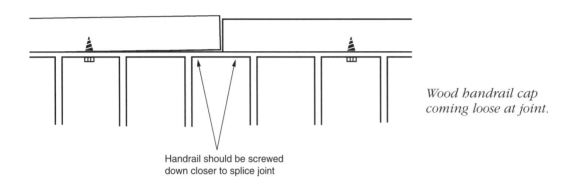

Wood handrail cap coming loose at joint.

☐ **994**

For a monument column at the entrance to a project, when there is a potted plant on the top of the column, place a saucer plate underneath the pot and consider sloping the column top slightly so that any water spilling over from the plant will not dribble over and stain the stucco and the project brass monument plaque, for example.

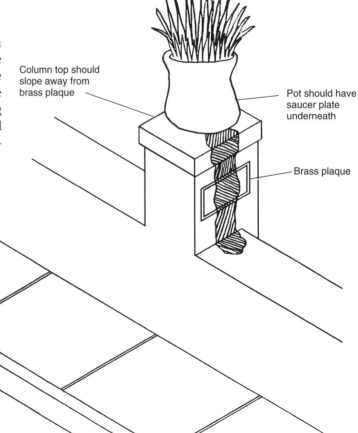

Irrigation water from pot stains a condominium project's brass monument plaque and surrounding stucco.

CHAPTER 32

Miscellaneous Exterior

☐ **995**
Sump pumps sometimes require a separate permit. The builder must produce a cut-sheet report showing a civil engineer's design, including the hydrology calculations, the recommended size of the sump pump motor, and the required switch gear; the civil engineer's wet-stamped seal and signature must appear at the bottom of the report. Consider this issue up front and include it in the sequence of construction so that this does not become a punch list item at the time of the final building inspection.

☐ **996**
For large sump pump pits with steel grate covers, consider splitting the steel grates into two or three smaller sizes to make them lighter and easier to handle.

Steel grate covers over sump pump pit

Steel grate over sump pump pit is in two pieces in order to reduce its weight and difficult handling.

☐ **997**

Consider including sump pumps at the bottoms of elevator pits, especially when landscaped planter boxes are designed on the exterior walls of an elevator shaft. This provides a way to discharge the water from the elevator pit if the water leaks through the planter box's waterproofing.

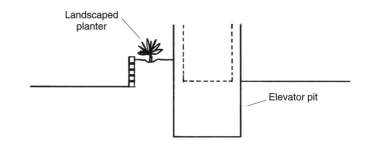

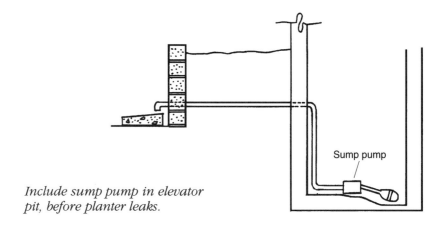

Include sump pump in elevator pit, before planter leaks.

☐ **998**

If you are considering a fountain on the project, ask whether the homeowners association will want to assume the cost of maintaining the fountain after the builder is done with the project. Fountains require the same maintenance support needed for swimming pools: pump motors, filters, chemicals, and a serviceperson.

☐ **999**

Check that the bottom-level basin for a fountain is large enough in diameter to contain the spillover from the fountain when the wind is blowing.

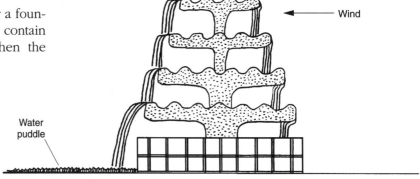

Water from fountain misses bottom pool.

☐ *1000*

Get the swimming pool plans for the recreation area approved by the city and/or county building department and health department early so that the swimming pool subcontractor can calculate the total gallons and the size and number of the risers coming up through the ground into the pool equipment room. Without this information the construction of the recreation building can be held up.

☐ *1001*

Lay out the swimming pool equipment, considering the ease of removal if a particular piece of equipment needs repair or replacement.

☐ *1002*

If a condominium, town house, or apartment project gets a direction map board, schedule its installation early in the project, at the time when the first homeowners move in.

☐ *1003*

Consider the use of photocell switches for all common area lights.

☐ *1004*

Check that the layout will not place temporary power poles in the way of the future locations of concrete walkways, driveways, underground utilities, and landscaping trees.

☐ *1005*

Check that all the temporary power pole outlets work and get any that are not working repaired before the tradespersons show up on the job site.

☐ *1006*

Check that the distances from the temporary power poles to the locations of second floors and roofs are reasonable. Tradespersons should not have to string out 200 feet or more of extension cord to reach the work areas.

☐ *1007*

Check that the landscape planting will not restrict airflow to the A/C condenser units and thus void their warranties.

☐ **1008**

For surface-mounted mailbox clusters, look for locations that are hidden and aesthetically pleasing in regard to the front elevation of the project. The locations of mailbox clusters should not be an afterthought.

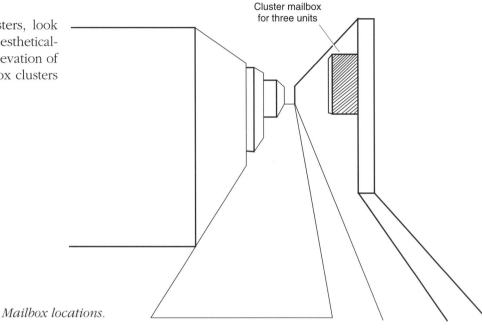

Mailbox locations.

☐ **1009**

For condominium and apartment projects with guest parking spaces and trash bin enclosure areas, check that a car parked in the space adjacent to a trash bin enclosure does not block access to the enclosure and that a car parked there also will not prevent the opening of the side door into the trash bin area.

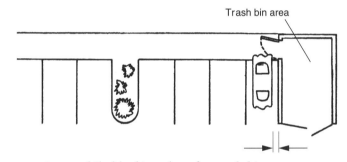

Automobile blocking door for trash bin area.

☐ **1010**

Make sure the building structure is not within the space envelope easement clearance requirements for a nearby telephone pole, including the yardarm at the top of the pole.

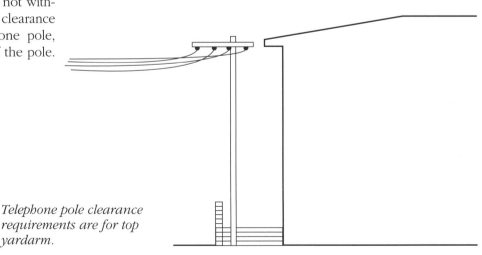

Telephone pole clearance requirements are for top yardarm.

☐ **1011**
Provide an adequate means for supporting exterior planter boxes so that when the soil, plants, and irrigation water are added, the planter boxes will not pull away from the exterior walls and sag downward.

☐ **1012**
Coordinate the elevation of exterior banding with a balcony deck so that the wrought-iron handrailing does not have to attach to or penetrate the banding, giving the appearance of a design and construction mistake.

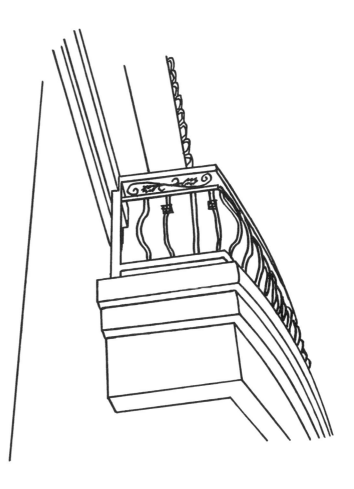

CHAPTER 33

Landscaping

☐ *1013*
In choosing a landscape architect, consider the type of project and match it with the expertise of a design professional. For example, sales models require an aesthetic visual impact, condominiums require maintenance economy and safety, and hillside projects require the proper selection of planting materials and irrigation to maintain the slopes.

☐ *1014*
Have the landscape architect who introduces irrigation water into an area for landscaping also design the drainage system to control the runoff.

☐ *1015*
Don't economize on landscaping in the project budget. Landscaping is a major ingredient in providing good curb appeal for a project.

☐ *1016*
Plan the landscaping design by starting with a realistic budget rather than with the client's "want list." This eliminates the need to spend time and money to come up with a landscape design the project cannot afford.

☐ *1017*
Get input and review from a landscaping contractor for the landscaping design, especially for condominium and apartment projects. Contractors sometimes have a better grasp of the maintenance costs involved with certain plants and trees and can make beneficial suggestions for alternative plants and trees that are easier and cheaper to maintain yet have equal visual impact.

☐ **1018**

Have the landscaping contractor maintain the landscaping for at least 60 to 90 days beyond the initial planting date. This allows the landscaping contractor to control the survival of the new trees and plants and provides time for the new landscaping to fill in to the point where a homeowners association can hire a landscape maintenance company.

☐ **1019**

For landscaping maintenance subcontracts for condominium and apartment projects, consider the possibility of contingency costs for dealing with snails, ants, and gophers. Include a line item in the budget for these unpredictable problems.

☐ **1020**

For a project with common area landscaping, get the project broken down into phases in terms of city or county approval. This allows the builder to move new home buyers in at each phase when the landscaping is completed and inspected for that phase only. It also prevents having to get the landscaping complete for twenty-five houses spread out over three streets, for example, just to move people into the first eight houses.

☐ **1021**

Select trees suited to a particular use: multi-branching for visual impact at focal points in the project, low branching to provide privacy, and canopy or umbrella at view locations and concrete walkways.

Multibranching   Low-branching   Canopy or umbrella

Three tree shapes.

Proper selection of canopy-type tree adjacent to street.

Landscaping 415

☐ **1022**

For condominium and apartment projects, come up with a custom lighting design that follows a hierarchy of lighting requirements for parking areas, stair steps, entries, landscaped areas, and streets and provides good lighting aesthetics for the project. Have this design reviewed and approved by the city or county before a uniform lighting requirement is imposed on the project from the city or county standards and is written into the project's conditions of approval.

☐ **1023**

Check that trees are not mistakenly designed directly below second floor balcony decks or cantilevered rooms.

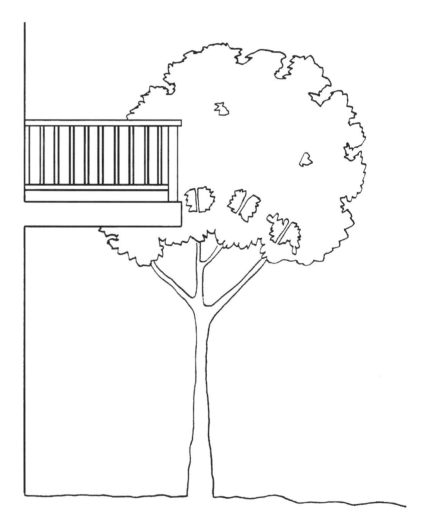

Tree designed at same location as balcony deck due to inadequate information shown on the preliminary base-sheet plans.

416   Chapter 33

☐ *1024*
Make sure trees are not placed on top of underground utilities.

Tree mistakenly designed and planted above underground utility pipe.

Landscaping

☐ *1025*
Check that landscaping at street corners will not block motorists' views of cross-traffic.

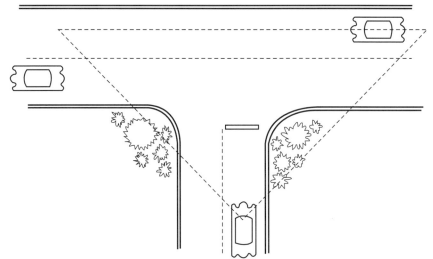

Trees and shrubs planted at street intersection block views of automobiles.

☐ *1026*
Do not place large plants, such as banana palms, in the landscaping areas immediately adjacent to garage doors. They may require continuous trimming to keep the leaves and branches from growing in front of garage doors and preventing them from opening.

Design correct plant type next to garage door.

☐ **1027**

For planter boxes, consider placing plants within pots inside the planter boxes. This removes the problem of plant roots penetrating through the planter waterproofing or clogging up the drainpipes. It keeps the plants isolated in the pots yet provides the visual aesthetics of landscaping.

Plants placed inside pots rather than the actual planter box.

☐ **1028**

Analyze the types of landscape irrigation sprinkler heads being used in terms of location and liability: pop-up-type heads at grass lawns and walkways and rigid or flexible riser-type sprinkler heads for ground cover planting.

☐ **1029**

Check that the landscape irrigation does not overspray onto sidewalks and streets and become a punch list item requiring correction that results in failing a landscaping inspection by the city or county.

☐ **1030**

Check that irrigation backflows, water meters, and sprinkler heads are not in the pathway of automobiles at curved driveways and cul-de-sac streets.

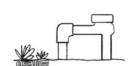
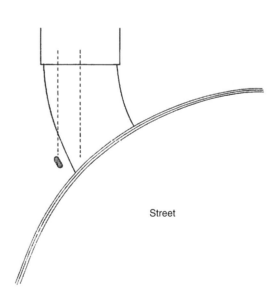

Irrigation backflow placed in line of car backing straight out of curved driveway.

Continued on next page

Landscaping 419

Continued from previous page

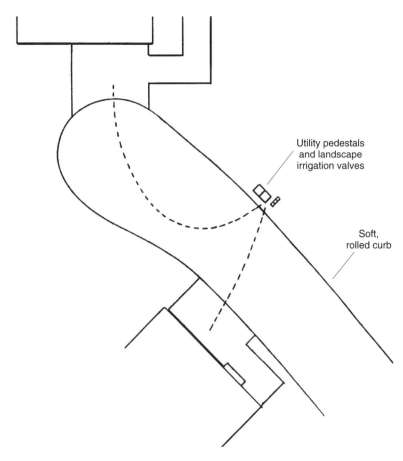

Cars backing out of driveways to turn around may hit irrigation valves and utility pedestals.

☐ **1031**
Check that traffic guardrails will not block the spray from irrigation sprinkler heads from reaching some of the landscape planting.

Traffic guardrail blocks irrigation spray from reaching upper-level plants.

420 **Chapter 33**

☐ **1032**

Find inconspicuous locations for irrigation meter pedestals with time clocks for sales models complexes and common area landscaping.

☐ **1033**

For a project with expansive soil, check that the landscaping does not prevent irrigation water from draining away from the building structures, thus minimizing the chance for subsurface water to create uplift underneath the foundations.

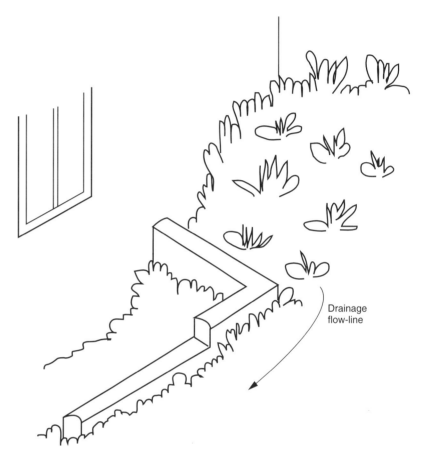

Dense landscape planting around back side of garden wall can block drainage flow.

☐ *1034*

Check that grass lawn areas surrounded by concrete walkways and driveways start with a subgrade that is at the correct elevation to allow water to sheet flow drain off the grass lawn areas. If water pools within these areas, a swamplike condition will exist with wet and soggy grass areas.

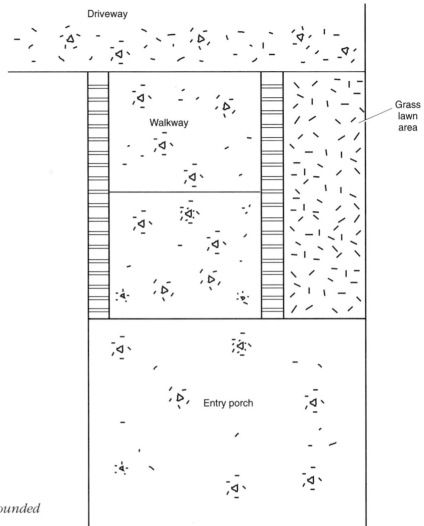

Grass lawn area surrounded by concrete.

☐ *1035*

Make sure downspout drains for planter boxes are placed at the low point of the planter boxes. Standing water in the bottom of planters may run down to a defect in the waterproofing and leak into the building structure.

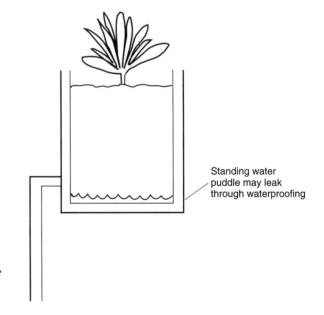

Standing water in bottom of planter due to downspout scupper placed too high.

422   **Chapter 33**

☐ *1036*

Install saucer plates underneath large plant pots. After the pots are filled with plants and soil, they may be too heavy to lift to slide the saucers into place. Irrigation water may then discolor the concrete walkways or stuccoed columns underneath the pots without saucer plates to collect and hold the water.

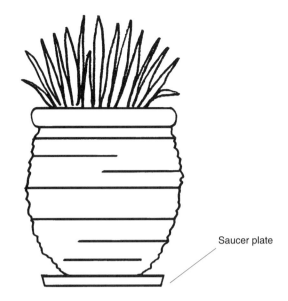

Saucer plate underneath plant pot collects excess irrigation water.

☐ *1037*

Include in the landscaping maintenance contract the provision that gardeners will clean out drainage ditches in the project that collect leaves, branches, grass, and debris as a result of the landscape maintenance cutting and trimming work or plants that grow over into drainage ditches and swales.

☐ *1038*

Consider using atrium-type drain covers at water collection points in drainage swale ditches to prevent the drains from being easily clogged by loose leaves and debris, causing a drainage ditch to overflow and erode adjacent areas.

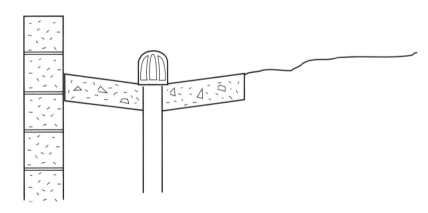

Round-type drain cover does not get clogged as easily by leaves and loose debris.

☐ *1039*

Take an active role in selecting "specimen" trees and plants for focal points in the project. Select the best looking trees and plants for visual impact.

☐ *1040*

When soil amendment material is delivered to the job site, check that the quantities are correct. After the material is mixed into the soil, it is too late to verify that the quantities invoiced by the landscaping subcontractor are what actually was installed.

☐ *1041*

Schedule the installation of large trees before doing the concrete flatwork on tight, high-density projects. This allows forklifts and backhoes to lift the trees and drive up next to the excavated holes to place the trees.

☐ *1042*

Use root barriers for trees and plants that are placed next to concrete flatwork.

☐ *1043*

Check that the landscapers properly compact the backfill into trenches for irrigation and drainage lines.

☐ *1044*

Check with the landscape architect and the soil engineer regarding the planting recommendations for palm trees. Expensive palm trees may die if sufficient drainage is not provided around the root balls.

☐ *1045*

Check with the landscape architect and the landscaping contractor regarding the recommended methods for maintaining the existing trees on the site during construction. For example, covering the root ball with soil and overwatering can kill oak trees.

☐ *1046*

For the landscape planting of trees around sales model complexes, consider planting the trees in their wooden box containers for the duration of the sales period. This removes the shock to the trees of being planted and then pulled up later to be planted elsewhere on the project. When this method is used, the trees remain in their boxes until a one-time-only planting.

Tree temporarily planted in a slope at sales model complex is left inside box for easier transporting and planting at later date.

☐ *1047*

When spraying herbicide for weed abatement before clearing and grubbing for landscape planting, for example, look for three or four days of dry weather. This allows the herbicide to penetrate into the soil and the roots of the weeds without being diluted by rainwater. After three or four days, irrigate the weed abatement areas to accelerate the spread of the herbicide from the roots up into the weed plants, which will kill the weeds.

☐ *1048*

Get the finish grading certified before starting the landscaping. If the landscaper adversely alters the drainage flow lines of the finish grading as a result of the landscape planting, any standing puddles of water complained about by the homeowners can be attributed to and repaired by the landscaping subcontractor.

☐ *1049*

Check that the finish grading for areas that will get grass lawns does not leave sharp berms or valleys. You don't want lawn mowers cutting the grass too short at berms or too long in valleys.

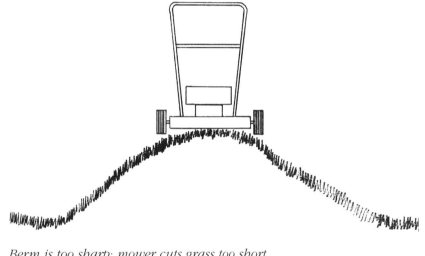

Berm is too sharp; mower cuts grass too short.

☐ *1050*

Don't allow painters and ceramic tile installers to clean up in soil areas that will get landscape planting later; designate concrete flatwork areas such as patios and driveways instead. Paint or tile grout that has been mixed into the soil can kill new landscape plantings.

☐ *1051*

Schedule the painting of wood or wrought-iron fencing before the landscape planting of trees, shrubs, and ground cover; landscaping will get in the way of the painting. Consider shop painting wrought-iron fencing before it is delivered and installed on the job site; it will require only a paint touch-up after installation.

☐ **1052**

For planter boxes next to interior building spaces, slope the bottoms of the planter boxes away from the building interiors to prevent standing water from leaking through the waterproofing and into the interior of the structure.

Build sloped ramp draining away from stairs at bottom of landscaped planter.

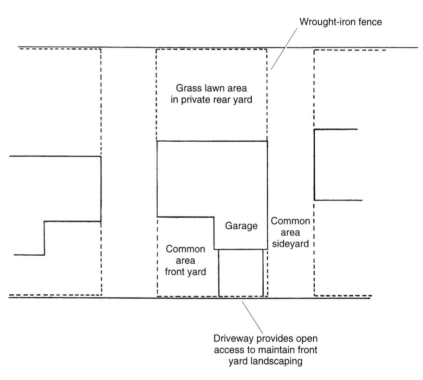

☐ **1053**

For single-family detached houses with zero lot lines and common area landscaping, if the rear yards are private exclusive-use areas with grass lawns enclosed by fencing, consider ahead of time how the homeowners will get their lawn mowers from the front yard garages to the rear yard lawns.

Wrought-iron fencing at zero lot lines prevents homeowners from getting lawn mower from front garage to private rear yard.

☐ **1054**

Provide the new homeowners with literature explaining the importance of maintaining the drainage contours and flow lines of the finish grading when they do their own landscaping.

426   **Chapter 33**

☐ **1055**

To mark the exact locations for the core drilling of holes through street curbs for landscape drainage pipes, hold a short length of plastic pipe with the same diameter as the core-drilled hole and then spray-paint the concrete street curb area contained within the pipe.

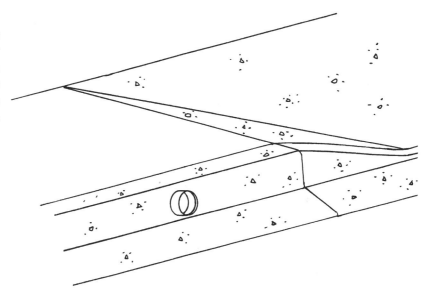

☐ **1056**

Consider ahead of time the fact that a typical 16-foot-wide driveway for a two-car garage takes up 40 percent of a 40-foot-wide lot when you are determining the locations for core-drilled holes through street curbs for landscape drainage pipes.

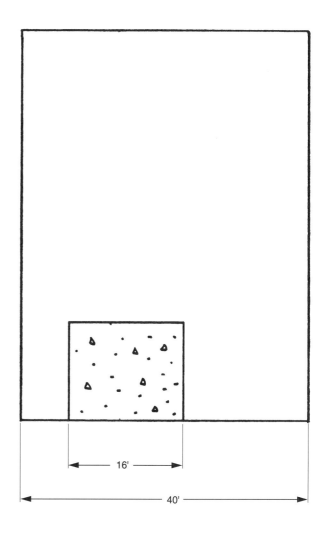

Typical 16-foot-wide driveway takes up 40 percent of 40-foot-wide lot.

☐ **1057**

For two houses that share a common, continuous driveway approach with a landscaped area between the two driveways, consider ahead of time the fact that a core-drilled hole through a street curb for a drainage line from this landscaped area must travel underneath one of the driveways; there is no street curb in a straight line directly out to the street because of the common driveway approach.

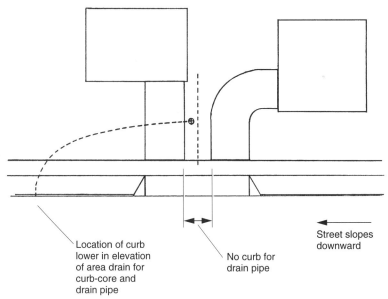

Area drain pipe must pass underneath driveway to reach downhill area of curb.

☐ **1058**

For condominium and apartment projects, consider ahead of time the number and diameter of the holes that must be core drilled through the street curb for landscape drainage pipes and include this cost in the project budget.

☐ **1059**

Make sure a correct diameter hole is core drilled through the street curb for a landscape drainage pipe or a discharge pipe from a sump pump pit. The general rule is that there should be at least 2 inches of concrete curb above the top of the pipe and at least 1 inch from the bottom of the pipe to the flow line in the gutter. The largest diameter hole core drilled through a 6-inch street curb should therefore be 3 inches. Check the public works standards.

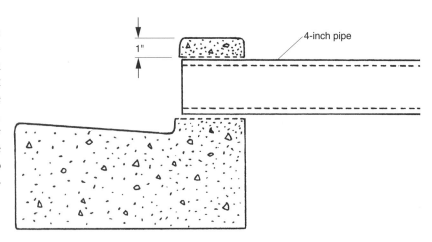

Side view of 4-inch drainage pipe inside a 6-inch curb.

Continued on next page

Continued from previous page

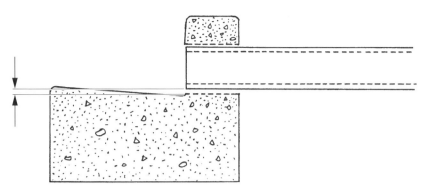

Street curb with hole for pipe core drilled too low.

☐ **1060**
Check the plans to ensure that trees and streetlights are not placed at the same locations required for curb cores for landscape drainage pipes. If they are at the same location, the drainage line may have to travel underneath driveways and walkways to the nearest available open street curb area down the street to maintain the proper vertical fall for the drainage pipe.

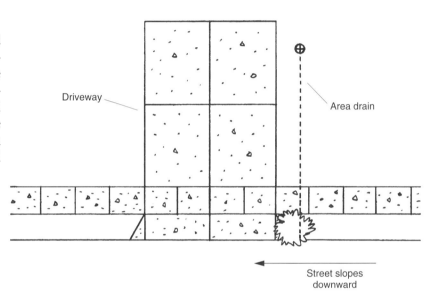

Tree blocking correct drainage of area drain in front yard.

☐ **1061**
For condominium and apartment projects with guest parking spaces, check that the various sets of plans do not place a tree or shrub in a landscaped area next to a parking space, preventing a car door from being opened.

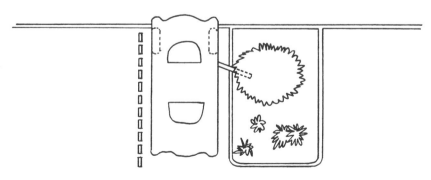

Car door hits tree placed too close to parking area.

Landscaping 429

☐ *1062*

For condominium and apartment projects with guest parking spaces without wheel stop bumpers, check that the various sets of plans do not place trees and streetlights at the center of the parking spaces.

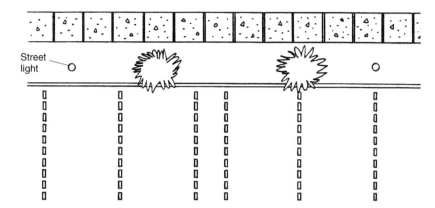

Trees and street light placement can interfere with parking spaces without wheel stop bumpers.

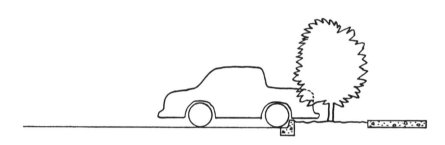

Front of automobile hitting tree before wheels hit curb.

☐ *1063*

For projects in which reclaimed water is used for landscape irrigation, check with the water utility company that the grade or quality level of the reclaimed water provided is allowable for "contact" areas within the project. "Secondary" grade reclaimed water may not be allowed in areas where the irrigation sprinklers could spray water on the tops of park benches or picnic tables, for example.

☐ *1064*

For condominium and apartment projects, locate dirt stockpiles that will be used for the backfilling of retaining garden walls and finish grading at locations on the project site that are close to the work for easier access and shorter haul routes.

☐ *1065*

For high-density condominium and apartment projects with dirt stockpiles used to backfill retaining garden walls and finish grading, consider ahead of time where to stockpile the needed dirt at the end of construction, when the nearly completed buildings will take up all the empty space on the project.

☐ *1066*

When the landscape drainage system is installed by one subcontractor and the landscape irrigation system is installed by another subcontractor, do not backfill over the tops of drainage pipes that are only a few inches below finish grade at the street curb until after the trenching for the irrigation system is complete. This prevents the landscaping subcontractor from hitting the shallow drainage pipes when trenching for the irrigation lines.

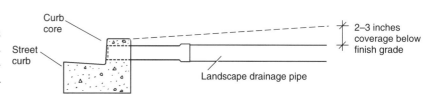

Dirt coverage above landscape drainage pipe at shallow depth.

☐ *1067*

Check the precise grading plans against the top of the footing elevation for the building structure to determine whether a side yard swale that drops in elevation from the rear to the front of the building will expose the top of the footing at the front of the building or cause the sides of the swale to become too steep to avoid exposing the footing.

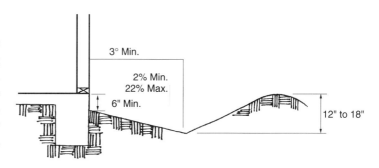

Comparison study of drainage swale elevations at both rear and front of building.

☐ *1068*

For projects in which trees must be removed because they are in the way of the new houses, check whether the city and/or county require new replacement trees to be planted elsewhere on the project and in what numbers and sizes.

☐ *1069*

When a project is required to replant new trees and/or do landscaping elsewhere on the project to replace existing trees or landscaping that had to be removed to make room for the new houses, check whether a revegetation monitoring plan is required and include the cost of that plan in the project budget.

☐ *1070*

For cities and counties that require that new trees be planted elsewhere within the project to replace existing trees that must be removed because they are in the way of the new construction, check whether some of the existing trees will be in the way of the trenching and installation of underground utilities. A 42-inch-wide trench for a 2-foot storm drainpipe, for example, may require a working space corridor of 16 feet to provide clearance for the equipment and a place to stack the spoils.

☐ *1071*

Have the following minimum information included in the precise grading plans:
1. The finish floor elevation of the structure
2. The finish floor elevation of the garage slab
3. The elevations of the rear patio stoop, front walkways, and driveway
4. The high point at the rear of the building lot
5. The swale flow lines around each side of the house
6. The catch basin top of grate, invert, and outlet elevations
7. The top of curb and flow line elevations at the street, preferably at intervals of at least 25 feet.
8. The fences and walls that might affect drainage.

CHAPTER 34

Construction Management

☐ **1072**
Create a system for identifying past and current construction problems and mistakes, using the following:
  ▷ Building inspection cards
  ▷ Punch lists
  ▷ Requests for information (RFIs)
  ▷ Redlined plans
  ▷ Home buyers' walk-throughs
  ▷ Customers service complaint letters
  ▷ Subcontractors' extras
  ▷ Field employee debriefings

☐ **1073**
Create a uniform, companywide comprehensive system applied equally to all projects to better define the objectives and balance out the performance.

☐ **1074**
Conduct periodic interviews with subcontractor company owners and field forepersons regarding construction problems and mistakes.

☐ **1075**
Have the architect demonstrate that all the dimensions on each page of the plans mathematically add up to the correct totals.

☐ **1076**
Have the architect or builder draw specific design areas in isometric three-dimensional drawings for improved clarity and to preanswer questions that will come up during the construction; use these drawings as cheat sheets during construction.

☐ **1077**
Cut up one set of plans and then paste the various door and window schedules and structural details on the appropriate pages of another set of plans for easier use in the field.

☐ *1078*
Try to minimize the number of plan revisions during the construction of the sales models to simplify the management of information before the issuance of the final revised set of building plans. Numerous cut sheets and addenda create an additional layer of information that must be incorporated into the construction.

☐ *1079*
Record the specific plan-checking corrections for each project during the city and/or county plan-checking review phase and then create a checklist that can be used to debug the plans for subsequent projects in terms of building code violations to speed the plan-checking process.

☐ *1080*
When product submittals must be included along with the building plans for the city and/or county plan check, review the product submittal package to make sure information given matches the building design. You don't want to submit wind load testing specifications for an 8-foot-wide sliding glass door, for example, when the master bedroom floor plan shows a 10-foot-wide sliding glass door. The product submittals must match the building plans.

☐ *1081*
Each time changes are made in the sales models, have the architect review the changes to find possible building code violations. Ungrading the size of a kitchen cooktop, for example, may not coordinate with the clearances to the combustible upper cabinets above the cooktop.

☐ *1082*
Have the architect remove crossed-out details on the plans that do not apply to the project. This simplifies the plans and may reduce the total number of pages.

☐ *1083*
Provide three-dimensional isometric drawings showing the start and stop lines for exterior elevation accent paint colors. The straight elevation view on the plans does not show where the accent colors start and stop at wall corners.

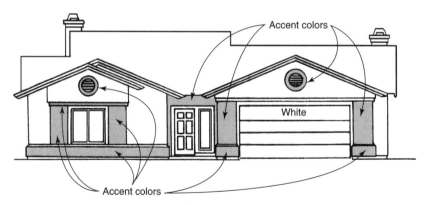

Elevation view does not show whether colors turn corners or stop at front edges.

Continued on next page

Continued from previous page

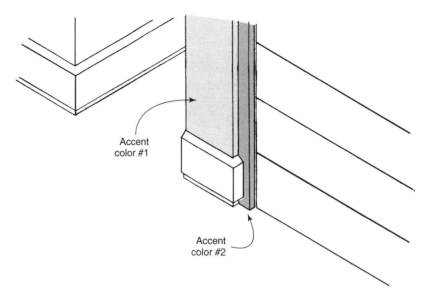

Accent colors shown in isometric side view show where colors stop (inside or outside corners).

☐ **1084**
Plan, monitor, and conserve time during the design phase to avoid placing the construction of the sales models and production units in a time-crunch mode.

☐ **1085**
Create a construction schedule with the correct number of activities and details to manage the work in the field, not a simplified milestone schedule suitable only for project management purposes.

☐ **1086**
Determine the number of tradespersons required for each activity to maintain the desired production rate. Then monitor the number of tradespersons on-site for each activity as a way of keeping the work on schedule. If the project needs fifteen drywall hangers to do two houses per day, for example, and there are only ten hangers on the job site, the work will run behind schedule.

☐ **1087**
Provide spread between the starting times of all the activities on the construction schedule and monitor and maintain the spread throughout the course of the project. This keeps each trade in its proper sequence and prevents it from overlapping and running into the trade directly ahead of it.

Construction Management

☐ *1088*
Analyze activities in the construction where the practice of piecework might slow the production rate. If twelve carpenters could be working on the setting of beams and floor joists on a condominium building but the piecework crew numbers only eight carpenters, some portion of the work will be late.

☐ *1089*
For large projects, consider hiring a full-time on-site deputy building inspector approved by the city or county to coordinate the building inspections with the work. This prevents the loss of blocks of time each day while you are waiting for the following day's inspection approval before proceeding to the next activity. With a deputy inspector, the next activity can start immediately after the completion and inspection of the previous activity.

☐ *1090*
On large projects, ask the city or county building department to assign the same building inspector to the project each day to provide continuity and uniformity in the inspection procedures and maintain a day-to-day history of the inspections.

☐ *1091*
On large projects, ask the city or county building department that a minimum length of time for inspections be reserved for the project each day. If the project needs at least 2 hours for inspections each day, a builder paying large permit and inspection fees should not have to compete with and get shortchanged by minor individual homeowner remodeling inspections, for example.

☐ *1092*
Consider concentrating more tradespersons in certain activities to accelerate the construction schedule.

☐ *1093*
For multiunit condominium sales model buildings that are subject to owners' changes involving wall layout, drop ceilings, soffits, and chases, for example, stop the work in all the units at the rough framing stage and do not begin the rough mechanical work until all the owner's changes have been made in the sales model units. This prevents having to tear out electrical work, plumbing, and HVAC work in the nonmodel units every time a change is made in the model units.

☐ *1094*
For a planned-unit-development project with common area landscaping and recreation areas that must be fully completed for the first groups of homeowners, plan, schedule, and coordinate the work so that the project is completed on time. The builder pays the portion of the common area maintenance costs for the unbuilt and unoccupied units until the project is completed.

CHAPTER 35

The Superintendent, the Construction Trailer, and Field Paperwork

☐ 1095
Analyze the plans before the start of the work and/or cut and paste areas of the construction that could benefit from single-page reduced cheat sheets. For example, reduce the floor plans to $8^1/2$ by 11 inches and show the shear panel locations on each floor level by using different colors. This material can be used during the shear panel inspections with the building inspector for large housing projects. This is easier and more efficient than carrying an entire set of plans around the job site and having to flip back and forth from one page to the next.

☐ 1096
The following cheat sheets can be used by the builder to check the work:
- ▷ Staking layout plan
- ▷ Building footings and trench layout
- ▷ Plumbing trenches
- ▷ Hardware embeds
- ▷ Utility sleeve layout
- ▷ Moment frames
- ▷ Structural posts and beams
- ▷ Shear panel
- ▷ Flush light head-outs
- ▷ Cutouts for mechanical vents
- ▷ Drop ceilings and soffits
- ▷ Fiberglass tub specs
- ▷ Master tub specs
- ▷ Tub and shower framing
- ▷ Door openings
- ▷ Window openings
- ▷ Medicine cabinets
- ▷ Towel bars and toilet paper holders
- ▷ Rough electrical work
- ▷ Rough HVAC work
- ▷ A/C condenser locations
- ▷ Mailbox locations
- ▷ Preliminary drywall

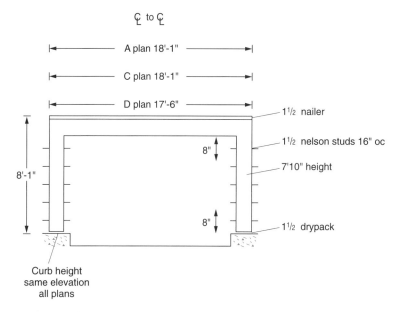

Sketch of steel moment frames at garage door openings. (This sketch was more informative than the detail shown in the plans and was used by the concrete contractor, steel fabricator, and framer.)

▷ Preliminary insulation
▷ Cabinets
▷ Intercom, vac, and security
▷ Appliance specs
▷ Sink cutouts
▷ List of roofjacks per house
▷ List of light fixtures per house
▷ List of hardware per floor plan

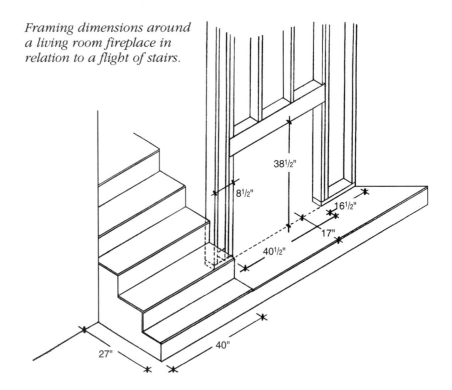

Framing dimensions around a living room fireplace in relation to a flight of stairs.

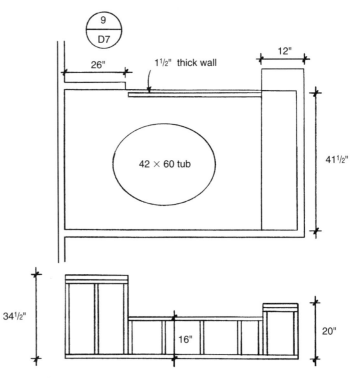

Framing around a bathtub, showing the dimensions of a wall between the tub and bath cabinets, dimensions of a tile seat or shelf, and a short backing wall for the bathtub front skirt. (This information used the dimensions from the specifications sheet of the bathtub purchased, which was not known when the plans were drawn.)

☐ *1097*

Develop a list, like the following, of construction activities that should be checked on every project to catch mistakes before they are covered up by the next activity:

- ▷ Building trench layout
- ▷ Plumbing trench layout
- ▷ Concrete forms
- ▷ Hardware embeds
- ▷ Utility sweeps layout
- ▷ Preslab
- ▷ Slabs ready for framing
- ▷ Steel moment frames and columns
- ▷ Slab humps and valleys
- ▷ Framing of the first floor
- ▷ Straightedge floor joists and beams
- ▷ Floor sheathing glue
- ▷ Stair framing
- ▷ Framing of the second floor
- ▷ Shear panel
- ▷ Facia
- ▷ Concrete chip and sack
- ▷ Door openings
- ▷ Closet door openings
- ▷ Entry door openings
- ▷ Slider openings
- ▷ Window openings
- ▷ Medicine cabinet openings
- ▷ Mechanical chases and cutouts
- ▷ Head-outs for flush lights
- ▷ Bathtub and shower framing
- ▷ Rough HVAC work
- ▷ Rough electrical work
- ▷ Rough plumbing work
- ▷ Sheet metal ducts
- ▷ Rough framing safety handrail
- ▷ Bathtub protectors
- ▷ Fireplaces
- ▷ Window reveals
- ▷ Home buyer options
- ▷ Preliminary insulation
- ▷ Preliminary drywall
- ▷ Drywall stocking
- ▷ Exteriors ready for lath
- ▷ Flashing
- ▷ Sight lath corneraid
- ▷ Pulling electrical wires out of exterior boxes before stucco application
- ▷ Stucco brown coat
- ▷ Plaster digs
- ▷ Stucco color correct type—check bags
- ▷ Tub scratches before tile installation
- ▷ Roofing
- ▷ Cabinets
- ▷ Finish carpentry
- ▷ Baseboards
- ▷ Utility trenches
- ▷ Utility services
- ▷ Tile
- ▷ Drywall prepaints
- ▷ Paint prep
- ▷ Window hack-outs
- ▷ Drywall finals
- ▷ Finish electrical, finish plumbing, and hardware
- ▷ Floor squeaks
- ▷ Toilet bases and fireplace hearths for cracks before carpeting
- ▷ Wood floors before installation of appliances and cleanup
- ▷ Appliances pigtailed
- ▷ Final punch list

☐ *1098*
Improve planning by making a daily activity list at the start of each workday.

Daily activity list.

thurs 9/26

(1) fix "D" plan powder bath floor jays
(2) walk paint pickup list
(3) call fiberglass tub — redo overflow patches & tub finals
(4) update sched
(5) make new drywall list
(6) rec bldg meter pedestal
(7) extend gas at rec bldg
   (a) correct irrigation
(8) constr. cleans bldg 16 non-models
(9) build tray fence models
(10) walk bldg 22
(11) sched scaffold bldg 22
(12) assign days to activities & plan to crunch
(13) c/s meeting 9:30
(14) call subs remaining pickup bldg 17
(15) ask framer complete bldg 18 by next friday
(16) call closet shelving start date
(17) check bldg 21 if entry columns & eyebrow roofs framed
(18) check backfill bldg 18 & sched entry block columns
(19) check "E" & "A" tongue platforms bldg 21
(20) sched prelim drywall bldg 21 for friday
(21) call tile lot 13 kitchen
(22) organize keys to rec/pool area

☐ *1099*
Make proactive problem solving part of the job site superintendent's role.

☐ *1100*
Establish quality levels that are reasonable and apply them uniformly to all projects.

☐ *1101*
Develop an empirical method for determining the number of job site superintendents on each project that is based on a listing of the daily, weekly, and monthly superintendent tasks and the estimated time period for completing each task.

☐ *1102*

Recruit new assistant superintendents from the ranks of tradespersons and forepersons. They already understand construction in terms of sequence and scheduling and are aware of the construction problems that can be solved through better supervision.

☐ *1103*

Develop in-house training for incoming, newly hired assistant superintendents to provide uniform comprehensive training for field employees that reduces the learning curve in the field.

☐ *1104*

Make a companywide employee register that lists all the office and field employees with photographs, titles, and brief job descriptions, including the owners and top managers, and give this register to each new employee. This reduces the time it takes field employees to learn the names and positions of people in the main office and on other job sites.

☐ *1105*

Have a filing expert from the main office set up the construction office's filing system in the construction trailer at the start of every new project and provide an orientation session for the field superintendents and customer service repairpersons on the job site.

☐ *1106*

Draw a flowchart that illustrates the steps involved in processing and paying invoices for better understanding on the part of field employees and subcontractors.

☐ *1107*

To avoid work stoppages or reduced labor power problems in the field resulting from subcontractor cash-flow problems, take a proactive approach in the accounts payable department to resolve subcontractor invoicing problems and mistakes.

☐ *1108*

Extra work order requests should be reviewed by the purchasing agent. Some extra work actually may be contractual work that is covered in the scope of work sections in the subcontracts.

☐ *1109*

Involve the purchasing agent in getting slow-acting subcontractors to perform pickup work or customer service repairs. Purchasing agents have a lot of influence when future work is being bid for upcoming projects.

☐ *1110*
Have a line item in the project budget for unassignable repair costs that cannot be back charged and use this line item for the costs to fix minor mistakes with subcontractors that are performing well. This way job site superintendents avoid back charging subcontractors that are otherwise performing well for minor repairs.

☐ *1111*
Consider the use of premade, in-house-generated invoices for some or all of the work. This simplifies the accounts payable process by taking the origination of invoices out of the subcontractor's hands.

☐ *1112*
Have a sign-out sheet in the construction trailer to record the number and types of plans used by the various subcontractors. This provides a method for tracking how many sets of plans are being used by a particular subcontractor and how many sets are returned after that subcontractor's work is completed.

☐ *1113*
Give copies of maps to the project to the various subcontracts and ask the subcontractors to make copies and distribute them to their material suppliers.

☐ *1114*
Spray-paint the ends of the various sets of plans different colors for easier identification. The ends of the grading plans might be spray-painted brown, the landscaping plans green, and so on.

☐ *1115*
Take extensive photographs of the sales model construction to use as a supplement to the building plans when you are explaining a portion of the construction to tradespersons at the start of a new phase of the work.

☐ *1116*
Provide the construction site with a trailer large enough to conduct a professional operation. There should be enough interior space for furniture and equipment and enough wall space to hang schedules.

☐ *1117*
Make a list of suggested office equipment for the construction trailer:
- ▷ Two-line speaker telephone
- ▷ Answering machine
- ▷ Fax machine
- ▷ Copier
- ▷ Drafting table, chair, and supplies
- ▷ File cabinets
- ▷ Bookshelves
- ▷ Plans rack
- ▷ Computer

☐ *1118*

Make a list of suggested office supplies:
- ▷ Address and telephone file
- ▷ Adhesive stick-on notes
- ▷ Ballpoint pens
- ▷ Building code books
- ▷ Business card holder
- ▷ Business cards
- ▷ Calendar
- ▷ Colored markers
- ▷ Colored pencils
- ▷ Copy machine toner
- ▷ Copy paper in standard and legal sizes
- ▷ Daily log book
- ▷ Drafting triangles and templates
- ▷ Erasers
- ▷ Fax machine paper and toner
- ▷ File folders
- ▷ File labels
- ▷ Fire extinguishers
- ▷ First-aid kit
- ▷ Hardware catalogs
- ▷ Job camera and film
- ▷ Keyed padlocks
- ▷ Key rings
- ▷ Liquid correction fluid
- ▷ Mechanical pencils
- ▷ Organizers
- ▷ Paper clips
- ▷ Pencil lead
- ▷ Pencils
- ▷ Pencil sharpener
- ▷ Plan holders
- ▷ Pushpins
- ▷ Round paper labels
- ▷ Scissors
- ▷ Scratch pads ($8^1/_2$ by 11 inches)
- ▷ Scratch pads, legal size
- ▷ Spray paint
- ▷ Stapler and staples
- ▷ Telephone answering message logbook
- ▷ Transparent tape holder and refills
- ▷ Trash cans and bags
- ▷ Upside-down spray paint
- ▷ Water dispenser and bottles of water

☐ *1119*

Consider having the construction trailer cleaned on a weekly basis, along with the sales office and sales models.

☐ *1120*

Consider placing gravel leading up to the construction trailer, a place mat at the entry door, and an awning or roof covering above the entry door to prevent people from tracking mud and dirt into the trailer.

☐ *1121*

Develop a policy regarding whether the trailer should be locked or unlocked during the workday when it is not occupied by the superintendents in terms of access to and use of telephones by subcontractors. Add a pay phone to the exterior of the construction trailer.

☐ *1122*

Keep the construction trailer close to the work to reduce the time spent by the superintendents and subcontractors walking from the work to the trailer and back. An old house that comes with the property may be converted into an excellent construction field office but may be too far from the construction.

☐ **1123**
When a field employee leaves a project, store that employee's set of keys for the project at the main office, not inside a desk drawer in the construction trailer. If the trailer is broken into and the keys are stolen along with other items, the integrity of the occupied units and the sales models complex will be compromised in terms of security, requiring that new keys be cut for the entire project.

☐ **1124**
When the project is enclosed by a chain-link fence, set the posts for the entrance gate in concrete and consider using 4×4 wood posts or steel I-beams so that the gates do not sag and end up dragging across the ground.

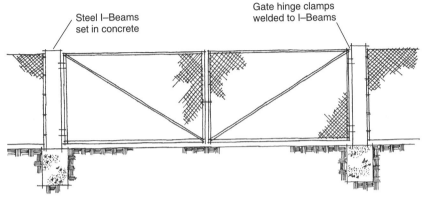

Chain-link fence with steel I-beams used for posts.

☐ **1125**
Analyze the project and select a location for the subcontractors' storage bins so that they will not have to be moved later.

☐ **1126**
Ask the various subcontractors ahead of time what they want done with the keys to storage bins as they are delivered to the job site before the start of their work. If keys are given to the superintendent, label and organize the keys until they are given to the subcontractors.

☐ **1127**
When a nylon windscreen is to be installed on chain-link fencing enclosing a project, consider setting the posts in concrete to prevent the fence from being blown over by the wind.

☐ **1128**
For materials such as stacks of lumber or drywall that must be placed temporarily at the curb of a public street, check with the city or county regarding the requirements for permits and procedures.

CHAPTER 36

The Final Preparation Phase

☐ **1129**
Getting the houses in nearly perfect condition at the time of the homeowner walk-through will create a positive first impression with new homeowners. After-the-fact customer service repairs should not be considered a positive but merely the resolution of a negative.

☐ **1130**
Quality requires a builder's prep crew to fine-tune the finished product after all the subcontractors' pickup repair work has been done. Subcontractor quality alone is not enough.

☐ **1131**
Have a superintendent or customer service prep crew person with a good eye for quality spend adequate time walking through each house to create the final prep punch list. This is the last pre-walk-through punch list and should duplicate exactly what the new homeowners will look at during the walk-through.

☐ **1132**
Don't allow the sales department to promise the completion and delivery of houses out of the scheduled sequence of construction. Jumping all over the project disrupts the construction and adversely affects quality.

☐ **1133**
After the carpeting is installed, keep the entry doors locked. This prevents tradespersons from thoughtlessly entering the units and tracking in dirt and/or mud. Locking the doors after carpeting has been installed provides the transition point from the construction mentality to the customer service mentality for the subcontractors.

☐ *1134*

Do not use the colored dot method to identify areas that need repairs. Use a punch list instead. Subcontractors may not see all the dots inside enclosed areas, such as cabinets and behind doors, and dots that are left behind will notify the homeowners of unfinished repairs during the walk-through or after moving in that they otherwise might not have noticed or complained about.

☐ *1135*

Test the keys and the garage door opener transmitters before giving them to the new homeowners. This prevents the problem of keys that were incorrectly labeled at the hardware store or by the hardware installer and transmitters that were incorrectly coded or labeled. You don't want the homeowner to show up on a Saturday morning with a moving van and have keys and transmitters that don't work.

☐ *1136*

When unsold production units are to be kept electrically energized under a "show and clean" rate, check the units after the meters are set to turn off any lights, fans, or appliances that were installed with their switches in the on position.

☐ *1137*

After the homeowner walk-through, keep the electricity on in the builder's name for a few more weeks to allow repairs to be made by the subcontractors and the builder's customer service crew before the homeowner moves in.

☐ *1138*

When fire extinguishers must be wall mounted inside individual garages on condominium projects, install the brackets on the walls but do not install the fire extinguishers. Place the fire extinguishers inside the lower cabinet underneath the kitchen sink at the end of the construction, just before the homeowner walk-through so that they can be installed by the homeowners after moving in. This prevents the fire extinguishers from being stolen during construction.

☐ *1139*

For production housing, schedule the installation of the central vacuum system tanks inside the garages after the construction is complete so that the tanks are not stolen. If the central vacuum company provides an orientation session with the new homeowners, that is a good time to have the tanks installed.

☐ *1140*

For two- and three-story condominium projects, check the painting, stucco, and roofing of buildings from the viewpoint of adjacent buildings. What looks good from ground level does not always look good from the vantage point of homeowners on the second or third floor of an adjacent building.

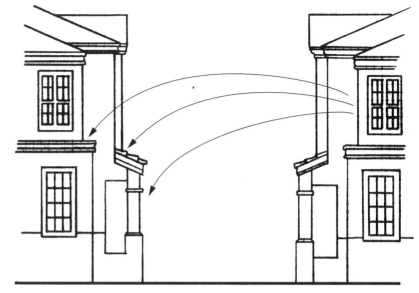

Can see accent color paint lines from 2nd floor of adjacent bldg.

Check exteriors of adjacent buildings.

☐ *1141*

When flooring is not installed in empty, unsold production units until each unit is sold and the flooring selections are made by the new home buyers, have in place a strategy for installing base shoe trim for the floor areas that get vinyl flooring. Have the finish carpenter cut and tack the base shoe in place ahead of time in probable vinyl flooring areas or have the customer service prep crew put in the base shoe after the flooring has been installed.

☐ *1142*

When the polishing of wood floors is part of the flooring subcontract, schedule this activity after the final interior cleaning phase but just before the homeowner walk-through. This gets the homeowners to see and accept the new flooring in its best condition at the time of the walk-through, before they have a chance to clean the floors themselves with the wrong cleaning substances or methods and thus damage the finish surface of the flooring. Otherwise the homeowners may blame the final cleaning subcontractor, the flooring subcontractor, or the builder.

CHAPTER 37

Customer Service

☐ *1143*
Don't view the quality level of customer service and the size of the customer service crew as a budgetary line item that should be economized on as if it were a competitive bid. This attitude produces customer service that is a compromise between service and cost, resulting in mediocre service with slow response times because of marginal crew sizes.

☐ *1144*
Come up with a monetary incentive program that rewards both the superintendents and the customer service crew for low-item walk-throughs and a minimal number of customer service complaint letters. This creates a quality team effort rather than a system in which one group passes off the responsibility and blame to the other group.

☐ *1145*
Make it a policy that homeowners send service complaint letters directly to the main office, not to the on-site customer service repair-persons. This allows the customer service department to establish a relationship with the homeowners and to be aware of and monitor the repairs being made in the field.

☐ *1146*
Have a clear policy regarding the twelve-month warranty period, for example, with a definite cutoff date for each house. When the builder continues to perform customer service repairs beyond the warranty period, homeowners think this applies to subcontractors as well. The plumbing subcontractor, for example, will get a telephone call from a homeowner for service in the fourteenth month of occupancy, beyond the subcontractor's twelve-month warranty period, because the builder is still doing customer service at homeowners' request.

☐ *1147*

Provide enough flexibility in the customer service department for the field representatives to say no to homeowner requests for customer service that are unreasonable, fall outside the scope of the warrantied work, or belong to the category of homeowner maintenance. Don't have a main office policy of saying yes to everything before it can be reviewed in the field; saying yes to nonwarrantied items is not equivalent to providing exceptional customer service.

☐ *1148*

For production tract housing, have a construction person conduct the homeowner walk-through so that quality and customer service questions can be answered up front. Unanswered questions regarding quality during the walk-through can end up becoming unreasonable service requests after the customer has moved in.

☐ *1149*

Keep a detailed record of the brand names, numbers, and colors of all the finish materials used in the construction of the production units and sales models to help the customer service repairpersons obtain materials and make repairs without having to spend time on the telephone.

☐ *1150*

For large projects of twenty houses or more, provide at least two repairpersons per project for customer service. Two people can do the work of three in terms of momentum and mutual support in dealing with homeowners' complaints.

☐ *1151*

During the walk-through explain to new homeowners that free access into the house by the customer service crew not only is more convenient for the homeowners, who will not have to take time off from work, but also gets the repairs done faster; the customer service crew will not have to coordinate the homeowner's and the subcontractor's schedules to make the repairs.

☐ *1152*

Provide new homeowners with a list of subcontractors they can contact directly. These subcontractors have their own service departments, are used to dealing directly with homeowners, and do repairs that usually are technically beyond the capacity of the builder's customer service crew:

▷ Electrical
▷ Plumbing
▷ Heating and air-conditioning
▷ Appliances
▷ Garage door openers
▷ Security alarm system
▷ Intercom system
▷ "Central vac"

☐ *1153*

Provide new homeowners with a "move-in repair kit" consisting of materials that will allow them to make their own minor repairs:
- ▷ Flat touch-up paint for interior walls
- ▷ Enamel touch-up paint for interior doors and trim
- ▷ One can of spackling paste
- ▷ One tube of latex caulking and a caulking gun
- ▷ One spray can of drywall texture
- ▷ One or two throwaway paintbrushes
- ▷ One tube of tile grout caulking to match the color of the bathroom tile grout
- ▷ One tube of tile grout caulking to match the color of the kitchen tile grout
- ▷ One can of spray lubricant, such as WD-40
- ▷ One can of paint remover, such as Goof-Off

☐ *1154*

Don't rob parts from completed production units to complete other units for homeowner walk-throughs. A missing doorknob can be removed and reinstalled five or six times in successive units until someone orders and receives a new doorknob.

☐ *1155*

When enamel paint touching up is done in occupied units while the homeowners are at work, place "wet paint" signs at the newly painted areas.

☐ *1156*

When homeowners are unsatisfied with the quality of the painting, give them colored, self-adhesive round paper dots and one opportunity to mark all the locations they think need touching up or repainting. This approach places the responsibility for identifying the paint repairs with the homeowner and limits the extent of the repair work for the painter to a one-time-only occurrence.

☐ *1157*

When wooden kitchen cabinets are stained and lacquered, write into the home buyer's manual and warranty the provision that the putty used to fill nail holes in the cabinets will not match exactly the varying colors of each individual piece of wood in the cabinets. This prevents an unreasonable home buyer from insisting that the nail putty be custom colored to match every shade on the cabinets.

☐ *1158*

When the job site superintendents and the customer service crew share a construction trailer, have separate phone numbers for construction and customer service. This allows the superintendents to avoid getting involved in customer service problems simply as a result of answering the telephone.

☐ **1159**

Periodically walk through unsold inventory units to check that the lights, fans, and appliances are not on and running.

☐ **1160**

Provide new homeowners with the names, numbers, and wattages for all the different types of light bulbs required in the interiors and exteriors of the houses.

☐ **1161**

Provide new homeowners with tubes of ceramic tile grout caulking that matches the color of their tile grout, along with instructions on how to caulk hairline grout cracks on countertops, bathtubs, and showers as part of the homeowner's maintenance. This saves the builder's customer service crew from having to caulk grout cracks throughout the twelve-month warranty period.

☐ **1162**

Provide new homeowners with appliances instruction booklets and warranty pamphlets arranged in some type of folder or binder to be more user-friendly and professional-looking.

☐ **1163**

For condominium projects with trash bin areas and a regularly scheduled number of trash pickups per week, budget for and schedule extra trash pickups during the move-in period after the completion of a building or phase. Otherwise, the boxes generated by homeowners when they are moving in will overflow the trash bin areas.

☐ **1164**

If the electrical outlet for the clothes dryer in the laundry room is 110 volts, inform the new homeowners before they move in. This will prevent complaints about a $250 or $300 extra cost to the homeowner to add a 220-volt outlet to match the dryer. This can be an unwelcome surprise to new home buyers.

☐ **1165**

Instruct the homeowners to keep the nut on the kitchen sink dish scrubber head tight as part of homeowner maintenance. This prevents water leaks from dribbling down underneath the kitchen sink and out onto the kitchen floor, possibly damaging the flooring.

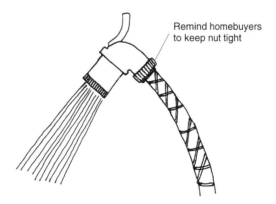

Tighten faucet scrubber head.

☐ *1166*
In condominiums with exterior light fixtures with photocell switches, have the customer service repairs covered by the HOA maintenance electrician instead of the builder's customer service crew. The maintenance electrician can clean the photocell glass, replace the photocell, and do minor electrical repairs as needed on the spot; if further electrical repairs are required, the builder can get involved.

☐ *1167*
Instruct new homeowners in condominium projects with exterior light fixtures that have photocell switches to clean the photocell glass periodically. This will prevent light fixtures from staying on during the daylight hours, causing the homeowners to think that the fixtures are defective.

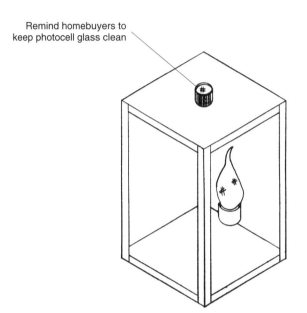

Clean photocells of outdoor lights.

☐ *1168*
If superintendents double as customer service repairpersons and keep tools and materials in a job site storage area, develop an equitable method for reimbursing these employees if their tools are stolen from the job site. Superintendents are not full-time customer service repairpersons with self-contained work trucks that can be moved between projects.

☐ *1169*
After window and roof leaks are repaired, water test the repairs with a garden hose to check that the leaks actually have been fixed. This eliminates the need to make the same interior repairs a second time for drywalling and painting if the leaks occur again.

☐ *1170*
Instruct new homeowners about the importance of maintaining the finish grading drainage slopes and flow lines when they have landscaping installed.

☐ *1171*
Order the following extra hardware for customer service use:
- ▷ Doorknobs
- ▷ Backsets
- ▷ Towel bars
- ▷ Towel rings
- ▷ Toilet paper holders
- ▷ Roller catches
- ▷ Flush bolts or surface bolts for double doors
- ▷ Pocket door latches

☐ *1172*
Order extra door stops: solid for entry doors and flexible or solid for interior doors.

☐ *1173*
Order extra electrical cover plates in all the sizes and types used on the project.

☐ *1174*
Order an assortment of extra light fixture parts so that light fixtures that have parts missing can be completed.

☐ *1175*
Order extra appliance instruction booklets and warranty pamphlets.

☐ *1176*
Order an assortment of extra window parts:
- ▷ Window latches
- ▷ Screen door latches
- ▷ Sliding glass door handles
- ▷ Painted screws
- ▷ Plastic inserts around screens
- ▷ Screen door rollers
- ▷ Spray touch-up paint

☐ *1177*
Order extra window screens.

☐ *1178*
Order spray touch-up paint for the prefabricated metal fireplace firebox.

☐ *1179*
Order spray touch-up paint for the metal thresholds at exterior doors.

☐ *1180*
Order extra tile grout and matching colored caulking.

☐ *1181*
Order extra colored door panel skin inserts for the dishwasher.

☐ *1182*
Order extra luminous ceiling plastic panels in the correct precut sizes.

CHAPTER 38
Sales Models

☐ *1183*
In scheduling the coordination of the sales models' grand opening date and the completion of construction, consider the need for electricity to be on in the models so that the light fixtures can provide lighting for the promotional photography shoot. Photographs for brochures and newspaper ads must be shot weeks before the actual grand opening.

☐ *1184*
For sales models opening during a hot weather period, consider that electricity must be on in the models to provide air-conditioning for the indoor plants that must be in place for the promotional photography shoot several weeks before the grand opening.

☐ *1185*
If the construction will not be completed to the point where the city building inspector can release the sales model complex to get electrical meters set and the power turned on in time for indoor plants and the promotional photography shoot, plan ahead to have a generator at the project.

☐ *1186*
Add several weeks of buffer time between the scheduled completion of the sales models and the grand opening date to deal with unexpected problems.

☐ **1187**
Order and install plastic covers over toilets in sales models to prevent people from using the toilets.

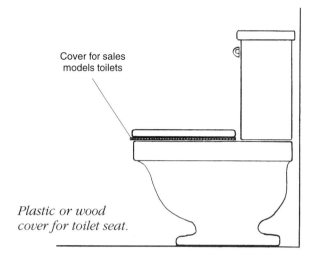

Plastic or wood cover for toilet seat.

☐ **1188**
Install plastic covers over the air-conditioning thermostats in sales models to prevent people from changing the temperatures inside the models by playing with the knobs.

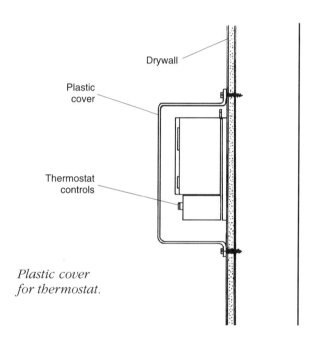

Plastic cover for thermostat.

☐ **1189**
Wire the electrical work inside the sales models so that the light fixtures are always on and cannot be operated manually by using the light switches. This prevents people from turning off the lights in hallways, stairways, walk-in closets, and bathrooms while walking through the models; there should be one switch operated by the sales staff only.

☐ **1190**
Disconnect the bath fans from the light switches in the sales models. This prevents people from turning on the fans and leaving them running. The bath fans can be reconnected to the switches after the project is sold out and the models are converted to occupiable units.

☐ *1191*

Do not plug in appliances in the sales models; people will push the buttons and turn the knobs if they get a response. If the appliances are not plugged in and there is no response, people will not fool with the appliances. This eliminates the need to replace electrical circuitry in the appliances after the models are converted and occupied by the new homeowners if the appliances do not work after a short period of use.

☐ *1192*

Consider ahead of time the number of copies of keys to the sales model complex that will be needed by the sales staff, construction superintendents, cleaning crew, indoor plant maintenance crew, etc. This prevents a last-minute crisis while you are preparing for the sales models' grand opening.

☐ *1193*

Reverse the doorknob locks on the inside doors between the house interiors and the garages in the sales models and keep those doors locked. This prevents the spring-loaded garage fire doors from closing behind unsuspecting home buyers after they enter the garages, trapping them inside a sales model's garage that is padlocked on the outside and has a fire door that is locked from the house's interior.

☐ *1194*

Discuss with the city building inspector ahead of time your intention not to install window screens on the sales models while they are open to view for the public to make the sales models' interiors as bright and sunny as possible. That way, this will not become a final building inspection issue. This prevents the builder from having to go through the senseless exercise of installing all the window screens to satisfy the final inspection requirements and then take them off immediately afterward.

☐ *1195*

When sales models are open beyond the daylight hours, consider ahead of time adequate exterior lighting for the pathway from the sales models' office trailer to the models and around the model complex for safety and liability reasons.

☐ *1196*

Consider placing the interior lighting of the sales models on a clock timer system that turns the electricity on and off from one location at preset times each day. This saves electricity and prevents the sales staff from having to walk through each model and manually turn the lights on and off each day.

☐ *1197*

For condominium projects with sales model buildings that have non-model production units that cannot be occupied until the project is complete and the sales models are converted to regular units, include in the project budget the cost to touch up or completely repaint these empty production units at the end of the project. Interior painted surfaces and painted wood doors and casings that are exposed to sunlight for months or years can turn yellow.

☐ *1198*

Include in the project budget a line item for periodic maintenance repairs for the sales models and sales office.

☐ *1199*

For sales models that have been open to the public for two or three years, consider ahead of time how to offer the typical twelve-month warranties for subcontractor work such as electrical, plumbing, and HVAC at the end of the project, when the sales models are converted to occupiable units and sold.

☐ *1200*

When the walls and the ceilings in sales models are painted different colors, determine ahead of time with the interior designer whether the vertical 12-inch "wall" area at a dropped ceiling soffit is a wall that will be painted with an accent wall color or a ceiling that will be painted white to blend in with the surrounding ceiling areas.

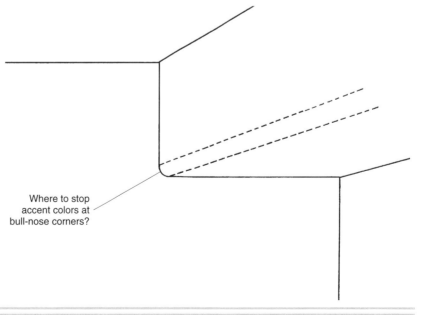

Is this space wall or ceiling when walls and ceilings are different colors?

☐ *1201*

When the walls and ceilings in sales models are painted different colors and the drywall has bull-nose corner bead, determine ahead of time with the interior designer where the accent colors will stop on the curvature of the bull-nose corners.

Where to stop accent colors at bull-nose corners?

Sales Models

ABOUT THE AUTHOR

Bart Jahn is a construction manager and a widely respected author on construction subjects. As a construction manager, he has worked for more than 20 years for several of Southern California's leading developers. He is also the author of *Practical Cost-Saving Techniques for Housing Construction,* *McGraw-Hill's Best Practices for Housing Construction,* and *Offsite Construction.*

INDEX

A

Accent colors, 434–435, 457
Access:
    to attic, 105, 167, 199
    electrical transformers, 13
    GFI receptacles, 176
    spa tub motors, 156
Address numbers:
    illumination of, 165
    size of, 297
    tile plaques for, 275
    visibility of, 298, 301
Adjacent buildings, exteriors of, 447
Adjacent windows, 85, 90, 196, 263
Adjustable shelves, 209
Air-conditioning, 43, 158–163, 175–176, 178, 284, 308–309
Air-conditioning condensers, 159, 163, 176, 188, 365, 385, 411
Alignment, 76, 246
Anchor bolts, 24–27, 31, 33, 35, 42, 52, 105, 107, 121–122, 356
Angle braces, 129
Angle stops, 217
Apartments, 5, 73, 75–76, 142, 154, 158, 164–165, 175, 188, 191–192, 203, 237, 280–282, 292, 296, 311, 346–347, 352–353, 356, 359, 364, 378, 392, 398, 411–412, 414–416, 428–431
Appliance instruction booklets, 451, 453
Appliances (*see specific appliances*)
Aprons:
    stairways, 98, 245
    windowsills, 90, 192–193, 239, 290
Archaeologists, 11
Arches:
    ceiling, 97
    entry, 118–119, 323
    recessed, 182
    window, 91, 230, 329
Architects, 50, 433–434
Architectural plans, 51, 87, 163, 188, 433–434, 437, 442
As-built drawings, 23
Asphalt, 132
Attics:
    access panels, 105, 167, 199
    drywall in, 199, 205
    lights, 176
Awnings, 186

B

Backfill/backfilling, 9–10, 14, 16–17, 43, 157, 365, 391, 424, 430–431
Backhoes, use of, 15, 424
Backing, 137–143
Backset screws, 292
Backside panels, cabinets, 213–214
Baffles, 284, 308
Balcony decks, 60, 66, 69, 83, 101, 126, 170, 185, 197, 277, 288, 302, 316, 322–324, 327, 330, 413, 416
Banding, 125–127, 351, 402, 413
Banjos, 283, 304
Bar cabinets, 267
Bar lights, 145, 177
Bar sinks, 252–253, 267
Barrel roof tiles, 86
Base plates, 356–357
Base shoes, 194, 215, 218, 220, 237, 287, 320, 332
Baseboard:
    as backing, 138
    in bathrooms, 135, 268, 332
    and cleanouts/vents, 151–152, 155
    and columns, 71
    in doorways, 81, 240
    and drywall, 194
    and fireplaces, 238
    and flooring, 333, 337, 339, 342
    during painting, 286
    and registers, 158, 160
    as trim, 229–230, 233–238, 240, 244
Batch numbers, paint, 282
Bathroom cabinets, 74, 146, 150, 152, 168, 177, 182, 208, 210, 217–218, 241, 257–259, 263, 283, 294, 296, 301–302, 313–314, 330, 334
Bathroom hardware, 291, 294–302
Bathrooms, 46, 48, 52, 54, 58, 74–75, 92–95, 119, 132, 134–136, 145–146, 150–156, 162, 168, 170, 175–177, 182, 201–202, 208, 210, 217–218, 227, 241, 247, 251–261, 263–270, 282–283, 285, 287, 304–307, 313–315, 320, 330–334, 336–337, 339, 342
Bathtubs, 48, 56, 58, 95, 132, 134–136, 150–151, 153–156, 176, 191, 201–202, 218, 251–254, 256, 258, 261, 264, 267–268, 282–283, 305–307, 320, 334, 342
Batt siding, 171, 179
Beam saddles, 124
Beams, 42, 59–68, 71, 97, 113–114, 121–124, 198, 444
Bedrooms, 151, 173, 241, 285, 311
Benches, curb, 9
Berber carpeting, 335–336
Berms, 8, 425
Bevel siding, 62, 178–179, 327
Bids, subcontractor, 2, 6, 441
Bifold doors, 45, 78, 152, 242, 295
Birds, prevention of perching by, 117
Bird's mouths, 281, 336
Blinds, 321
Block columns (*see* Masonry block columns)
Block footing, 27
Block walls (*see* Masonry block walls)
Bolts, 61–62
    anchor, 24–27, 31, 33, 35, 42, 52, 105, 107, 121–122, 256
    breakaway, 15
    lag, 408
    striker, 207
    threaded, 123
    throw, 292
Bonds:
    encroachment, 8
    improvement, 9, 21
Bottom guides, 78
Bottom plates, 67, 77, 138
Bottom steps, 100–101, 103
Boxes:
    fireplace, 108–109, 112
    utility, 32, 57, 140
Braces:
    angle, 129

let-in, 57, 111
Brackets, 53–54, 142, 232, 234–235, 287, 293
Breadboards, 217, 284
Breakaway bolts, 15
Breaker switches (*see* Circuit breakers)
Bricks, 153
Buckets, 61–62, 113, 122–123
Building footings, 24, 27–31, 37, 50
Building pads, 2–3, 6, 8
Building paper, 256
Built-out walls, 240
Bulkhead doors, 75
Bull-nose corners, 81, 102–103, 197–199, 207, 220, 249–250, 261, 267, 270, 314, 322
Bull-nose tiles, 155, 186, 259, 266, 270
Bumper jambs, 138, 231, 233
Bumper pads, 216, 219
Bumper posts, 286
Bumpers, wheel stop, 430
Buried pipes (*see* Subdrain pipes: Underground utilities)
Butt-joint splices, 236

C

Cabinets, 213
    bar, 267
    bathroom cabinets, 74, 146, 150, 152, 168, 177, 182, 208, 210, 217218, 241, 257–259, 263, 283, 294, 296, 301–302, 313–314, 330, 342
    electrical outlets in, 312
    hardware for, 294, 296, 302
    island, 208, 363
    kitchen, 47, 57, 69–70, 74, 83, 87–88, 131, 145, 166, 168–169, 180–181, 187, 208–209, 211–213, 215–217, 219, 222–225, 284, 312, 320–321, 342, 450
    laundry room, 222, 290
    and light fixtures, 165–166, 210
    linen, 48, 214
    medicine, 52, 182, 285, 296, 302, 313–314
    painting of, 283–285, 287, 290
    pantry, 54, 209, 215, 224
    tiles on, 251, 257, 262, 265–266
    and windows, 247
Cable:
    direct burial, 13–14
    television, 177, 189
Cambered members, 33–34, 59
Canopy windows, 92, 262
Cantilevering, 60, 104, 185, 278, 368, 416
Canvas awnings, 186
Caps:
    handrail, 138, 239–240, 242–243, 245–246, 338, 406–408
    for pipes, 44
    post, 113
    tile, 155, 186, 259, 266, 270
Carpentry phase, finish, 227–249
Carpeting, 72, 76, 100, 102, 194–195, 228, 231, 233, 236, 238, 242, 245, 306, 332, 334–339, 341, 343–344
Casement windows, 403
Cash-flow problems, 441
Casings:
    door, 72–76, 80–81, 148, 155, 160, 171, 200, 207–208, 221, 227, 229–230, 233, 235, 240–241, 259, 269, 287, 340
    window, 90, 94–95, 247, 250, 265, 268
Catches, magnetic, 142
Caulking, 55, 89, 94, 110, 112, 182, 206, 228, 230, 242, 252–253, 266, 283, 285, 290, 309, 320, 333, 370, 451, 453
Cedar shingles, 327
Ceiling arches, 97
Ceiling backing, 137, 139
Ceiling fans, 154, 174, 312
Ceiling joists, 63, 96, 115–116, 137, 139, 149, 163, 174, 185
Ceilings:
    air-conditioning registers on, 178
    attic access panels, 105
    dropped, 70
    framing of, 59–72
    and roofs, 114–116
    sloped, 71, 161, 226, 270
    spring lines, 149
    and stairways, 96–97
Centered arches, 97
Centered windows, 84
Ceramic tiles (*see* Tiles)
Certification, 3
Chain-link fences, 20, 444
Chases, 124

fireplace flues, 108
mechanical ducts, 214
walk-in closets, 46
Cheat sheets, 24, 298, 433, 437–438
Checklists, 437–439
Chimneys, 111–112, 116
Circuit breakers, 13, 57, 164, 311
Civil engineers, 3–4, 6–8, 373, 409
Clay roof tiles, 279–281
Clay soil, 16
Cleanouts, 13, 151–152, 155, 371
Cleanup phase, 133, 445–447
Clear/grub, 2, 8
Clearances:
    air-conditioning condensers/furnaces, 159, 163
    archways, 323
    baseboard, 71
    bathtub spouts, 305
    cabinet drawers, 342
    and cantilevering, 278
    cleanouts, 155
    countertops, 223
    domestic/reclaimed water, 12
    doors, 72, 75, 80, 152, 160, 165, 167, 178, 182, 211, 221–223, 228–229, 241, 263, 296, 354
    driveways, 368–369
    dryer vents, 161
    electrical meters, 49, 164
    electrical outlets, 186, 188
    electrical transformers, 13
    exterior lights, 186
    fire hydrants, 15
    fireplaces, 108, 195, 255
    floors/ceilings, 72
    hose bibs, 153
    microwaves, 321
    mirror doors, 296
    range ducts, 169
    roofs, 118, 329
    stairways, 97, 101
    telephone poles, 412
    toilets, 265, 304
    trenching equipment, 30
    walls, 45–47, 49, 52, 58
    water heater vents, 304
    water meters, 23
    weep screeds, 125
    windows, 92, 94, 247, 265, 329, 403
    (*See also* Elevations)
Cleats, 209, 215, 232, 283

Closet doors, 45, 47, 49, 73, 78, 167, 231, 294, 299
Closets:
   backing for, 138–139
   coat, 73
   framing of, 45–49, 53–54, 130
   laundry, 78, 152
   pantry, 224
   walk-in, 46, 167
   wardrobe, 53, 130, 138–139, 231–233, 241, 299
Clothes poles, 46, 53, 232, 241
Coat closets, 73
Colored bathtubs, 253
Colored cement/concrete, 281, 362, 366, 371, 374
Colored stucco, 275–276, 288, 290
Colored tiles, 258, 262, 338
Colors, paint, 282, 288–289, 434–435, 457
Columns:
   bases of, 323
   exterior, 127, 408
   masonry block, 383, 395–396, 400–402
   and moment frames, 121–124
   pop-outs, 94, 298
   prefabricated, 55–56, 71, 325, 353–354
   structural steel, 25, 27–28, 31
   stucco, 271
Common area fencing, 405
Common area landscaping, 392–393, 415, 421, 426, 428, 436
Compaction, 3, 10, 16–17, 43, 157, 365–366, 391
Complaints, homeowner, 448–449
Compression, 60
Concrete:
   colored, 281, 362, 366, 371, 374
   excess, 31, 34, 133, 362
Concrete driveways, 10, 22, 117, 281, 345–378, 385–388, 419–420, 422, 427–429
Concrete foundations, 24–36
Concrete handrails, 277, 322, 324, 330, 406–407
Concrete roof tiles, 279–281
Concrete slabs, 27, 29–31, 33, 35, 38, 40–41, 72, 273, 344, 354, 360, 372, 374–375
Concrete slough, 31, 34
Concrete slurry mixes, 17

Concrete trim, 324, 329–330
Concrete walkways, 5, 10, 43, 117, 156–157, 281, 290, 345–378, 387–388, 422–423
Condensers, air-conditioning, 159, 163, 176, 188, 365, 385, 411
Condominiums, 5, 20, 73, 75–76, 142, 154, 158, 164–165, 175, 188, 191–192, 203, 237, 276, 280–282, 292, 296, 306, 311, 345–347, 352–353, 356, 359, 364, 372, 375, 378, 380, 392, 398, 405, 408, 411–412, 414–416, 428–431, 436, 447, 451–452, 457
Conduit pipes, 13–14, 18, 32–33
Conduit sleeves, 32–33, 57, 366, 370
Conduit sweeps, 40
Construction management, 433–436
Construction schedules, 11, 41–42, 58, 88, 123, 136, 142, 204, 213, 276, 282, 285, 291, 298, 306, 314, 433, 435–436, 445, 454
Construction traffic, 20
Construction trailers, 442–444
Control valves, 151
Corbels, 125, 129, 185, 290, 368
Corian marble, 88, 257
Corner beads, 81, 197–199, 207, 220, 249–250, 261, 267, 270, 314, 322
Corners (see Wall corners)
Corridors (see Hallways)
Cosmetic boxes, 170, 296
Countertops, 47, 57, 74, 87–88, 91, 94, 131, 181–182, 184–185, 202, 208, 213, 219, 223–225, 241, 247, 249–250, 253–254, 256–257, 259, 261–263, 265–267, 269270, 283, 330
Coupling, pipe, 33
Coved vinyl flooring, 215
Cover plates, 168–169, 171–172, 182, 186, 206, 213, 244, 262, 311–312
Cripples, 52, 91
Cross-sectional lumber, 128
Cross-slopes, 348–350
Crown molding, 70, 212, 220–221, 223–226, 239
Curb benches, 9
Curbs, 8–9, 12, 15, 17–18, 22–23, 350, 362, 376–378, 389, 427–429
Curved driveways, 374, 419
Curved stairways, 96, 99, 236, 350

Curved walkways, 350
Curved walls, 237, 241
Customer service, 445, 448–453
Cut and fill maps, 3
Cut sheets, 13, 25
Cutout holes, for furnaces, 133
Cuts, 2, 4, 6–8

D

Daily activity lists, 440
Dams, shower, 132, 135, 251
Deadbolt locks, 79, 292
Decks (see Balcony decks)
Demolition, 2
Design/raw cut, 2
Design phase, 43, 71, 101
Differential displacement, 374–375
Dining rooms, 173
Direct burial cable, 13–14
Dirt:
   as backfill, 43
   excess, 2–3, 6
   under curbs, 9
   (see also Soil)
Dirt fingers, 5
Dirt ramps, 6
Dirt stockpiles, 430–431
Dishwashers, 180, 211, 305, 312
Domestic water, 12
Door casings, 72–76, 80–81, 148, 155, 160, 171, 200, 207–208, 221, 227, 229–230, 233, 235, 240–241, 259, 269, 287, 340
Doorbells, 174–175
Doorjambs, 76, 78–80, 132, 201, 207, 227–231, 235, 293, 296, 300, 333
Doorknobs, 142, 207, 230, 291–292, 294–296, 300, 303
Doors, 194, 201
   balcony deck, 302
   bathroom, 58, 75, 263, 299–300
   bedroom, 285
   bifold, 45, 78, 152, 242, 295
   cabinet, 165, 208, 211, 216, 219, 221–223, 225, 257, 290
   closet, 45, 47, 49, 73, 78, 167, 231, 294, 299
   entry, 5, 58, 101, 167, 238, 254, 288, 291–292, 295–298, 302, 316, 321, 325, 346, 352, 354–355, 359–360
   fire, 142, 297
   framing of, 73–81

garage, 58, 64–65, 67, 80, 129, 132, 196, 297, 316, 372, 374, 418
glass, 83, 87, 94, 101, 121–122, 148, 170, 203, 230, 237, 252, 274, 276, 297, 300–301, 324
hardware for, 291–303
installation of, 227–231, 235, 238, 240–242
mirror, 231
pocket, 78, 195, 228-229, 256, 293, 300, 343
shower, 252, 297, 300–301
swinging, 293
Doorstops, 299–300
Double joists, 62–63
Downspouts, 326, 422
Downward settlement, 60
Drainage:
driveways, 371–372, 375–376, 427–429
landscape, 379–381, 383, 386, 389, 414, 421–424, 426–429, 431–432
walkways, 360–361, 377–378
windowsills, 272
Drainage pipes, 10, 389–390, 419, 423–424, 427–429, 431–432
Drains, 3, 69, 307
Drawers, cabinet, 208, 211, 218, 223, 257, 284, 290, 301, 342
Drip kerfs, 68
Driveways, 10, 22, 117, 281, 345–378, 385–388, 419–420, 422, 427–429
Drop beams, 61, 71
Dropped ceilings, 70
Dryer vents, 31, 152, 159, 161–162, 274, 362–363
Dryers, 45, 78, 152, 161, 181, 364, 451
Drywall:
access panels, 199
in attics, 199, 205
and backing, 137–138, 142
bulges, 148, 197
and cabinets, 219–220, 225, 285, 294, 296, 330
and countertops, 131, 219
and doors, 73, 76, 81, 193–195, 203, 207, 293
and electrical wiring, 166–167, 172–175, 178, 193, 311
excess, 133, 193, 201

and fireplaces, 109–110, 195, 289
and floors/ceilings, 63, 67–69, 71–72, 206
and foundations/slabs, 28–30, 35
in garages, 196–197
and handrails, 198, 242
installation of, 191–207
nailing/screwing, 200
and plumbing, 136, 150, 155, 191–192, 201–202
preliminary, 191–192
repairs to, 276, 315
and roofs, 116
and shear panels, 106
and shelves, 130, 196, 232, 284
and stairways, 96, 98, 102, 195, 203, 245
stocking, 190–191
taping mud, 195, 200–204
textures, 200–204, 206, 213
and tiles, 249, 256, 259, 261, 267, 270
and walls, 47–48, 52, 57, 192, 194–195, 202, 206, 272
and water heaters, 130–131, 192
and windows, 88–89, 94, 132, 192–193, 196, 199–200, 272
Dual-glazed doors/windows, 83, 92
Ducts, 161, 169, 175, 214, 284
Dust shelves, 224
Dust-control measures, 6
Dusting, before painting, 282, 286–287

E

Eaves, 118–119, 275, 288, 290, 330
Edges, painting of, 290
Efflorescence stains, 390
Electrical meters, 49, 164, 188, 312, 352
Electrical outlets, 168–174, 176–183, 185–189, 193, 202, 204, 213, 262, 309, 311–312, 327, 451
Electrical panel boxes, 32, 57, 140, 164, 311
Electrical phases:
finish, 304–312
rough, 150–189
Electrical receptacles (*see* Electrical outlets)
Electrical switches (*see* Light switches)
Electrical transformers, 13–14

Electrical wires, 13–14, 40, 174–176, 192, 216, 311–312
Electricians, 179–180
Elevations:
backings, 138
banding, 125, 413
building pads, 2–3
countertops, 91, 261
doors, 78, 359
driveways, 371–372, 422
electrical outlets, 178, 185
fireplaces, 112
floors/ceilings, 72, 188, 368, 372, 378
garages, 368, 372
headers, 87
and paint colors, 434
plumbing, 151
roofs, 86, 114
stair handrails, 248
street subgrade, 9
swales, 375, 431
tiles, 86, 218
underground utilities, 12
vents, 362–363
walkways, 345, 348, 422
walls, 38–40, 138, 379–381, 386–390
window arches, 91
windows, 86–87, 90–91, 262–263
windowsill stools, 90
yards, 354, 392, 431
(*See also* Clearances)
Elevators, 237, 410
Employee registers, 441
Encroachment permits/bonds, 8
Engineering improvement plans, 9
Entertainment centers, 189
Entry arches, 118–119, 323
Entry doors, 5, 58, 101, 167, 238, 254, 288, 291–292, 295–298, 302, 316, 321, 325, 346, 352, 354–355, 359–360
Entry handrails, 148, 174, 291, 346–347, 352, 356–359, 361, 364, 397
Entry stairs, 345–351, 355–359, 364
Environmental impact reports, 11
Erosion, 383, 386
Escutcheon cover plates, 244
Expansive soil, 8, 43, 359, 375, 383, 421
Export yardage, 2–3, 6

Exposed factory edges, 260
Extension cords, 216, 305, 309, 312, 315, 411
Exterior doors (*see* Entry doors)
Exterior framing, 125–129
Exterior lights, 167, 186, 309, 311, 328, 452
Exterior walls, 327

F

Face frames, 208–210, 221–223
Faces, fireplace, 109–110, 289, 322
False rafters, 116
Fans:
    ceiling, 154, 174, 312
    range, 213
Fascia joints, 119, 154, 232, 288
Faucet aerators, 92
Faucet handles, 296, 305
Faucet scrubber heads, 451
Faucets, 305–306
Fees, 436
Fences:
    construction site, 20, 444
    residential, 141, 353, 359, 393–401, 405, 425–426, 444
Fiberboard, 195
Fiberglass bathtubs/showers, 135–136, 150, 154, 191, 201
Field paperwork, 437–444
Filing systems, 441
Final preparation phase, 445–447
Finger cups, 299
Finish, 2
Finish carpentry phase, 227–249
Finish electrical phase, 304–312
Finish graded slopes, 7–8
Finish HVAC phase, 304–312
Finish plumbing phase, 304–312
Fire alarm systems, 18, 308
Fire doors, 142, 297
Fire extinguishers, 446
Fire hydrants, 12, 15, 19
Fire sprinklers, 12, 154, 156, 305, 307–308, 405
Fire walls, 106, 191–192
Fireplaces, 106
    backing for, 143
    baseboard near, 238
    and carpeting, 337
    drywall/fiberboard for, 195
    finishing of, 317–319, 322
    framing of, 108–112

    log lighters, 153
    painting of, 287, 289
    prefabricated, 318–319, 322
    and roofs, 116
    tiles for, 109–110, 253–255
Fixed shelves, 209
Flag lots, 19
Flange/kerf connections, 66
Flanges, 114, 123, 136, 295, 326, 356–357
Flat roofs, 120
Flex conduits, 166, 169, 189
Floor humps, 59–60, 71
Floor joists, 63, 65–66, 68, 99, 110, 114, 125, 191–192
Floor plans, 51, 58, 73, 84, 87, 95, 101, 108, 151, 188, 191, 211, 298–299, 310, 336, 373, 437, 441
Floors:
    balcony decks, 126
    and cabinets, 210, 215, 218, 221
    elevation of, 72, 188, 368, 372, 378
    framing of, 59–72
    installation of, 331–334
    marble, 333, 338–339, 342
    and sinks, 306, 315
    split-level, 29–30, 35, 102, 138, 221, 325
    tile, 56, 332–334, 338, 341–343
    vinyl, 56, 215, 218, 227, 255, 287, 320, 332–333, 335, 337–338, 340, 447
    and walls, 147
    wood, 102–103, 147, 221, 245, 306, 332–334, 338–339, 447
    (*See also* Carpeting; Subfloors)
Flowcharts, use of, 441
Flowmeters, 305, 307–308
Flues, 108, 112
Fluorescent light fixtures, 166, 210
Flush beams, 61, 71
Flush lights, 63, 115, 154, 163, 165, 182, 185, 204, 212
Foam bedding, for bathtubs, 135
Foam polyseal, 132
Foam trim, 118–119, 170, 274, 327–330
Footings:
    buildings, 24, 27–31, 37, 50
    walls, 31, 37–40, 42
Forced-air units, 45, 133, 176
Forklifts, use of, 22, 190, 200, 424

Formboards, 24–25, 31, 33–34
Foundation vents, 140, 273
Foundations, 24–36, 362–363, 421
    (*See also* Slabs)
Fountains, 410
Frames:
    moment, 27, 31, 121–124
    window, 88–90, 132, 192–193, 272, 330
Framing, 14
    of bathtubs, 134–136
    of doors, 73–81
    exterior, 125–129
    of fireplaces, 108–112
    of floors/ceilings, 59–72
    of roofs, 113–120
    miscellaneous, 130–133
    of stairs, 96–103
    of walls, 45–58, 104, 106, 122, 124, 137, 140–142, 148, 175, 178, 183, 190–192, 195, 205, 344
    of windows, 83–95, 260
Furnaces:
    air-conditioning, 163
    heating, 45, 133, 286, 304
Furring, 26, 61, 63, 98, 104, 122, 136
Fuse-primer paints, 288
Future construction, 18, 23

G

Gables peaks, 84
Galvanized nails, 128
Garage door openers, 297, 370, 446
Garage doors, 58, 64–65, 67, 80, 129, 132, 196, 297, 316, 372, 374, 418
Garages, 27, 31–33, 38, 41, 57, 107, 133, 140–141, 164, 187, 196–197, 200, 272, 286, 304, 306, 316, 323, 346–346, 353–355, 366, 369–376, 385–386
Garbage disposals, 305–307, 309, 312
Garden walls, 13, 382, 387–390, 421
    (*See also* Masonry block walls)
Gas meters, 156, 158
Gas pipes, 12, 15, 34, 112, 152–153, 158
Gaskets, 309
Gates, 141, 352–353, 359–360, 383, 396–397, 400–401, 404
Geologic tests, 4
GFI receptacles, 176

Index   *463*

Glass, tempered, 83
Glass doors, 83, 87, 94, 101, 121–122, 148, 170, 203, 230, 237, 274, 276, 324
Globe light fixtures, 165, 167, 178, 309
Glue, wallpaper, 316
Glue laminated lumber, 59–61, 63, 129
Grab bars, 140–141
Grading phase, rough, 2–11
Grading plans, 12, 163, 375, 432, 442
Grading stakes, 7
Grain patterns, 128–129
Granite countertops, 223, 257, 262, 270
Granular backfill, 43
Grass lawns, 356, 422, 425–426
Grates, 409
Gravel, as backfill, 43
Greenhouses, 92, 262
Grid lines, 50
Grout, 42, 253–256, 272, 289, 322–323, 332, 338, 366, 391, 451, 453
Grub/clear, 2, 8
Guardrails, 420
Guest parking spaces, 19, 375, 412, 429–430
Gutters:
    roof, 326, 329
    street, 8–9, 12, 17–18, 377–378

H

Half-hot outlets, 311–312
Hallways, 48, 73, 100, 104, 120, 122, 137, 142, 147, 165, 224, 240–241, 245–246, 341, 344
Handicap ramps, 348–349
Handrail caps, 138, 239–240, 242–243, 245–246, 338, 406–408
Handrail pickets, 400, 407
Handrail spindles, 55–56, 242–243, 316–317, 338
Handrails, 271–272, 408
    backing for, 138, 142
    balcony decks, 126, 277, 288, 316, 322, 324, 327, 330, 413
    and drywall, 198, 242
    entry, 148, 174, 291, 346–347, 352, 356–359, 361, 364, 397
    height/spacing of, 407
    second-floor, 240, 326–327
    stairways, 98–99, 162, 180, 198, 234–235, 242–243, 245–246, 248, 287, 317, 326–327, 338–341, 395, 402–403, 406
Handsaws, use of, 77
Hangers, 61–64, 113–114, 116
Hanging light fixtures, 165–166
Hardware, 291–303, 453
Hardwood floors (*see* Wood floors)
Haul routes, 6
Headers:
    garage doors, 58, 64–65, 67, 129, 132, 196, 316, 372
    house doors, 75, 77, 193, 229, 231, 233
    windows, 87, 91
Hearths, 238, 254–255, 317, 319, 337
Heating, 158–163
Herbicides, 425
Hickeys, 178
Hillside lots, 3, 7, 11
Hinge butts, 298
Hinges, 404
Hip hangers, 113–114, 116
Hip roofs, 119
Hips, 281
Hold-downs, 24–26, 35, 52, 107
Hoods, range, 169, 213, 215–216, 223, 286
Hook strips, 232
Horizontal rebars, 25
Hose bibs, 153–154, 156, 308
Hot mopping, 132
Hot sticks, 13
HVAC phases:
    finish, 304–312
    rough, 150–189
Hydrants, fire, 12, 15, 19

I

I-beams, 60, 62–64, 66, 121–123, 444
I-joists, 62
Import yardage, 2–3, 6
Improvement bonds, 9, 21
Improvement plans, 12
Inspectors/inspections, 6, 8, 21, 25, 107, 153, 165, 204, 281, 419, 436
Instruction booklets, appliances, 451, 453
Insulation, 58, 175, 271–278
Intercoms, 173, 183–184
Invoices, 441–442

Irrigation (*see* Landscape irrigation)
Island cabinets, 208, 363
Isometric drawings, 434–435

J

Jacks:
    roof, 280, 288
    stair, 96, 98, 100
    telephone, 173
Jacuzzi bathtubs, 136
Jamb stops, 229
Jambs:
    bumper, 138, 231, 233
    door, 76, 78–80, 132, 201, 207, 227–230, 235, 293, 296, 300, 333
Jigs, 31
Joist blocks, 61
Joist tails, 61
Joists, 59, 61–66, 68, 96, 99, 110, 114–116, 125, 137, 139, 149, 163, 174, 185, 191–192

K

Kerf/flange connections, 66
Keys, 297, 446, 456
King rafters, 113
King studs, 138
Kitchen cabinets, 47, 57, 69–70, 74, 83, 87–88, 131, 145, 166, 168–169, 180–181, 187, 208–209, 211–213, 215–217, 219, 222–225, 284, 312, 320–321, 342, 450
Kitchens, 47, 57, 69–70, 74, 83, 87–88, 91–92, 94, 131, 145, 153, 166 168–169, 173, 175, 180–181, 184–185, 187, 208–209, 211–213, 215–217, 219, 222–225, 227, 249–250, 253, 255, 258, 262, 268, 270, 278, 284, 287, 290, 305–307, 309, 312, 320–321, 333, 340, 450
Knobs:
    cabinet, 211, 218
    door, 142, 207, 230, 291–292, 294–296, 300, 303
Knockdown texture, 206
Knots, in wood flooring, 126

L

Lacquer, 242, 245, 284, 287
Ladder trusses, 205
Lag bolts, 408

Laminated veneer lumber, 61, 63–65, 129
Landings, 96–97, 99, 240, 245–246, 326–327, 346, 348, 352, 354–355, 366
Landscape architects, 414, 424
Landscape design, 414–416
Landscape irrigation, 10, 12, 18, 366, 383, 400, 414, 419–424, 430–431
Landscape lighting, 18, 366, 416
Landscaping:
    common area, 392–393, 415, 421, 426, 428, 436
    and electrical transformers, 13
    and erosion, 383, 386
    excess dirt from, 3
    plants for, 415–426
    of slopes, 278, 392
    and walls, 39, 421
Large-dimension lumber, 60–61
Latches, 292–293
Lateral connection points, for utilities, 13
Lateral terminus points, for utilities, 12–13
Lath, 61, 80, 111, 197, 201–202, 256, 271–278
Laundry rooms/closets, 45, 78, 152, 161, 175, 181, 222, 290, 364
Laundry valves, 183
Lawns, 356, 422, 425–426
Layout, 41
    of doors, 73–81
    of walls, 45–58
    of windows, 83–95
Layout sheets, 24
Leaks (*see* Water leaks)
Let-in braces, 57, 111
Lever doorknobs, 292, 294, 303
Light bulbs, 311, 451
Light fixtures:
    fluorescent, 166, 210
    globe, 165, 167, 178, 309
    hanging, 165–166
    location of, 178–179, 310–311
    strip, 213
    wall sconce, 167, 178
Light switches, 171, 180, 183, 298, 312, 411
Lighting:
    of address numbers, 165

    exterior, 167, 186, 309, 311, 328, 452
    landscape, 18, 366, 416
    of sales models, 456
    street, 21, 430
Lights:
    attic, 176
    bar, 145, 177
    closet, 167
    flush, 63, 115, 154, 163, 165, 182, 185, 204, 212
Lightweight concrete, 35
Line-of-sight studies, 3
Linear footage intervals, soil tests, 11
Linen cabinets, 48, 214
Loader buckets, amounts in, 2
Locks, 79, 292, 297, 300, 456
Log lighters, 153
Lot numbers, 19, 58
Louvered doors, 45, 78

M
Magnetic catches, 142
Mailboxes, 412
Mantels, 109, 143, 195, 318
Maps, 3, 442
Marble baseboard, 339
Marble countertops, 88, 225, 257, 262, 264–265, 269
Marble faces, 109–110, 322
Marble floors, 333, 338–339, 342
Marble sheet panels, 268
Marble splashes, 268
Masonite veneer, 294
Masonry block columns, 383, 395–396, 400–402
Masonry block walls, 6–8, 13, 31, 37–44, 272, 357–358, 365–366, 376–392, 396–397, 404
Mass grading projects, 4, 10–11
Medicine cabinets, 52, 182, 285, 296, 302, 313–314
Metal brackets, 142, 287, 293
Metal detectors, 11
Metal fireplaces, 110, 255
Metal flanges, 114
Metal flues, 112
Metal grab bars, 140–141
Metal shoes, 404
Metal straps, 50, 68, 197
Metal surfaces, painting on, 288
Metal tracks, 139

Metal trim, 196, 271, 335, 347
Metal window frames, 88
Meter rooms, 14, 40, 49, 188, 405
Meters:
    electrical, 49, 164, 188, 312, 352
    gas, 156, 158
    water, 22–23, 168, 305, 307–308, 421
Microlaminated beams, 64–65
Microwave ovens, 216, 222–223, 305, 307, 320–321
Mirror doors, 231
Mirrors, 93, 144–146, 168, 170, 177, 247, 259, 289, 296, 313–315, 330
Mitred splices, 236
Mobilization, 2
Models (*see* Sales models)
Moisture content of soil, 10, 16
Molding, 70, 74, 212, 214–215, 220–221, 223–226, 239, 247
Molly-bolt fasteners, 294–295
Moment frames, 27, 31, 121–124
Monument columns, 408
Mortar splay, 391
Mortises, 292
Mudsills, 105, 121–122

N
Nail dimples, 200
Nail guns, use of, 233, 236
Nail holes, sanding of, 283
Nail shiners, 117
Nailers, 121
Nailing:
    of backing, 137–139, 142
    of baseboard, 236
    of drywall, 195, 198, 200
    of particleboard, 233
    of plywood, 191
    spacing patterns, 24, 104–105
    of studs, 190
Nailing flanges, 136
Nails, 128
Nelson studs, 121, 123
Newel posts, 242–243, 246
Nosings:
    tread, 102–103, 334–336
    trim, 221
Numbers:
    address, 165, 275, 297–298, 301
    lot, 19, 58
Nuts, 26, 35, 61–62, 121

O

Oak trees, 424
Off-site construction, 12
Office equipment, 442–443
Ogee trim, 303
Outlets:
   electrical, 168–174, 176–183, 185–189, 193, 202, 204, 213, 262, 309, 311–312, 327, 451
   plumbing, 153
   telephone/television, 173
Outlookers, 117
Oval windows, 265
Overhangs (see Roof overhangs)

P

Paint colors, 282, 288–289, 434–435, 457
Painting, 55, 94, 142, 154, 182, 201, 203–204, 206, 213, 227–228, 230–232, 235, 237, 275, 280, 282–290, 425, 450, 453
Paleontologists, 11
Palm trees, 418, 424
Pancake outlet boxes, 178
Panel boxes, 32, 57, 140, 164, 311
Paneling, 237, 271
   (See also Shear panels)
Pantry cabinets/closets, 54, 209, 215, 224
Paper-backed insulation, 271
Paperwork, field, 437–444
Parking areas/spaces, 9, 19, 387, 412, 429–430
Particleboard, 233–234, 283, 290
Party walls, 25, 104–106, 191–192
Pass-through bar tops, 270
Patios, 290, 353
Pedestal sinks, 54, 306, 315
Peepholes, 295
Permanent slopes, 3
Permits, 8, 436, 444
Photocell switches, 411, 452
Photographs, use of, 16, 18, 442
Pickets, 400, 407
Picture-frame fencing, 393–394
Pie-shaped steps, 96, 99, 236, 350
Pigtail extension cords, 305, 309, 312
Pipe coupling, 33
Pipes:
   conduits, 13–14, 18, 32–33
   drainage, 10, 389–390, 419, 423–424, 427–429, 431–432
   gas, 12, 15, 34, 112, 152–153, 158
   plumbing, 25, 28, 33, 38, 47, 52, 57, 154, 192, 214, 244, 306
   sewer, 15, 306
   subdrain, 11–14, 42, 44, 380–381, 390
   water, 12, 15, 17, 22, 28–29, 34, 169, 306–308
Pitch roofs, 115
Plant pots, 423
Planter boxes, 413, 419, 422, 426
Planting of trees, 3, 415–418, 423–425, 432
Plantons, 90, 109, 127, 154, 231, 277
Plaster (see Stucco)
Plaster bedding, for bathtubs, 135
Plastic covers, 455
Plastic trim, 320, 327–330
Plumbers, 22, 24
Plumbing angles, 217
Plumbing cleanouts, 151–152, 155, 371
Plumbing phases:
   finish, 304–312
   rough, 150–189
Plumbing pipes, 25, 28, 33, 38, 47, 52, 57, 154, 192, 214, 244, 306
Plumbing trenches, 24
Plumbness, 76, 78, 93, 110–111, 146, 155, 162, 229, 309, 314
Plywood, 65, 204
   for canopy windows, 92
   for exterior framing, 125–127
   for fireplaces, 111
Plywood doors, 49
Plywood furnace platforms, 133
Plywood pot shelves, 130
Plywood rough tops, 217
Plywood scraps, 92, 111, 133
Plywood shear panels, 24, 48, 51, 104–107, 126–127, 150, 172, 191
Plywood sheathing, 111
Plywood siding, 109–110
Plywood stair treads, 99–100
Plywood subfloors, 68, 72, 132, 255, 331, 335, 342
Plywood templates, 27, 31, 66
Plywood trim, 240
Plywood water heater platforms, 130–131
Plywood window arches, 91
Pocket doors, 78, 195, 228–229, 256, 293, 300, 343
Polyseal, 132
Pony walls, 47–48, 98–99, 102, 131, 197, 245, 317, 322, 324
Pop-out banding, 125–127
Pop-outs:
   column, 94, 298
   fireplace, 109, 195
Porches, 323, 346–347, 354, 360–361, 365–366
Post anchors, 24
Post caps, 113
Post footings, 28
Posts, 26–29, 35, 50, 62, 65, 99, 113–114, 116, 122–123, 180, 191, 242–243, 246, 325, 353–354, 356, 444
Pot shelves, 130, 196, 284, 290
Power poles, 411
Predrilled holes, 139, 142
Prefabricated columns, 55–56, 71, 325, 353–354
Prefabricated fireplaces, 318–319, 322
Presaturation of soil, 8
Private streets/parking areas, 9, 21
Production tract housing, 73, 83, 106, 200, 209, 282, 291–292, 302, 305–306, 446
Property lines, 373
Pullman tops, 257
Punch lists, 21
Putty, 283

Q

Quality levels, 440

R

Radial-arm cutoff saws, use of, 52
Radio panels, 173, 184
Rafters, 113–117, 149, 290, 326
Railing (see Handrails)
Rain gutters, 326, 329
Ramps:
   handicap, 348–349
   interior, 333
   temporary, 6, 325
Random length wood veneer kits, 237
Ranges, 153, 168–169, 181, 213, 215–216, 223, 286, 305, 307, 321, 363

Raw/design cut, 2
Rebar stakes, 22
Rebars, 25, 31, 37
Receptacles (*see* Outlets)
Recertification, 3
Recessed arches, 182
Reclaimed water, 12, 430
Recreation buildings, 297
Refrigerators, 211
Registers (*see* Return-air registers)
Removal, 3, 8
Removal of trees, 8, 431–432
Repairs, 4, 204, 213, 253, 272, 274, 276, 315, 442, 445–446, 450, 452
Reports:
   environmental impact, 11
   soil, 10
Restaking, 12
Retaining walls (*see* Masonry block walls)
Retesting, 10
Return-air registers, 158, 160–162, 178, 239, 283–284, 308–309, 342
Reveals:
   drywall, 52, 57, 69, 71, 155, 207, 249, 259
   window frame, 88–91, 132, 192–193, 260
Ridges, 113–115, 119, 281
Rise, 113
Risers, 100, 102, 335
Rodents, 45
Roof gutters, 326, 329
Roof jacks, 280, 288
Roof load, 149
Roof overhangs, 75, 93, 117–118, 154, 275, 288, 326
Roof paper, 276, 280
Roof tiles, 86, 93, 276, 279–281
Roof vents, 280
Roofing, 279–281
Roofs:
   air-conditioning furnaces on, 163
   flat, 120
   framing of, 113–120
   gable peaks, 84
   hip, 119
   painting of, 288, 290
   pitch, 115
   sloping, 86, 113
   solar panels on, 151
Rosettes, clothes pole, 46, 241

Rough electrical phase, 150–189
Rough grading phase, 2–11
Rough HVAC phase, 150–189
Rough plumbing phase, 150–189
Rough-sawn plantons, 231
Round doorknobs, 292, 295, 300, 303
Round windows, 199
Rubber baseboard, 237
Rubber gaskets, 309

S
S-Shaped roof tiles, 86
Saddles, 61–62, 122, 124
Sales models, 12, 211, 276, 310–311, 421, 424, 434–436, 442–444, 449, 454–457
Sand, as backfill, 43
Sandblasting, 288
Sanding, 230, 257, 283, 290
Sashes, 88, 250
Saucer plates, 423
Saw kerfs, 233
Saws, use of, 52, 77
Scaffolding, 272, 276–277
Schedules (*see* Construction schedules)
Sconces, 167, 178
Scrap lumber, 92, 111, 133
Screwdrivers, use of, 218
Screwing:
   of backing, 140–142
   of cabinet knobs, 218
   of drywall, 195, 200
   of support brackets, 235
Screws, for hardware, 292, 294–296
Scribe molding, 74, 214–215, 241
Scupper drains, 69
Separation gaps, 388
Separations, required, 12
Service connections, 15, 22–23
Service request, 448–449, 452
Setters, cabinets, 216, 218
Sewer laterals, 15, 157
Sewer pipes, 15, 306
Sewers, 3, 12–13, 15
Shampoo shelves, 254
Shear panels, 24, 48, 51, 104–107, 126–127, 150, 172, 191
Sheathing, 111, 117, 192
Shelves:
   bathroom, 134, 266–267
   cabinet, 152, 168, 187, 209–210, 214, 216, 224, 320

   closet, 46, 53–54, 232–233, 241
   dust, 224
   greenhouse, 92
   pot, 130, 196, 284, 290
   shower, 254
   tiled, 266–267
   windowsill, 87
Shims, 79, 215, 218
Shingles, 327
Shiplap siding, 289
Shoes (*see* Base shoes)
Shower dams, 132, 135, 251
Shower doors, 252, 297, 300–301
Showerheads, 92, 150
Showers, 95, 132, 135–136, 150–151, 153–155, 191, 201–202, 251–252, 254, 263, 265, 268, 270, 307
Shrubs, 418, 423–425, 429
Shut-off valves, 17, 23
Shutters, 232, 239
Sidewalk parkways, 9, 157
Sidewalks, 15, 345, 355, 367–369, 388, 419
Siding:
   batt, 171, 179
   bevel, 62, 178–179
   plywood, 109–110
   shiplap, 289
   for windows, 128
   wood, 62, 88, 128, 171–172, 178–179, 185, 289, 327, 332
Sign-out sheets, 442
Sills (*see* Windowsills)
Single-family houses, 23, 426
Sinks:
   bar, 252–253, 267
   bathroom, 54, 150, 217, 257, 305–306, 315
   kitchen, 83, 92, 180, 253, 262, 306–307, 312
   laundry, 222
Site preparation, 2, 14
Site walls, 3
Skip-trowel finish, 276
Skirtboard, 98, 100, 149, 204, 233–236, 244–245, 286, 326–327
Skylights, 92
Slabs, 27, 29–31, 33, 35, 38, 40–41, 72, 273, 344, 354, 360, 372, 374–375
(*See also* Foundations)
Sleepers, 33–34
Sleeves, conduit, 32–33, 57, 366, 370

Sliding glass doors, 83, 87, 94, 101, 121–122, 148, 170, 203, 230, 237, 274, 276, 324
Sliding wood doors, 299
Sloped ceilings, 71, 161, 226, 270
Sloped driveways, 367–368, 370–372, 377, 385–386, 388
Sloped roofs, 86, 113
Sloped walkways, 345, 347–350, 361
Slopes:
   finish graded, 7–8
   landscaped, 278, 383, 386, 392
   permanent, 3
   repairs to, 4
   ribbon gutters, 378
   sidewalks, 367
   subdrain pipes in, 11
   temporary, 7–8
   vertical, 6, 43, 380–381
   yard, 377
Slump-stone walls, 390
Slurry mixes, 17
Soffits, 55, 69–72, 124, 212, 223, 225, 239, 296, 323
Soil:
   amendment material, 424
   clay, 16
   excess, 2–3, 6
   expansive, 8, 43, 359, 375, 383, 421
   moisture content, 10, 16
   presaturation of, 8
   (*See also* Dirt)
Soil engineers/technicians, 4, 6, 8–11, 16, 383, 424
Soil reports, 10
Soil tests, 4, 9–11
Solar panels, 151
Sound insulation, 92
Spa tubs, 156, 176, 253–254, 306
Spackling, 230, 283
Spindles, 55–56, 242–243, 316–317, 338
Splashes, 94, 145–146, 168, 173, 180, 182, 184–186, 202, 218, 247, 253, 259, 262–263, 267–268, 330
Splice joints, 219, 236, 408
Split-level block walls, 40
Split-level floors/slabs, 29–30, 35, 102, 138, 221, 325
Split-level houses, 6

Spouts, bathtub, 134, 305
Spray-paint, use of, 18, 58, 331, 442
Spring lines, ceiling, 149
Sprinklers:
   fire, 12, 154, 156, 305, 307–308, 405
   water, 419–420, 430
Squeaks, floors, 331
Stain, 128–129, 242, 245, 284, 287–288
Stainless steel nails, 128
Stair handrails, 98–99, 162, 180, 198, 234–235, 242–243, 248, 287, 317, 326–327, 395, 402–403, 406
Stair jacks, 96, 98, 100
Stair landings, 96–97, 99, 240, 245–246, 327–328, 346, 348, 352, 354–355, 366
Stair risers, 100, 102, 335
Stair skirtboard, 233–234, 236, 244–245, 286, 326–327
Stair stringers, 96, 98
Stair treads, 99–100, 102–103, 286, 334–336
Stairways, 122, 195, 224, 308
   and bulkhead doors, 75
   carpeting of, 334–339, 344
   entry, 345–351, 355–359, 364
   framing of, 96–103
   installation of, 242–246, 248
   near electrical transformers, 13
   texturing, 203
   tiling of, 341–343
   walls, 149, 383
Stakes/staking, 7, 12, 22, 30, 41
Standpipes, 405
Steel beams, 60–66, 71, 121–123, 444
Steel columns, 25, 27–29, 31, 121–124
Steel grates, 409
Steel hangers, 61–64, 113–114, 116
Steel jigs, 31
Steel moment frames, 27, 31, 121–124
Steel nails, 128
Steel posts, 27–29, 35, 113–114, 122–123
Steel saddles/buckets, 61–62
Steel wire, around subdrain pipes, 11
Steps:
   bottom, 100–101, 103
   pie-shaped, 96, 99, 236

   top, 101
Stiles, 208–211, 224, 291
Stirrups, 113, 197
Stools, windowsill, 90, 239, 250, 268, 290
Stoops, 366
Storage bins, 256, 444
Storm drains, 3
Straightedge, 144–149
Straps, 50, 68, 197
Street curbs (*see* Curbs)
Street improvements, 9
Street sections, 3
Streetlights, 21, 430
Streets, 9, 21, 345, 419
Striker bolts, 207
Striker plates, 296
Stringers, stair, 96, 98
Strip light fixtures, 313
Structural engineers, 25, 50, 62
Structural plans, 50–51
Stub-outs, 153, 156–158
Stucco, 61–62, 67, 87–88, 111, 118, 125, 127, 129, 133, 141, 148, 170, 174, 176, 197, 200–202, 232, 256, 271–278, 288, 290, 347, 386, 388–389, 408, 423
Studs, 53–54, 61, 65, 67, 98, 109, 121, 123, 125, 136, 138, 140, 142, 148, 168–169, 174, 183, 190, 198, 202, 294
Subcontractor bids, 2, 6, 441
Subcontractors, dealing with, 433, 441–447
Subdrain pipes, 11–14, 42, 44, 380–381, 390
Subfloors, 67, 72, 132, 155, 211, 255, 287, 320, 331, 335, 342
Subsurface conditions, testing for, 4, 9–11
Subterranean garages, 353, 370, 386
Sump pumps, 409–410, 428
Superintendents, 10, 437–444, 452
Support brackets, 53–54, 234–235, 287
Surrounds, 92
Surveyors, 7, 12, 30
Swales, 371, 375, 389, 423, 431
Swimming pools, 297, 392, 400, 411
Swinging doors, 293
Switches (*see* Circuit breakers; Light switches; Photocell switches)

T

T-bar tracks, 199
Tack strips, 333–334, 336
Taping, drywall, 195, 197, 199–204
Tar, 132
Telephone lines, 14
Telephone outlets, 173
Telephone poles, 412
Television lines/cable, 14, 177, 189
Television outlets, 173, 189
Tempered glass windows, 83
Templates, 27, 31, 66, 216–217
Temporary fences, 20
Temporary plantings, 424
Temporary power poles, 411
Temporary ramps, 6, 325
Temporary slopes, 7–8
Tension, 60
Testing, for subsurface conditions, 4, 9–11
Texture spraying, 202–203
Thermostats, 183
Threaded bolts, 121, 123
Thresholds, 292, 297, 316, 325
Throw bolts, 292
Tiles:
   address numbers, 275
   bathroom, 95, 134–135, 150, 155, 202, 218, 241, 247, 251–261, 263–270, 332
   countertop, 91, 94, 213, 223, 241, 249–250, 253–254, 256–257, 259, 261–263, 265–267, 269–270
   decorative, 262
   entry, 247–348, 355
   fireplace, 109–110, 253–255, 337
   floor, 56, 332–334, 338, 341–343
   installation of, 249–270
   kitchen, 249–250, 253, 255, 258, 262, 268, 270
   roof, 86, 93, 276, 279–281
   splash, 94, 145–146, 168, 173, 180, 182, 184–186, 202, 218, 247, 253, 259, 262–263, 267–268
   wall, 95, 145, 150, 155, 218, 241, 247, 249–251, 258–261, 265, 269–270, 332
Toe kicks, 208, 210, 334
Toilet paper holders, 291, 294–295, 298, 301, 306
Toilets, 46, 75, 150–151, 156, 244, 258, 261, 265, 269, 283, 304, 306, 331–332, 336–337, 339

Top plates, 114–115, 137
Top steps, 101
Torpedo levels, use of, 309
Touch-ups, painting, 287, 450, 453
Towel bars, 291, 294–295, 297–299, 301, 306
Towel rings, 291, 294–295, 298
Tracks, 139, 199, 231, 233, 313
Tract housing (*see* Production tract housing)
Traffic, construction, 20
Trailers, construction, 442–444
Training, in-house, 441
Transformers, electrical, 13–14
Trash bins, 352, 370, 404, 412, 451
Trash chutes, 47
Trash pickup schedules, 451
Travel expenses, of soil technicians, 10
Treads, stair, 99–100, 102–103, 286, 334–336
Trees, 3, 8, 13, 415–418, 423–425, 429–432
Trenches/trenching, 3, 9–10, 13–18, 24, 28–30, 32, 34, 41, 157, 391–392, 424, 431–432
Trim, 64, 81, 90, 94–95, 118–119, 129, 131, 139, 154, 170–171, 192, 196, 200, 202, 207, 214–215, 219–221, 225–226, 229–234, 239–240, 243–247, 250, 265, 268–269, 271, 274, 277, 285, 288, 290, 303, 316, 320, 324, 327–330, 335, 347
Trimmer return, 138
Trimmers, 75, 77, 79, 193
Truck beds, amounts in, 2
Truss joints, 59, 62–63
Trusses, 205
Tudor-style architecture, 277
Turnarounds, 19, 420
Turning radius, 369, 385
Twist straps, 68

U

Underground utilities, 3, 9, 12–23, 28–29, 417
Upholstered carpeting method, 335
Upward movement, 60
Utilities, underground, 3, 9, 12–23, 28–29, 417
Utility panel boxes, 32, 57, 140, 164, 188

Utility rooms, 14, 40, 49, 164, 188, 405
Utility sleeves, 32–33
Utility vaults, 371

V

Vacuum tanks, 187, 446
Valances, 210, 219
Valve keys/wrenches, 17
Valves:
   control, 151
   laundry, 183
   shut-off, 17, 23
Vault boxes, 23
Vaults, 371
Veneer, 213, 237, 294
Ventilation, 158–163
Vents, 288
   bathroom, 162
   dryer, 31, 152, 159, 161–162, 274, 362–363
   foundation, 140, 273
   garage, 366
   range hood, 169, 223, 286
   roof, 280
   water heater, 304
Vertical chases, 46, 108
Vertical cuts, 4, 6–8
Vertical slopes, 6, 43, 380–381
View lots, 3, 7
Vinyl flooring, 56, 215, 218, 227, 255, 287, 320, 332–333, 335, 337–338, 340, 447

W

Wainscots, 139, 269, 330
Walk-in closets, 46, 167
Walk-throughs, homeowner, 445–450
Walkways, 5, 10, 43, 117, 156–157, 281, 290, 345–378, 387–388, 422–423
Wall backing, 138–139
Wall corners, 70–71, 75, 80, 85, 90, 92, 100, 102–103, 124, 126, 135, 146, 160–161, 170, 180, 212, 240, 242, 269–270, 274, 313, 318–319, 330
Wall sconces, 167, 178
Wallpapering, 202, 311, 315–316
Walls:
   bar light, 145
   bolts/nuts inside of, 26, 61–62

bowed, 148, 197
built-out, 240
and centered windows, 84
curved, 237, 241
exterior, 327
fire, 106, 191–192
and floors, 147
framing of, 45–58, 104, 106, 122, 124, 137, 140–142, 148, 175, 178, 183, 190–192, 195, 205, 344
garden, 13, 383, 387–390, 421
masonry block, 6–8, 13, 31, 37–44, 272, 357–358, 365–366, 376–392, 396–397, 404
mirror, 144
and moment frames, 122, 124
near electrical transformers, 13
party, 25, 104–106, 191–192
pony, 47–48, 98–99, 102, 131, 197, 245, 317, 322, 324
shear, 104–107
site, 3
tiles on, 95, 145, 150, 155, 218, 241, 247, 249–251, 258–261, 265, 269–270
(*See also* Drywall)
Wardrobe closets, 53, 130, 138–139, 231–233, 241, 299
Warranties, 448–450, 453
Washers, 26, 35
Washing machines, 45, 78, 364
Water, reclaimed/domestic, 12, 430
Water heater platforms, 130, 192
Water heaters, 131, 169, 286, 304, 306
Water jetting, 16
Water leaks, 22, 75, 87, 281, 307, 390–391, 410, 422, 426, 451–452
Water main shutoff valves, 17, 23
Water meters, 22–23, 168, 305, 307–308, 421
Water pipes, 12, 15, 17, 22, 28–29, 34, 169, 306–308
Water pressure, 305, 307
Water sprinklers, 419–420, 430
Waterfall carpeting method, 335
Waterproofing, 41, 75, 171–172, 197, 390, 410, 419, 422–423, 426
Weather stripping, 75, 292, 297, 316
Wedges, 77
Weed abatement, 425
Weep holes, 272
Weep screeds, 125, 133, 347

Wheel stop bumpers, 430
Window blinds, 321
Window casings, 90, 94–95, 247, 250, 265, 268
Window frames, 88–90, 132, 192–193, 272, 330
Window sashes, 88, 250
Window shutters, 232, 239
Windows, 77, 190, 227, 272
adjacent, 85, 90, 196, 263
arched, 91, 230, 329
canopy, 92, 262
casement, 403
centered, 84
and exterior light, 328
framing of, 83–95, 260, 330
greenhouse, 92, 262
oval, 265
round, 199
and siding boards, 128
skylights, 92
tempered glass, 83
and tiles, 249–250
view of roof from, 118–120
and wall framing, 51–52
Windowsills, 87–88, 90–91, 132, 192–193, 199–201, 220, 239, 250, 260–261, 263, 268, 272, 290
Windscreens, use of, 444
Wires (*see* Electrical wires)
Wood aprons, 192–193
Wood beams, 42, 63–68, 71, 122–123, 198
Wood caps, 138, 239–240, 242–243, 338, 408
Wood carpet returns, 343
Wood casings, 94–95, 230, 250, 268
Wood cleats, 215, 232
Wood corbels, 125, 129, 185, 290
Wood countertops, 215
Wood doors, 288
Wood faces, 109
Wood fences, 393–397, 425
Wood flooring, 102–103, 126, 221, 245, 306, 332–334, 338–339, 447
Wood gates, 396–397
Wood handrails, 142, 234–235, 242–243, 316, 408
Wood jambs, 79
Wood mantles, 143
Wood nailers, 121
Wood paneling, 237

Wood posts, 26, 50, 113, 180, 325, 353–354, 444
Wood shims, 79, 215, 218
Wood shingles, 327
Wood shutters, 232, 239
Wood siding, 88, 109, 128, 171–172, 178–179, 185, 289, 327, 332
Wood sleepers, 33–34
Wood surrounds, 92
Wood trim, 64, 81, 90, 94–95, 129, 131, 139, 154, 171, 192, 200, 202, 207, 214–215, 219, 225–226, 229–232, 239–240, 243–247, 265, 268, 277, 285, 288, 290, 303, 316, 335
Wood wedges, 77
Work orders, 441
Working space, 6, 14, 23, 30, 41, 52, 116, 169, 278–279, 307, 364
Wrought-iron fences, 398–401, 405, 425–426
Wrought-iron gates, 352, 359–360, 400–401, 404
Wrought-iron handrails, 138, 174, 198, 240, 271–272, 317, 326–327, 340–341, 346–347, 352, 356–359, 361, 364, 395, 397, 399–400, 402–403, 408, 413

Y
Yardage, 2–3, 6
Yards, 353–354, 377, 379–380, 385, 392, 426, 431

Z
Zero lot lines, 30, 426